Performance Tuning of Scientific Applications

Edited by
David H. Bailey
Robert F. Lucas
Samuel W. Williams

CRC Press
Taylor & Francis Group
Boca Raton London New York

CRC Press is an imprint of the
Taylor & Francis Group, an **informa** business

A CHAPMAN & HALL BOOK

CRC Press
Taylor & Francis Group
6000 Broken Sound Parkway NW, Suite 300
Boca Raton, FL 33487-2742

First issued in paperback 2019

© 2011 by Taylor and Francis Group, LLC
CRC Press is an imprint of Taylor & Francis Group, an Informa business

No claim to original U.S. Government works

ISBN-13: 978-1-4398-1569-4 (hbk)
ISBN-13: 978-0-367-38330-5 (pbk)

Library of Congress Cataloging-in-Publication Data

Performance tuning of scientific applications / David H. Bailey, Robert
    F. Lucas, Samuel W. Williams, editors.
       p. cm. -- (Chapman & Hall/CRC computational science)
    Includes bibliographical references and index.
    ISBN 978-1-4398-1569-4 (hardback)
    1. Science--Data processing--Evaluation. 2. Electronic digital computers--Evaluation.
3. System design--Evaluation. 4. Science--Computer programs. I. Bailey, David H. II.
Lucas, Robert F. III. Williams, Samuel Watkins. IV. Title. V. Series.

Q183.9.P47 2011
502.85--dc22   2010042967

Visit the Taylor & Francis Web site at
http://www.taylorandfrancis.com

and the CRC Press Web site at
http://www.crcpress.com

Performance Tuning of Scientific Applications

Chapman & Hall/CRC
Computational Science Series

SERIES EDITOR

Horst Simon
Deputy Director
Lawrence Berkeley National Laboratory
Berkeley, California, U.S.A.

AIMS AND SCOPE

This series aims to capture new developments and applications in the field of computational science through the publication of a broad range of textbooks, reference works, and handbooks. Books in this series will provide introductory as well as advanced material on mathematical, statistical, and computational methods and techniques, and will present researchers with the latest theories and experimentation. The scope of the series includes, but is not limited to, titles in the areas of scientific computing, parallel and distributed computing, high performance computing, grid computing, cluster computing, heterogeneous computing, quantum computing, and their applications in scientific disciplines such as astrophysics, aeronautics, biology, chemistry, climate modeling, combustion, cosmology, earthquake prediction, imaging, materials, neuroscience, oil exploration, and weather forecasting.

PUBLISHED TITLES

PETASCALE COMPUTING: ALGORITHMS AND APPLICATIONS
Edited by David A. Bader

PROCESS ALGEBRA FOR PARALLEL AND DISTRIBUTED PROCESSING
Edited by Michael Alexander and William Gardner

GRID COMPUTING: TECHNIQUES AND APPLICATIONS
Barry Wilkinson

INTRODUCTION TO CONCURRENCY IN PROGRAMMING LANGUAGES
Matthew J. Sottile, Timothy G. Mattson, and Craig E Rasmussen

INTRODUCTION TO SCHEDULING
Yves Robert and Frédéric Vivien

SCIENTIFIC DATA MANAGEMENT: CHALLENGES, TECHNOLOGY, AND DEPLOYMENT
Edited by Arie Shoshani and Doron Rotem

INTRODUCTION TO THE SIMULATION OF DYNAMICS USING SIMULINK®
Michael A. Gray

INTRODUCTION TO HIGH PERFORMANCE COMPUTING FOR SCIENTISTS
AND ENGINEERS, **Georg Hager and Gerhard Wellein**

PERFORMANCE TUNING OF SCIENTIFIC APPLICATIONS, **Edited by David H. Bailey,
Robert F. Lucas, and Samuel W. Williams**

Contents

Contributor List

Mark Adams
APAM Department,
Columbia University
New York, New York

David H. Bailey
Lawrence Berkeley National Lab
Berkeley, California

Vincent Beckner
Lawrence Berkeley National Lab
Berkeley, California

John Bell
Lawrence Berkeley National Lab
Berkeley, California

Laura Carrington
San Diego Supercomputer Center
San Diego, California

Jonathan Carter
Lawrence Berkeley National Lab
Berkeley, California

Jacqueline Chame
University of Southern California
Information Sciences Institute
Los Angeles, California

Kaushik Datta
University of California at Berkeley
Berkeley, California

Bronis R. de Supinski
Lawrence Livermore National Lab
Livermore, California

Jack Dongarra
University of Tennessee,
Oak Ridge National Lab
Knoxville, Tennessee

Erik W. Draeger
Lawrence Livermore National Lab
Livermore, California

Stéphane Ethier
Princeton Plasma Physics Lab,
Princeton University
Princeton, New Jersey

Mary Hall
University of Utah School of
Computing
Salt Lake City, Utah

Jeffrey K. Hollingsworth
University of Maryland
College Park, Maryland

Khaled Ibrahim
Lawrence Berkeley National Lab
Berkeley, California

Jesus Labarta
Barcelona Supercomputing Center
Barcelona, Spain

Byounghak Lee
Texas State University, San Marcos
San Marcos, Texas

Allen D. Malony
University of Oregon
Eugene, Oregon

John Mellor-Crummey
Rice University
Houston, Texas

Juan Meza
Lawrence Berkeley National Lab
Berkeley, California

Shirley Moore
University of Tennessee
Knoxville, Tennessee

Leonid Oliker
Lawrence Berkeley National Lab
Berkeley, California

Erik Schnetter
CCT, Louisiana State University
Baton Rouge, Louisiana

Martin Schulz
Lawrence Livermore National Lab
Livermore, California

John Shalf
Lawrence Berkeley National Lab
Berkeley, California

Hongzhang Shan
Lawrence Berkeley National Lab
Berkeley, California

Sameer S. Shende
University of Oregon
Eugene, Oregon

Allan Snavely
University of California, San Diego
San Diego, California

Erich Strohmaier
Lawrence Berkeley National Lab
Berkeley, California

Daniel K. Terpstra
University of Tennessee
Knoxville, Tennessee

Ananta Tiwari
University of Maryland
College Park, Maryland

Lin-Wang Wang
Lawrence Berkeley National Lab
Berkeley, California

Harvey Wasserman
Lawrence Berkeley National Lab
Berkeley, California

Vincent M. Weaver
University of Tennessee
Knoxville, Tennessee

Samuel W. Williams
Lawrence Berkeley National Lab
Berkeley, California

Patrick H. Worley
Oak Ridge National Lab
Oak Ridge, Tennessee

Katherine Yelick
Lawrence Berkeley National Lab
Berkeley, California

Zhengji Zhao
Lawrence Berkeley National Lab
Berkeley, California

About the Editors

David H. Bailey is the chief technologist of the Computational Research Department at Lawrence Berkeley National Laboratory (LBNL). Prior to coming to LBNL in 1998, he held a position with the NASA Advanced Supercomputing Center at NASA's Ames Research Center. Bailey's 140 published papers and three books span studies in system performance analysis, highly parallel computing, numerical algorithms, and computational mathematics. He has received the Sidney Fernbach Award from the IEEE Computer Society, the Gordon Bell Prize from the Association for Computing Machinery, and the Chauvenet and Merten Hesse Prizes from the Mathematical Association of America.

Robert F. Lucas is the Computational Sciences Division director of the Information Sciences Institute, and also a research associate professor of computer science in the Viterbi School of Engineering at the University of Southern California. Prior to coming to USC/ISI, he worked at the Lawrence Berkeley National Laboratory and the Defense Advanced Research Project Agency. His published research includes studies in system performance analysis, real-time applications, heterogeneous computing, and system simulation.

Samuel W. Williams is a research scientist in the Future Technologies Group at the Lawrence Berkeley National Laboratory (LBNL). He received his PhD in 2008 from the University of California at Berkeley. Dr. Williams has authored or co-authored 30 technical papers including several award-winning papers. His research includes high-performance computing, automatic performance tuning (auto-tuning), computer architecture, and performance modeling.

Chapter 1

Introduction

David H. Bailey

Lawrence Berkeley National Laboratory

1.1 Background

The field of performance analysis is almost as old as scientific computing itself. This is because large-scale scientific computations typically press the outer envelope of performance. Thus, unless a user's application has been tuned to obtain a high fraction of the theoretical peak performance available on the system, he/she might not be able to run the calculation in reasonable time or within the limits of computer resource allocations available to the user. For these reasons, techniques for analyzing performance and modifying code to achieve higher performance have long been a staple of the high-performance computing field.

What's more, as will become very clear in the chapters of this book, understanding why a calculation runs at a disappointing performance level is seldom straightforward, and it is often even more challenging to devise reasonable changes to the code while at the same doing avoiding doing violence to the structure of the code.

Thus over the past few decades as scientific computing has developed, the field of performance analysis has developed in tandem. Some familiar themes include:

- *Benchmark performance and analysis.* Comparing the performance of a scientific calculation on various available computer systems or studying the performance of a computer system on various scientific applications.

- *Performance monitoring.* Developing hardware/software techniques and tools that can be used to accurately monitor performance of a scientific application as it is running—floating-point operation counts, integer operations, cache misses, etc.

- *Performance modeling.* Developing techniques and tools to encapsulate the performance behavior of applications and computer system into relatively simple yet accurate models.

- *Performance tuning.* Developing semi-automatic techniques and tools to make necessary code changes to optimize the run-time performance of a scientific application (or class of applications).

The purpose of this book is to introduce the reader to research being done in this area as of the current date (summer 2010). The various chapters are written by some of the most notable experts in the field, yet the book is not targeted primarily to experts in the performance field. Instead, the target audience is the much broader community of physicists, chemists, biologists, environmental scientists, mathematicians, computer scientists, and engineers who actually use these systems. In other words, the intent of this book is to introduce real-world users of computer systems to useful research being done in the field of performance analysis.

The hope is that the exposition in this book will assist these scientists to better understand the performance vagaries that may arise in scientific applications, and also to better understand what practical potential there is to actually improve the performance—what changes need to be made; how difficult is it to make these changes; what improvement can be expected if such changes are made; what the are prospects of semi-automatic tools to make these changes, etc.

In general, it should be clear from the scope of research work discussed in this book that the field of performance analysis is a relatively sophisticated field. Indeed, it is now a full-fledged instance of empirical science, involving:

- *Experimentation.* One of the first steps in understanding the performance characteristics of a given application is to perform some carefully conducted measurements, collecting data that can be used to confirm or refute various conjectures, probe anomalies, develop accurate models, and compare results.

- *Theory.* Once a set of validated experiments have been run, performance researchers typically attempt to make sense of this data by developing analytic models that accurately represent this data, and which lead to further insights, both general and specific, that will be of use to other researchers.

- *Engineering.* One important activity in the field of performance analysis is the development of tools and techniques to make more accurate measurements, develop more accurate models, and to assist users in making requisite changes in codes.

Indeed, efforts in this field have led to a significantly increased emphasis on thoroughness and rigor in conducting performance experiments (see next section), and in drawing conclusions from performance data. In this regards, the field of performance analysis is on a trajectory that is typical of almost every other field of science—after some initial work, researchers must develop and adopt significantly more rigorous, reliable methods to further advance the state of the art. We can expect this trend to continue into the future.

1.2 "Twelve Ways to Fool the Masses"

The need for increased sophistication and rigor in the field can be most clearly (and most amusingly) seen in an episode from the early days of parallel computing, which will be briefly recounted here.

Until about 1976, most scientific computing was done on large general-purpose mainframe computers, typically in the central computing centers of universities and research laboratories. Some of the most widely used systems include the IBM 7030, 7040 and 7090 systems, and the CDC 6600 and 7600 systems. Some minicomputers, notably the PDP-11 and DEC-10, were also used for scientific computation.

The Cray-1 made its debut in 1975, designed by legendary computer architect Seymour Cray. This system featured a vector architecture, which proved to be very efficient for many types of scientific applications at the time, because it could sustain large fractions of the system's peak performance while accessing multi-dimensional arrays along any dimension (not just the first), provided that some care was taken in coding to avoid power-of-two memory strides. Many scientists achieved dramatic performance improvements, compared with what they could achieve on other systems available at the time.

As a result of this initial success, systems manufactured by Cray Research dominated the world of scientific computing for at least a decade. The Cray X-MP, introduced in 1982, featured multiple processors. The Cray-2, introduced in 1985, dramatically increased the memory available to scientific programs. A few years later, the Cray Y-MP continued the X-MP architectural line. Several large Japanese firms, including NEC, Fujitsu and Hitachi, also introduced competing systems based on vector designs.

Beginning in 1988, several distributed memory parallel systems became commercially available. These included the Connection Machine, from Thinking Machines, Inc., and systems from Intel and other vendors that were constructed entirely from commodity microprocessors and network components. Some initial work with these systems achieved remarkably high performance rates. These results generated considerable enthusiasm, and soon it became clear that a paradigm shift was underway.

However, some researchers in the field began to grow concerned with what appeared to be "hype" in the field—in numerous cases, scientists' descriptions of their work with these highly parallel architectures seemed indistinguishable from the sort of salesmanship that one might expect to hear from computer vendor marketing personnel. While some degree of enthusiasm was perhaps to be expected, there was concern that this "enthusiasm" was leading to sloppy science and potentially misleading results. Some researchers noted examples, even in presumably peer-reviewed literature, of questionable research work. The most common example was questionable comparisons between highly parallel systems and Cray vector systems.

In 1991, one of the present editors thought that it was time to publicly express concern with these practices. This was done in the form of a humorous tongue-in-cheek article entitled "Twelve Ways to Fool the Masses When Giving Performance Reports on Parallel Computers" [39], which was published in the now-defunct publication *Supercomputing Review*. This article attracted an astonishing amount of attention at the time, including mention in the *New York Times* [229]. Evidently it struck a responsive chord among many professionals in the field of technical computing who shared the author's concerns.

The following is a brief summary of the "Twelve Ways:"

1. Quote only 32-bit performance results, not 64-bit results, and compare your 32-bit results with others' 64-bit results.

2. Present inner kernel performance figures as the performance of the entire application.

3. Quietly employ assembly code and other low-level language constructs, and compare your assembly-coded results with others' Fortran or C implementations.

4. Scale up the problem size with the number of processors, but don't clearly disclose this fact.

5. Quote performance results linearly projected to a full system.

6. Compare your results against scalar, unoptimized, single processor code on Crays [prominent vector computer systems at the time].

7. Compare with an old code on an obsolete system.

8. Base megaflops operation counts on the parallel implementation instead of on the best sequential implementation.

9. Quote performance in terms of processor utilization, parallel speedups or megaflops per dollar (peak megaflops, not sustained).

10. Mutilate the algorithm used in the parallel implementation to match the architecture. In other words, employ algorithms that are numerically inefficient in order to exhibit artificially high megaflops rates.

11. Measure parallel run times on a dedicated system, but measure conventional run times in a busy environment.

12. If all else fails, show pretty pictures and animated videos, and don't talk about performance.

Subsequently, the present editor previously wrote a more detailed paper, which quoted actual passages and graphs taken from peer-reviewed work [40]. To avoid undue public embarrassment of the scientists involved, which was not the purpose of the article, the source articles were not explicitly given in that paper. This material was also presented in a session at the 1992 Supercomputing conference. These actions caused considerable consternation in some quarters at the time (and the present editor does not recommend repeating this experiment!), but many in the field appreciated these much-needed reminders of the pressing need for greater professional discipline in the field.

1.3 Examples from Other Scientific Fields

We do not need to look very far to see examples in other fields where well-meaning but sloppy practices in reporting and analyzing experimental results have led to distortion. One amusing example is the history of measurements of the speed of light [105]. Accurate measurements performed in the late 1930s and 1940s seemed to be converging on a value of roughly 299,760 km/s. After World War II, some researchers made more careful measurements, and thereafter the best value converged to its present-day value of 299,792.458 km/s (which is now taken as a standard, indirectly defining the meter and the second). No researcher seriously believes that the speed of light changed between 1930 and 1950. Thus one is left with the sobering possibility that sloppy experimental practices and "group-think" led earlier researchers to converge on an incorrect value.

Difficulties with experimental methods and dispassionate assessment of results have also been seen in the social sciences. In his book *The Blank Slate: The Modern Denial of Human Nature* [277], Harvard psychologist Steven Pinker chronicles the fall of the "blank slate" paradigm of the social sciences, namely the belief that heredity and biology play no significant role in human psychology—all personality and behavioral traits are socially constructed. This paradigm prevailed in the field until approximately 1980, when a wave of indisputable empirical evidence forced a change of paradigm. The current consensus, based on latest research, is that humans at birth universally possess sophisticated facilities for language acquisition, visual pattern recognition, numerical reasoning and social life, and furthermore that heredity, evolution and biology are major factors in human personality—some personality traits are as much as 70% heritable.

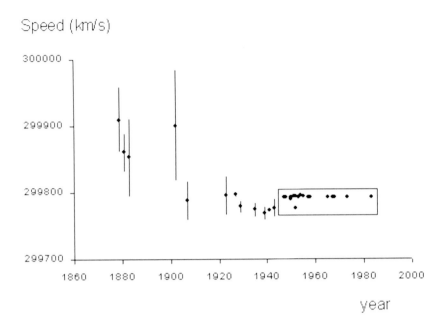

FIGURE 1.1: History of measurements of the speed of light.

Along this line, anthropologists, beginning with Margaret Mead in the 1930s, painted an idyllic picture of primitive societies, notably South Sea Islanders, claiming that they had little of the violence, jealousy, warfare, and social hangups that afflict the major Western societies. But beginning in the 1980s, a new generation of anthropologists began to re-examine these societies and question the earlier findings. They found, contrary to earlier results, that these societies typically had murder rates several times higher than large U.S. cities, and death rates from inter-tribe warfare exceeding those of warfare among Western nations by factors of 10 to 100. What's more, in many of these societies, complex, jealousy-charged taboos surrounded courtship and marriage. For example, some of these societies condoned violent reprisals against a bride's family if she was found not to be a virgin on her wedding night. These findings represented a complete reversal of earlier claims of carefree adolescence, indiscriminate lovemaking, and peaceful coexistence with other tribes.

How did these social scientists get it so wrong? Pinker and others have concluded that the principal culprit was sloppy experimental methodology and wishful thinking [277].

1.4   Guidelines for Reporting High Performance

When the pressures of intense marketplace competition in the computer world are added to the natural human tendency of scientists and engineers to be exuberant about their own work, it should come as little surprise that sloppy and potentially misleading performance work has surfaced in scientific papers and conference presentations. And since the reviewers of these papers are themselves in many cases caught up in the excitement of this new technology, it should not be surprising that they have, in some cases, been relatively permissive with questionable aspects of these papers.

Clearly the field of technical computing does not do itself a favor by condoning sloppy science, whatever are the motives of those involved. In addition to fundamental issues of ethics and scientific accuracy, the best way to ensure that computer systems are effective for scientific or engineering applications is to provide early feedback to manufacturers regarding their weaknesses. Once the reasons for less-than-expected performance rates on certain problems are identified, improvements can be made in the next generation.

Virtually every field of science has found it necessary at some point to establish rigorous standards for the reporting of experimental results, and the field of high-performance computing is no exception. Thus the following guidelines have been suggested to govern reporting and analysis of performance [40]:

1. If results are presented for a well-known benchmark, comparative figures should be truly comparable, and the rules for the particular benchmark should be followed.

2. Only actual performance rates should be presented, not projections or extrapolations. For example, performance rates should not be extrapolated to a full system from a scaled-down system. Comparing extrapolated figures with actual performance figures, such as by including both in the same table or graph, is particularly inappropriate.

3. Comparative performance figures should be based on comparable levels of tuning.

4. Direct comparisons of run times are preferred to comparisons of Mflop/s or Gflop/s rates. Whenever possible, timings should be true elapsed time-of-day measurements. This might be difficult in some environments, but, at the least, whatever timing scheme is used should be clearly disclosed.

5. Performance rates should be computed from consistent floating-point or integer operation counts, preferably operation counts based on efficient implementations of the best practical serial algorithms. One intent here is to discourage the usage of numerically inefficient algorithms, which may exhibit artificially high performance rates on a particular system.

6. If speedup figures are presented, the single processor rate should be based on a reasonably well tuned program without multiprocessing constructs. If the problem size is scaled up with the number of processors, then the results should be clearly cited as "scaled speedup" figures, and details should be given explaining how the problem was scaled up in size.

7. Any ancillary information that would significantly affect the interpretation of the performance results should be fully disclosed. For example, if performance results were for 32-bit rather than for 64-bit data, or if assembly-level coding was employed, or if only a subset of the system was used for the test, all these facts should be clearly and openly stated.

8. Due to the natural prominence of abstracts, figures and tables, special care should be taken to ensure that these items are not misleading, even if presented alone. For example, if significant performance claims are made in the abstract of the paper, any important qualifying information should also be included in the abstract.

9. Whenever possible, the following should be included in the text of the paper: the hardware, software and system environment; the language, algorithms, the datatypes and programming techniques employed; the nature and extent of tuning performed; and the basis for timings, flop counts and speedup figures. The goal here is to enable other scientists and engineers to accurately reproduce the performance results presented in the paper.

1.5　Modern Performance Science

In the chapters that follow, we will present a sampling of research in the field of modern performance science and engineering. Each chapter is written by a recognized expert (or experts) in the particular topic in question. The material is written at a level and with a style that is targeted to be readable by a wide range of application scientists, computer scientists and mathematicians. The material is organized into these main sections:

1. *Architectural overview* (Chapter 2). This chapter presents a brief overview of modern computer architecture, defining numerous terms, and explaining how each subsystem and design choice of a modern system potentially affects performance.

2. *Performance monitoring* (Chapters 3–5). These chapters describe the state-of-the-art in hardware and software tools that are in common use

for monitoring and measuring performance, and in managing the mountains of data that often result in such measurements.

3. *Performance analysis* (Chapter 6). This chapter describes modern approaches to computer performance benchmarking (i.e., comparative studies of performance), and gives results exemplifying what insights can be learned by such studies.

4. *Performance modeling* (Chapters 7–9). These chapters describe how researchers deduce accurate performance models from raw performance data or from other high-level characteristics of a scientific computation.

5. *Automatic performance tuning* (Chapters 10–13). These chapters describe ongoing research into automatic and semi-automatic techniques to modify computer programs to achieve optimal performance on a given computer platform.

6. *Application tuning* (Chapters 14–16). These chapters give some examples where appropriate analysis of performance and some deft changes have resulted in extremely high performance, in some cases permitting scientific applications to be scaled to much higher levels of parallelism, with high efficiency.

1.6 Acknowledgments

Writing, compiling and reviewing this book has involved the efforts of many talented people. In addition to the authors of the respective chapters, the editors wish to thank officials at Taylor and Francis for their encouragement and support. Also, this work was supported by numerous research grants, mostly from the U.S. Department of Energy, but also from other U.S. and Spanish research agencies. The particular agencies and grants are summarized in each chapter.

David Bailey is supported in part by the Performance Engineering Research Institute, which is supported by the SciDAC Program of the U.S. Department of Energy, and by the Director, Office of Computational and Technology Research, Division of Mathematical, Information and Computational Sciences, U.S. Department of Energy, under contract number DE-AC02-05CH11231.

Chapter 2

Parallel Computer Architecture

Samuel W. Williams

Lawrence Berkeley National Laboratory

David H. Bailey

Lawrence Berkeley National Laboratory

In this chapter we provide a brief introduction to parallel computer architecture. The intent of this material is to provide a survey of the principal features of computer architectures that are most relevant to performance, and to briefly define concepts and terminology that will be used in the remainder of this book. Obviously there is substantial literature on the topic of computer architectures in general, and even on the subtopic of computer architectures within the realm of high-performance computing. Thus, we can only provide a sampling here. For additional details, please see the excellent references [104, 164, 268].

We commence this chapter with a discussion of the fundamental forces enabling advances in architecture, proceed with a high-level overview of the basic classes of parallel architectures, and then provide a retrospective on the challenges that have driven architecture to date. Along the way, we note particular challenges that face present-day designers and users of computer systems, such as the recent re-emergence of heterogeneous architectures.

2.1   Introduction

2.1.1   Moore's Law and Little's Law

In a seminal 1965 article, little noticed at the time, Intel founder Gordon Moore wrote

> The complexity for minimum component costs has increased at a rate of roughly a factor of two per year. ... Certainly over the short term this rate can be expected to continue, if not to increase. Over the longer term, the rate of increase is a bit more uncertain, although there is no reason to believe it will not remain nearly constant for at least 10 years. That means by 1975, the number of components per integrated circuit for minimum cost will be 65,000. [247]

With these sentences, Moore stated what is now known as Moore's Law, namely that advances in design and fabrication technology enable a doubling in semiconductor density every 12 months (a figure later modified to roughly 18 months). Astonishingly, Moore's Law has now continued unabated for 45 years, defying several confident predictions that it would soon come to a halt. This sustained and continuing exponential rate of progress is without peer in the history of technology. In 2003, Moore predicted that Moore's Law would continue at least for another ten years [18]. As of the writing of this volume (summer 2010), it appears that at least an additional ten years are assured.

It should be emphasized that not all parameters of computer processor performance have improved. In particular, processor clock rates have stalled in recent years, as chip designers have struggled to control energy costs and heat dissipation. But their response has been straightforward—simply increase the number of processor "cores" on a single chip, together with associated cache memory, so that aggregate performance continues to track or exceed Moore's Law projections. And the capacity of leading-edge DRAM main memory chips continues to advance apace with Moore's Law.

Beyond the level of a single computer processor, computer architects capitalize on advances in semiconductor technology to maximize sustained performance, productivity, and efficiency for a given application (or set of applications), within constraints of overall system cost. Indeed one over-arching design objective of computer architecture is to ensure that progress in the aggregate computational power and memory storage of computer systems advance, over time, at a rate that at least on a par with Moore's Law, if not faster.

Related to Moore's Law is "Bell's Law of Computer Classes." This is an observation by computer pioneer Gordon Bell, originally made in 1972, that roughly every ten years or so a new class of computer systems emerges that is essentially constructed from components of the previous class. The best-known instance of this law is the emergence of scientific supercomputers consisting of large clusters of personal computer modules and commodity network components. These clusters first appeared about 1995; now they constitute a large fraction of the systems on the Top500 list. Today, we may be at a similar precipice where embedded processors and technologies are use to create large-scale, efficient parallel systems. Some additional background on Bell's Law, and trends for the future, are presented in a recent article by Gordon Bell [47].

Another important principle in the study of computer architecture is "Little's Law" [41], which is actually a result of queuing theory. Let N be the average number of waiting customers, let A be the average arrival rate, and let T be the average wait time. Then Little's Law is the assertion that

$$N = AT \qquad (2.1)$$

Little's Law can be seen as a consequence of Fubini's theorem, namely the result from mathematical integration theory that under appropriate conditions, one can integrate to compute the area under a curve along either of two axes and obtain the same result [41]. In other words, Little's Law is a very basic principle that applies to a wide range of phenomena.

In the context of a single processor computer system, one application of Little's Law is as follows. Assume for point of discussion that a given computation deals exclusively with data in main memory—there is little or no cache or register re-use, and there is a one-to-one ratio between floating-point operations and data fetches. Assume also that the system has a balanced design: the length of vector registers, in a vector system; the size of cache lines, in a cache-based system; or the maximum number of outstanding memory references, in a multi-threaded system, is comparable to the memory latency as measured in clock periods.

By Little's Law, the average number of words in transit between memory and the processor is given by the product of the memory bandwidth and the memory latency (assuming the memory bandwidth is being fully utilized). Now if we assume a balanced design as above, then the number of words in transit at a given time is comparable to the degree of concurrency inherent in the processor-memory design. Thus we can write the (approximate) relation

$$C = BL, \qquad (2.2)$$

or, in other words, concurrency equals bandwidth times latency measured in seconds. This is the form that Little's Law is most often used in discussions of computer architecture. It can be applied to numerous aspects of a computer system, including floating-point units, memory, and interprocessor network.

2.2   Parallel Architectures

At the present time, a large-scale scientific computation typically involves many trillions of floating-point computations. Unfortunately, the performance of individual processing elements is limited to the frequency at which they may be clocked. For this reason, the performance of a scientific computation (even setting aside the question of whether there is enough memory to hold the necessary data) is limited to basically the reciprocal of the clock period. However, since virtually all computations exhibit at least a modest degree of parallelism, this means that at a given level of technology, the only means of increasing performance is to employ parallel processors.

There are many styles of parallel architectures, but all present-day designs consist of processing elements, memory, and an interprocessor network. The primary differentiating factor is whether computation is instruction-driven or data-driven. In this chapter, we will restrict ourselves to the former, namely architectures whose processing elements are based, in a fundamental sense, on a classical von Neumann design, originally sketched by John von Neumann. Within this class, there are two principle types of parallel architectures: shared-memory and distributed-memory. These two styles are illustrated in Figure 2.1.

We will discuss processor architecture, memory architecture, and network architecture in the following sections. As mentioned earlier, we will keep this discussion at a rather high-level, since parallel processing may be realized across a wide variety of scales, from multicore chips to supercomputers. For additional details, see [104, 164, 268], or any of many articles on the topic, many of which are accessible via the Internet.

2.2.1   Shared-Memory Parallel Architectures

Figure 2.1(a) illustrates a *shared-memory* parallel architecture. In this architecture, the von Neuman architecture is extended by adding additional processing elements (each with their own program counters) yet keeping one (conceptually) memory that is now shared among processing elements. Instantiation of a network between processing elements and memory allows any processor to reference any memory location.

Due to physical and economic constraints, the architecture is often partitioned into "nodes." Each node can contain one or more processing elements, a partition of the memory, and a network interface.

Although a shared-memory design permits extremely low overhead com-

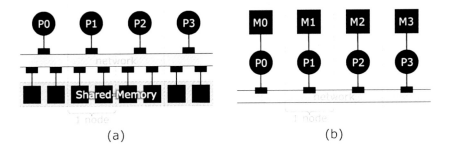

FIGURE 2.1: Parallel Architectures. (a) Shared-Memory, (b) Distributed-Memory.

munication of data (load and store instructions), it does create a major programming challenge: resolution of data dependencies. To that end, a number of instruction-set (atomic instructions) and programming model paradigms (bulk synchronous) have been invented to simplify this problem.

Caches may filter bandwidth and reduce memory access time. To maintain correct behavior, caches must be kept "coherent" with each other. That is, there is always one correct value of a memory location. Unfortunately, global cache coherency is a major impediment to the scalability of shared-memory architectures. When caches are placed on the processor side of the network (where a processor may locally cache any memory location), a cache miss necessitates a global inquiry as to the location of the current value of a memory address. In a snoopy protocol, all caches must be queried to ascertain if they possess the latest copy—a undesirable solution as it necessitates a quadratic increase in network performance. Directory-based protocols create a directory that specifies which processors have cached the data. Requesting processors simply query the fixed directory location, and then query the caching agent. Although this solution requires additional overhead per cache line, when the number of caching processors is low, the number of network transactions is also low.

An alternate solution is to place the caches on the memory-side of the network. In such an architecture, no coherency protocol is required as only one cache may ever cache a particular memory address. This solution is appropriate when memory latency exceeds network latency.

2.2.2 Distributed-Memory Parallel Architectures

An alternative to a shared-memory design is a *distributed-memory* architecture, as depicted in Figure 2.1(b). At its inception in the late 1980s, the distributed-memory architecture was a rather radical departure from classical shared-memory designs. Since processors no longer share memory, they must communicate via rather bulky network operations, instead of lightweight loads and stores. On the plus side, processors need only cache local data. Thus, cache coherency is not an impediment to scalability on distributed-memory systems. This is exemplified by machines like IBM's Blue Gene system, an architecture that has reached unprecedented scales of parallelism (hundreds of thousands of processors).

To simplify the communication challenges, two major communication models (for the purposes of controlling parallelism in a user's program) have emerged: two-sided message passing and one-sided partitioned global address space (PGAS).

In the two-sided message passing arena, an abstraction layer is applied to the communication interface. The abstraction layer distills communication into send and receive operations, which in turn can be blocking or non-blocking. In two-sided message passing, it is the programmer's responsibility to pair a send on one processor with a receive on another processor—if he/she fails to do this properly as required, data could be lost, resulting in strange intermittent errors plaguing an otherwise valid program.

Moreover, in two-sided message passing, it is often necessary, for performance reasons, for the programmer to aggregate data into larger messages as this helps amortize latency and overhead.

In the one-sided PGAS world, the abstraction layer conceptually unifies the distributed memories into one shared memory. Additionally, the PGAS model exposes two very useful primitive operations: put and get. For example, a processor may "put" local data to a remote memory, which in essence copies local data to a remote memory location.

There is often confusion on the difference between send/receive semantics (in two-sided message passing) and put/get semantics (in one-sided message passing). The send-receive protocol is not only two-sided but also tag-based. This means that the receiver must match the tag of an incoming message to a tag associated (by a previous receive operation) with a buffer. In contrast, put/get commands are one-sided and based on addresses. In essence, there is no level of indirection to determine the relevant remote memory address with put/get semantics.

2.2.3   Hybrid Parallel Architectures

At the present time, there are many legacy shared-memory applications, particularly in the corporate sector. Partly for this reason, there is a continuing and substantial economic incentive to build mid-range shared-memory parallel architectures. In the 1990s, the scientific community began to leverage these commodity shared-memory multiprocessors (SMPs) as components of large supercomputers. In effect, these machines combine both the shared-memory and distributed-memory paradigms. That is, the shared-memory paradigm is restricted to a local node (a few processors) of some commodity-based SMP, but the distributed-memory paradigm is applied to connect hundreds of thousands of nodes into a large highly parallel system.

In the past, shared-memory nodes have been built mainly from discrete single-core processors. Today, shared-memory nodes are almost invariably built from multicore chips. This has resulted in a paradigm shift of sorts, in which the concurrency per node has grown from a few processors per shared-memory node to potentially hundreds. At the present time, many programmers of large scientific programs ignore the hybrid nature of machines of this class and simply program them with a flat message passing programming model. Such codes usually work at reasonably good performance rates, but there is growing concern that they miss out on the full performance potential of emerging systems consisting of multicore or manycore nodes. Thus some users have converted their programs to employ two levels of parallelism—simple shared-memory parallel commands at the local node level, and message-passing (either one-sided or two-sided) at the global level.

2.3   Processor (Core) Architecture

Virtually every processor used in scientific computing today shows a clear evolutionary lineage from the simple clock-synchronized classical von Neumann architecture. A *von Neumann* architecture is an instruction-controlled architecture that comprises a single (conceptually) memory array, a control unit, and an execution unit (ALU). Programs are comprised of a sequence of atomic (user perspective) instructions that are fetched from memory (addressed via a program counter or instruction pointer) and executed. Typically, execution of an instruction proceeds by first fetching the requisite operands from memory, then performing the computation, and finally committing the results to memory. The latches of the processor (and thus these phases) are synchronized by a processor clock. Each phase may require one or more clocks to complete.

Over the course of the last 50 years, computer architects have exploited Moore's Law to evolve these architectures to improve performance and programmer productivity. Initially, due to the relative performance of memory and processors, architects have primarily focused on inventing architectural paradigms to exploit instruction- and data-level parallelism. To that end, in this section, we examine the evolution of processor core architectures and emerging challenges for programmers of such systems.

Pipelined Processors: In their simplest (non-pipelined) form, processors complete the execution of one instruction before commencing the execution of the next. By Little's Law (latency=1, bandwidth=1), we see that such architectures demand no instruction-level parallelism (ILP) to attain peak performance. Although instructions appear atomic to programmers, their execution is clearly composed of a number of stages including fetching the instruction word from memory, decoding it in the control unit, fetching the relevant operands, performing the computation, and committing the results of the instructions. Architects quickly realized that the phases of sequential instructions could be overlapped. In doing so they could "pipeline" their execution: at the same instant the first instruction is committing its results, the second instruction is performing its computation, the third, is fetching operands, etc. As such, rather than completing one instruction every five cycles, the processor's peak throughput has been increased to one instruction per cycle. Over the years, architects have continually reduced the functionality of each stage. In doing so, they can substantially increase the clock frequency ("superpipelining") and thereby increase peak performance.

As dependent instructions are now typically overlapped, stall cycles must be inserted to ensure proper execution. To ensure the programmer is not unduly burdened by resolving these resulting data dependencies, virtually every processor uses an interlock system to transparently insert the appropriate number of stall cycles.

Although hardware will ensure correct execution, the performance of the architecture is highly dependent on the dependencies in the instruction stream. By Little's Law, we see that the ILP (concurrency) demanded by a pipelined architecture is equal to the number of stages (latency). Data forwarding from the execution stage to prior ones limits the demanded concurrency to the latency of the execution unit. To achieve peak performance, the programmer (or compiler) is responsible for explicitly and continually expressing this ILP in the instruction stream. When ILP is deficient, performance is correspondingly reduced.

Out-of-Order Processors: Unfortunately, the compile-time analysis of data dependencies and latencies has often proved unwieldily and limits optimized binaries to one microarchitecture. To that end, system architects have endeavored to extend the processor to dynamically inspect the instruction sequence and discover instruction-level parallelism. The resulting "out-of-order" architecture can be implemented several ways. Generally, additional components are added to the standard pipelined architectures: a register renaming unit, reservation stations, and a reorder buffer. Upon decoding an instruction, a temporary "rename" of the target register is created. The renamed register effectively becomes the real register for purposes of resolving data dependencies for subsequent instructions. Instructions are queued in reservation stations until their operands are available. After computation is completed, results are queued into a reorder buffer. In effect, execution is decoupled from the rest of the processor. Although programs must still exhibit instruction-level parallelism to attain peak performance, dedicating reservation stations for different classes of instructions (functional units) allows out-of-order processors to overlap (and thus hide) instruction latencies.

Superscalar Processors: Once out-of-order processors were implemented with multiple functional units and dedicated reservation stations, system architects attempted to process multiple instructions per cycle. Such a bandwidth-oriented approach potentially could increase peak performance without increasing frequency. In a "superscalar" processor, the processor attempts to fetch, decode, and issue multiple instructions per cycle from a sequential instruction stream. Collectively, the functional units may complete several instructions per cycle to the reorder buffer. As such, the reorder buffer must be capable of committing several instructions per cycle. Ideally, this architecture should increase peak performance. However, the complexity involved in register renaming with multiple (potentially dependent) instructions per cycle often negatively impacts clock frequency and power costs. For this reason, most out-of-order processors are limited to four-way designs (i.e., four instructions per cycle). Moreover, by Little's Law, the latency–bandwidth product increases in proportional to the number of instructions that can be executed per cycle — only exacerbating the software challenge.

VLIW Processors: Discovering four-way parallelism within one cycle has proven to be rather expensive. System architects can mitigate this cost by adopting a very-long instruction word (VLIW) architecture. The VLIW paradigm statically (at compile time) transforms the sequential instruction stream into N parallel instruction streams. The instructions of these streams are grouped into N-instruction bundles. On each cycle, a VLIW architecture fetches a bundle and resolves any data dependencies with previous bundles. As the instructions within a bundle are guaranteed to be parallel, the complexity of wide issue is limited. VLIW architectures may be out-of-order (to mitigate unpredictable latencies) or in-order to maximize efficiency. Like superscalar, VLIW simultaneously improves peak performance while challenging the ability to attain it (because of Little's Law).

SIMD/Vector Processors: Superscalar and VLIW processors increase instruction-level parallelism. Scientific applications often apply the same instruction to multiple data elements (a vector). As such, architects re-designed instruction sets to be data-parallel. The resultant single instruction, multiple data (SIMD) or vector instruction sets apply the same operation to multiple data elements — Figure 2.2. Doing so reduces the per-operation instruction fetch and decode overhead, since the SIMD paradigm mandates all operations in a SIMD instruction be data-parallel. One should note that the number of operations expressed by a SIMD instruction need not match the number of execution units. For example, if a SIMD instruction expresses 16-way data level parallelism is executed on a machine with four FPUs, the instruction will require four cycles to complete. When the number of cycles a SIMD instruction requires to complete exceeds the latency of the functional units, the programmer need only express data-level parallelism to fully utilize a processor.

There are two camps within the SIMD design world. In the first, the length of vector operations is encoded in the instruction set. All such SIMD instructions always express N-way parallelism. Such an approach requires instruction-set changes (longer SIMD instructions) to improve processor performance. In the second, the length of the vector operations is implementation-dependent. That is, two different processors implementing the same vector in-

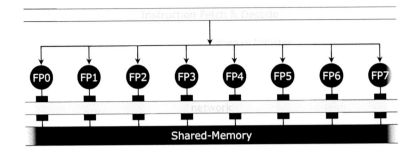

FIGURE 2.2: Single Instruction, Multiple Data (SIMD) architecture.

struction set may have not only different numbers of functional units, but also different vector lengths. By encoding the implementation-specific maximum vector length (MVL) into a register, binary compatibility may be maintained. A program simply reads this register, and strip-mine loops based on the vector length. This approach allows vector designs to gracefully expand to handle ever longer vectors.

SIMD architectures are only effective when the compiler infrastructure can encode vast data-level parallelism into the instruction stream. The Fortran-90 (and above) standard includes array syntax that potentially facilitates such SIMD operations. For example, the Fortran code

```
double precision a(10:10), b(10:10), c(10:10)
...
c = cos (a) + sin (b)
```

performs the indicated operations over the entire 100 element-long arrays. As such, SIMD parallelism has been made explicit to the compiler. Nevertheless, effective "SIMDization" has been a challenge for other languages.

Multithreading: Superscalar processors are extremely power-hungry and require abundant implicit parallelism in the instruction sequence. Conversely, although VLIW and vector architectures are typically much more efficient, they require substantially more software infrastructure to generate high-quality (i.e., highly efficient) code. To that end, delivered performance on these architectures often suffers. Multithreaded architectures virtualize hardware so that it appears from the programmer's perspective to be N distinct processors. In reality, there is one core, but N thread contexts that hardware (rather than software) time multiplexes between. Clearly, multithreading is a style of parallel architecture. Almost invariably, multithreading is implemented using the shared-memory parallel architecture paradigm.

There are three main styles of multithreading that are principally delineated by the granularity of how often hardware switches tasks. In coarse-grained multithreading, the processor only switches to different threads after issuing a long-latency operations. Between such events, the full capability of the processor is dedicated to one thread. Long latency is a sliding scale and over the years has meant page faults, floating-point divides, or memory operations. Regardless, the justification is that while that long-latency operation is being processed, the processor may rotate through all other thread contexts. As such, latency is hidden, and forward progress is made. There are two types of fine-grained multithreading. In vertical multithreading (VMT), the processor switches threads every cycle. On that one cycle, all issue slots are available to only one thread. Before selecting a thread, the processor will determine if it's next instruction is available to execute. In simultaneous multithreading (SMT), a superscalar processor divvies up its issue width among a number of threads. For this reason, SMT often makes more efficient utilization of superscalar processors, since, with enough threads, the processor is usually able to fill all issue slots. It should be noted that VMT and SMT, instructions are fetched, decoded, and buffered prior to the determination of which thread to run.

The advantage of fine-grained and simultaneous multithreading is that these designs are extremely friendly to programmers and the current state of compiler technology. With sufficient threads, the performance characteristics of each thread looks like the unpipelined processors of yesterday. That is, although hardware may still demand a sizable operation concurrency (ILP), the concurrency demanded per thread ($\frac{ILP}{TLP}$) can be quite low. For this reason, programmers, or more often compilers, are rarely required to express any instruction-level parallelism. The clear downside of this design decision is that programmers are required to create thread- or process-parallel programs.

Multicore: Multithreading increases efficiency and programability, but it does not necessarily increase peak performance. Recently, architects have endeavored to exploit Moore's Law by simply placing multiple processors on-chip. Just as we saw for multithreading, almost universally the shared-memory parallel architecture is used to interconnect the cores of a multicore processor. From a programming perspective, these multicore processors are indistinguishable from their SMP brethren. Although the processing cores themselves are no faster than their discrete counterparts, optimized parallel applications need only communicate on chip and may thus improve performance. However, using de-facto supercomputing programming paradigms, multicore only delivers better efficiency and performance per unit cost, not better raw performance per core.

Single Instruction, Multiple Threads (SIMT): Recently, NVIDIA, a graphics processing unit (GPU) manufacturer, attempted to merge multicore and SIMD in what they call single instruction, multiple threads (SIMT). Unlike the traditional SIMD architecture, in which instructions express SIMD semantics, the NVIDIA G80 and newer architectures group multiple cores into multiprocessors. In such an architecture, cores, and the threads they execute are nominally independent. However, there is only one fetch and decode unit per multiprocessor. As such, when threads execute independently (decoupled), they stall seven out of eight cycles, because of contention for the instruction fetch unit. However, through some software and programming model help, when all threads wish to execute the same instruction, hardware may broadcast that instruction to all cores and they may attain peak performance. Like SIMD, SIMT amortizes the hardware overhead for instruction fetch and decode. However, in the presence of scalar or non-SIMD code paths, SIMT may execute more efficiently.

2.4   Memory Architecture

Moore's Law has enabled a level of integration in computer memory devices that would have been unimaginable fifty years ago. The same technology that has enabled multiple processors to be integrated on a chip has enabled multigigabit memory chips. Unfortunately, Moore's Law says nothing about latency, the relative performance of DRAM and SRAM circuits, or the chip pinout technology. Indeed, memory technology vendors compete principally on the basis of density and capacity, not speed. As a result, the clock frequency of CPUs has greatly outpaced that of capacity storage technologies like DRAM. Measured in terms of processor clock cycles, the latency to main memory has increased from one cycle to nearly 1000 core cycles. Moreover, the bandwidth into a chip (pins and frequency) has not kept pace with CPU performance. Over the last 20 years, the resulting memory wall has become the largest impediment to sustained processor performance. Thus, architects have toiled unceasingly to devise clever means of mitigating it. In doing so, they have both transformed the abstract machine representation and attempted to exploit alternate paradigms to ensure memory performance does not mask potential core performance.

2.4.1   Latency and Bandwidth Avoiding Architectures

In a world where performance is increasingly limited by the discrepancy in CPU and memory performance, we may dramatically improve performance by trying to avoid or eliminate both memory latency and the volume of memory traffic. Architects have continually evolved memory architectures to that end. **Accumulators and Specialized Registers:** When on-chip (SRAM) memory was at a premium, but off-chip memory was (relatively) fast and cheap, most processors implemented very few registers. Some early architectures instantiated only an accumulator register. Such architectures typically implemented a register/memory architecture, in which computations involved the accumulator and a memory location (acc←acc OP memory). Over time, additional registers were added, although often they were used for specialized operations (memory addressing, loop counts, stack pointers, condition codes, etc.). In a similar way, instructions were often encoded to use one particular register. Today, we see the vestiges of these specialized registers in AX/BX/CX/SP registers of Intel-x86-based processors, although Moore's law has allowed them to grow from eight-bit registers to 64-bit registers.

General Purpose Register Files: Eventually, most early architectures standardized their registers into what is today known as a general purpose register file. In essence, this is a small on-chip memory of perhaps 32 words, which functions as an immediately addressed memory. Programs (via compilers) may use it to exploit temporal locality of the variables, and thereby eliminate some of the stall cycles for main memory accesses. Instruction sets have evolved to so that they may reference any one, two, or three of these registers as operands or destinations for computation, address calculation, or stack management. In so doing, they created both an opportunity and challenge for programmers and compilers, namely register allocation. After decades of research, compilers do a pretty good job scheduling registers and instructions.

Some architectures (MIPS, POWER, x86) chose to split their register files into floating-point/multimedia registers and (integer) general purpose registers, since it is relatively rare integer and floating-point data are used for the same computation. A more profound change was the adoption of load-store architectures. No longer can computational instructions refer to memory locations. Rather there is a clean separation between memory operations designed to transfer data between main memory and the register file, and computational instructions that operate on data in the register files.

Caches: As the gap between register and main memory latencies and bandwidth continued to widen, it became apparent that a new paradigm was required to bridge the memory–processor gap. The random and unpredictable memory access patterns effectively prevented compilers from exploiting locality on granularities larger than a basic block. The solution was to exploit Moore's law to create a hardware-controlled, demand-based cache of main memory. The cache would create copies of requested memory addresses and keep them on-chip. On subsequent references, the cache would intercede and return its copy of the data rather than generating a memory transaction. As capacity is finite, hardware must manage which address it "caches" and which which should be returned to main memory. Architects have endeavored to create extremely efficiency caches, including multi-level caches and victim caches, as well as selecting the optimal associativity and memory block size.

In general, caches act as a filter on main memory accesses. Hardware subsumes the responsibility of managing and exploiting spatial and temporal locality. In doing so, they eliminate some of the latency and memory traffic. This reduces the number of stalls seen by the processor core thereby keeping it busy performing computation. Broadly speaking, caches have been successful in enhancing productivity and performance. No programming effort is required to exploit them, and they almost always provide at least some benefit, often a substantial benefit. Unfortunately, as commodity cache architectures are optimized for consumer workloads, many scientific programs, with their giant working sets and power-of-two problem sizes, generate pathological cases and deliver suboptimal performance. For this reason, programmers in these fields are forced to work around the hardware's heuristics in an attempt to elicit better cache behavior.

Local Stores: Periodically, architects have experimented with local stores or scratchpad memories. In essence, these are disjoint address spaces or an address range carved out of the physical address space. In some cases, such as the Cell architecture from Sony, Toshiba and IBM, which was first employed in the Sony Playstation 3, each core has its own private local store. In other systems, such as various GPUs, a local store may be shared by a cluster of cores. Because the size of these memories is sufficiently small, they may be placed on chip using SRAM or embedded DRAM. Thus, rather than having the programmer try and change loops to elicit better cache behavior, he/she simply blocks his/her code from the beginning. Data for which temporal locality is desired is copied from main memory to the local store from where it may be referenced for computation. Unlike caches where there is a long latency tail associated with access time, the average memory access time for a local store is tightly bound. Moreover, as the programmer controls its contents, local stores do not suffer from cache capacity or conflict misses. The peak energy efficiency and hardware overhead per byte stored for local stores is much better than caches. Nevertheless, local stores are a disruptive technology, since they require programmers to rewrite code to explicitly exploit spatial and temporal locality.

2.4.2 Latency Hiding Techniques

Thus far, we've examined paradigms architects have employed to avoid memory latency or traffic. However, existing and novel paradigms can be used to simply hide latency rather than eliminate it. These methods must fill these latency cycles with other useful work. Unfortunately, today, with main bandwidth figures exceeding 20 Gbyte/s and latencies in approaching 200 ns, Little's Law demands processors express 4 Kbytes of concurrency or memory-level parallelism (MLP) to the memory subsystem. No one should underestimate the consternation this has produced among architects and programmers alike.

Out-of-Order Execution: Out-of-order processors can keep dozens of memory instructions in flight. Unfortunately, each memory instruction only loads or stores one word (64 bits). As a result, the total concurrency exploited by out-of-order execution may be less than 400 bytes — only 10% of that typically demanded by Little's Law. Today, out-of-order execution can only hide the latency to caches.

Software Prefetching: Often programs and thus programmers know the memory addresses long before the data is actually needed. Using an orphaned load instruction to warm up the cache may limit the instruction-level parallelism, but will not satisfy Little's Law, since the out-of-order memory unit has limited concurrency. However, architects can create instructions that fall right through the memory unit and immediately request data. These software prefetch instructions have no target register and thus do not occupy slots in the reservation stations or reorder buffers. In essence, their execution is decoupled from the rest of the core. By properly scheduling them (number and location) programs may completely hide memory latency. In other words, the cache is used to satisfy the concurrency demanded by Little's Law.

Unfortunately, in practice, software prefetch have limited applicability. It is difficult to express this level of MLP in the more challenging programs. As such, their integration is a very user-intensive and unproductive task.

Hardware Stream Prefetching: Architects quickly realized that they could incorporate this prefetching concept directly into hardware. After detecting a sequence of cache misses, the resultant hardware stream prefetcher would speculatively begin to prefetch cache lines that the miss sequence suggests will be demand requested in the future. By prefetching sufficiently far ahead, memory latency can be hidden. Therefore, hardware prefetching has been gracefully incorporated into virtually every high-end processor and readily exploited by most scientific computations. However, there are a few caveats that programmers should adhere to in order to maximize their effectiveness. First, being outside the core, prefetchers are not privy to address translations. For this reason, they may only prefetch data until they cross a page boundary (where an additional translation would be required). Second, until recently, prefetchers could only detect unit- (or, more properly, sub-cache line) strides. Thus, such systems perform best on long unit-stride or low-stride accesses.

DMA: Since some programs are not amenable to long unit-stride access patterns, software prefetch has been resurrected as direct memory access (DMA). DMA will copy a block of memory from one address to another. With sufficiently long DMAs, Little's Law is easily hidden. Recently, processors like Cell have evolved DMA to include gather-scatter support. In essence, the programmer can specify a list of DMA's (as large as 16 KByte or small as a 16 bits) to perform. Unfortunately, like so many similar methods, DMA requires substantial programming effort to exploit. For the simplest cases, it provides no better performance or efficiency than hardware prefetching. For the more complex cases, we see that multicore and multithreading can more productively satisfy Little's Law.

Multithreading and Multicore: Placing multiple cores (each with multiple threads) on chip is common today. With respect to Little's Law, multithreading and multicore is very friendly, since 4 KByte (or 512 double-precision words) data can now be distributed among a dozen out-of-order CPU cores or a thousand in-order GPU threads. For this reason, multicore processors can be quite adept at saturating memory bandwidth.

2.5   Network Architecture

Thus far, we have addressed two of the three components shown in Figure 2.1: processor architecture and memory architecture. The third component, namely network architecture, is a description of how the elements of the system are interconnected and details of how they communicate. Indeed, the term *network architecture* encompasses all levels of communication, from global connections (between nodes in a supercomputer), to the inter-processor network on a node, to the network-on-a-chip that connects cores in a multi-core processor. Just as differences in scaling technologies have pushed memory architecture to the forefront today, the advent of manycore and billion-way parallel machines are likely to raise interconnect and communication systems to be the pre-eminent challenge facing the high-performance community in the coming years.

Broadly speaking, network architecture can be categorized by topology and switching technology. Networks are composed of interface chips (NICs) connected by links. Processors (and memory in the case of shared-memory architectures) connect to the interface chips. Links can be cables (optical or copper), wires on a printed circuit board, or metal routes on a chip. Network transactions, or messages, are sent to the NIC which then arbitrate and send them to their destinations. In the case of shared-memory architectures, upon receipt of a load message, memory will respond (via its interface to the network) with the desired data.

2.5.1   Routing

There are two major types of networks: point-to-point networks and packet-switched. The principal distinction is that the former results in systems in which communication between arbitrary processors is managed in software by transmitting data between directly connected processors while in the latter, data and an "address" of the destination processor are packaged into packets. Hardware then manages the delivery to a processor indirectly connected. In doing so, hardware "routes" these packets along a series of links between connected processors making a "hop" at each step. Among historical computer systems, the Illiac IV is a notable example of a point-to-point chordal network. Today, such point-to-point networks are increasingly relegated to ever small, on-chip scales.

Packet-switched networks are more common at the present time, both for Internet-type applications and also for interprocessor networks in high-performance computer systems. One key figure of merit is how many "hops" are required, on average, for communication between nodes. Since each "hop" entails some delay, the average number of hops is a good index of average delay.

2.5.2   Topology

The term *network topology* describes the graph connectivity of nodes. In considering what topology to employ, system architects must weigh the costs and benefits associated with each possible design to determine which is appropriate for the desired target scale and performance.

As mentioned above, the simplest network is a fully connected network in which each node is directly connected to every other node. Although only one hop is required to reach any node, the number of links grows quadratically with the number of nodes (in particular, $n(n + 1)/2$ links). This relegates such designs to very small scales—beyond roughly eight nodes these networks become impractical.

An alternate design is based on a *hypercube*. In this design, nodes sit at the vertices of an n-dimensional hypercube, which can be thought of as an n-dimensional generalization of a cube. The *dimension* of a hypercube is $log_2 P$, where P is the number of nodes. One advantage of a hypercube architecture is that although the number of hops is equal to the dimension, the number of links is only $P/2 \cdot log_2(P)$. This cost grows at a fairly modest rate, and thus reasonable network performance can be achieved even for fairly large configuration. One difficulty with the hypercube topology is that the two-dimensional planar geometries of chips and the three-dimensional universe we live in complicate the manufacture of high-dimensional hypercubes. Thus while they were quite popular for many years, today they are fading in favor of other topologies.

Today, *mesh* and *toroidal* geometries are the most commonly used topologies for connecting multiple processors. In either topology, nodes fill an n-dimensional rectahedral volume. For example, 2048 processors might be contained in a 4-D mesh of dimensions $4 \times 8 \times 8 \times 8$. Each processor (or, more accurately, network interface card) is connected with eight neighbors—left, right, above, below, in front, behind, as well as those in the fourth dimension.

A torus differs from a mesh in how nodes on the boundary are connected. In a mesh, nodes on the boundary lack one or more neighbors. In a torus, links wrap around to the far side of the network. In other words, a torus topology has a periodic boundary. In either case, the number of links is proportional to the product of dimensionality and nodes, but the number of hops grows in proportion to the n-th root of the number of nodes. This flexibility in performance versus cost allows system architects to tailor the dimensionality and often the dimensions themselves to achieve an optimal design.

Two parameters that are often associated with a network are *diameter* and *bisection bandwidth*:

- Diameter: This is the maximum distance, in hops, between any two nodes in the system.

- Bisection bandwidth: This is the minimum bandwidth between two halves of the system, where the minimum is taken over all ways of cutting the system into two pieces. This parameter is particularly important

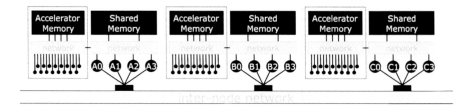

FIGURE 2.3: Accelerator-based Parallel Architecture.

for applications such as those dominated by global fast Fourier transform operations, since such codes typically involve numerous all-to-all exchanges of data.

2.6   Heterogeneous Architectures

Over the years, we have observed that disruptive technologies emerge periodically. These technologies cannot be directly incorporated into existing products due to technological, economical, or programming challenges. So architects and integrators have offered heterogeneous solutions that pair existing solutions with these novel accelerator architectures. If these architectures that employ novel accelerators prove valuable, then over some period of time, they find their way into standard CPU architectures.

Recently, we have observed processors like GPUs and Cell being integrated into large scale supercomputers using this accelerator model. Figure 2.3 shows what these machines often look like. Interestingly, at this granularity, both the Cell (Roadrunner at Los Alamos) and GPU accelerator architectures look quite similar. We see a conventional four-processor CPU node has been paired with a 16-processor accelerator. Invariably, these accelerators have their own private (and often high-cost/performance) memory. That is, the CPU cores (A0..A3) may access their memory via load and store instructions, the accelerator cores may access their memory via load and store instructions, but CPU cores cannot directly load from accelerator memory. In practice, to exploit this accelerator, data must be copied (via a DMA) from the CPU (node) shared memory to the accelerator memory where the accelerator cores may access it. As the overhead of this DMA is high, programmers must explicitly manage data transfers to attain high temporal locality in accelerator memory lest the performance potential be squandered.

Generally, only the CPU cores may initiate inter-node communication. This produces rather complex communication patterns in which data might need to be transferred from accelerator to CPU, from CPU to a remote CPU, and finally from that CPU to its accelerator. Thus, programmers must amortize this tripling in overhead via larger messages and coarser parallelization.

2.7 Summary

We have seen that modern high-performance computer systems span a wide range of designs. Currently systems in widespread usage range from custom-built clusters of personal computers, using inexpensive commodity networking components for communication, to systems involving networks of commercial servers, to custom-designed systems such as the IBM Blue Gene system. At the same time, several large heterogeneous system have been deployed that employ relatively conventional components, such as personal computer modules, together with game- and graphics-based accelerators.

At the same time, some very large systems are being deployed. The high-performance computing world has already passed the 1 Pflop/s mark (i.e., 10^{15} floating-point operations per second). By the time this book appears in print, it is likely that some 10 Pflop/s will already be deployed, and exascale systems (1 Eflop/s, i.e., 10^{18} floating-point operations per second) are already being sketched.

The convergence of these two trends—peta- and exascale computing together with an unprecedented proliferation of disparate architectures— portends a future in which the task of extracting optimal performance from a scientific application will be more challenging than ever before.

In the chapters that follow, we will see numerous ways in which state-of-the-art work in the performance arena is meeting this challenge with new tools, techniques and research.

2.8 Acknowledgments

This work was sponsored by the Performance Engineering Research Institute, which is supported by the SciDAC Program of the U.S. Department of Energy, and by the ASCR Office in the DOE Office of Science, under contract number DE-AC02-05CH11231, Microsoft (Award #024263), Intel (Award #024894), and by matching funding through U.C. Discovery (Award #DIG07-10227).

2.9 Glossary

ALU: Arithmetic and Logic Unit is an execution (functional) unit that performs the basic integer or bit-vector operations.

CPU: Central Processing Unit is a somewhat ambiguous term that may refer to a processor core, processor chip, or an entire SMP. Typically CPUs are optimized for general purpose computing.

DLP: Data-Level Parallelism is a measure of the number of independent data items one operation may be applied to.

DMA: Direct Memory Access is an architectural paradigm designed to maximize memory-level parallelism by expressing memory operations as decoupled bulk transfers.

GPU: Graphics Processing Unit is a processor optimized for graphics computing.

ILP: Instruction-Level Parallelism is a measure of the number of independent (no data hazards) instructions in one instruction stream.

ISA: Instruction Set Architecture is a specification of programmer-visible registers, address spaces, and the instructions that operate on them.

MLP: Memory-Level Parallelism is a measure of the number of independent memory operations that may be injected into the memory subsystem. This may be a quantified as maximum, minimum, or average as calculated over a window of the instruction stream

Multithreading: is an architectural paradigm in which hardware efficiently virtualizes one core into several.

MVL: Maximum Vector Length is maximum number of data items an operation can be applied to in a SIMD archtiecture. This vector length may be encoded in the ISA or in an implementation-specific register.

NIC: Network Interface Chip.

PGAS: Partitioned Global Address Space is a programming paradigm in which the shared memory semantics are realized on a distributed memory architecture.

SIMD: Single-Instruction, Multiple-Data is the common architectural paradigm for exploiting data-level parallelism.

SIMT: Single-Instruction, Multiple-Threads is NVIDIA's approach to efficient exploitation of data-level parallleism while maintaining productive exploitation of thread-level parallelism.

SMP: Shared Memory Parallel is a parallel architecture paradigm in which multiple processing elements shared a common address space.

SMT: Simultaneous Multithreading is a multithreading architectural paradigm in which hardware partitions each cycle's instruction issue bandwidth among two or more threads. This is also known as horizontal multithreading.

TLP: Thread-Level Parallelism is a measure of the number of parallel instruction stream a program could support.

Vector: A SIMD-based architectural paradigm designed to exploit data-level parallelism. A purest definition of vector does not require the vector length to be fixed by the ISA.

VLIW: Very Long Instruction Word is an architectural paradigm that explicitly encodes instruction-level parallelism as a bundle of independent instructions.

VMT: In Vertical Multithreading, each cycle's instruction issue width is entirely allocated to one thread. As such, if visualized with time vertical and issue width horizontal, threads appear to be vertically stacked.

Chapter 3

Software Interfaces to Hardware Counters

Shirley V. Moore

University of Tennessee

Daniel K. Terpstra

University of Tennessee

Vincent M. Weaver

University of Tennessee

3.1   Introduction

Hardware performance counters are registers available on modern CPUs that count low-level events within the processor with minimal overhead. Hardware counters may also exist on other system components, such as memory controllers and network interfaces. Data from the counters can be used for performance evaluation and tuning. Each processor family has a different set of hardware counters, often with different names even for the same types of events. Models in the same processor family can also differ in the specific events available. In general similar types of events are available on most CPUs.

Hardware counters are a powerful tool, but widespread adoption has been hindered by scant documentation, lack of cross-platform interfaces, and poor operating system support. Historically hardware counters were used internally by chip designers and interfaces were not always documented or made publicly available. As application and performance tool developers began to discover the usefulness of counters, vendors started supporting the interfaces (SGI's perfex interface is an early example). These interfaces were platform-specific; an application or tool developer had to use a different interface on each platform. Typically special operating system support is needed to access the counters, and for some popular operating systems (such as Linux) this support was missing.

Linux support for hardware counters has been substandard for years. Until recently the default kernel did not support access to the counters; the kernel had to be specifically patched. Various counter patch sets were available, including the perfctr patches (developed by Mikael Pettersson of Uppsala University) and the perfmon and perfmon2 [129] projects (developed by Stéphane Eranian). In 2009 performance counter support was finally merged into the Linux kernel, as the Performance Counters for Linux (PCL) project (since renamed to simply "perf events").

The divergent nature of performance counters across various platforms, architectures, and operating systems has led to the need for a layer of abstraction that hides these differences. This is addressed by the Performance API (PAPI) [64] project at the University of Tennessee, which has developed a platform-independent interface to the underlying hardware counter implementations.

The remainder of this chapter explores the features, uses, and limitations of hardware performance counters, and describes operating system interfaces and the PAPI interface.

3.2 Processor Counters

The hardware counters available on CPUs expose the underlying architecture of the chip for program analysis. Various types of events are available, but care should be taken when extrapolating conclusions from the results.

Counters are a limited resource; in general only a few (typically from 2-5) events can be counted at a time. On modern CPUs there may be hundreds of available events, making it difficult to decide which events are relevant and worth counting.

Commonly available events include the following:

- Cycle count

- Instruction count

- Cache and memory statistics

- Branch predictor statistics

- Floating point statistics

- Resource utilization

Counts can be collected as aggregate totals, or else sampled. Sampling helps when counting more events than available counters, although this introduces some degree of error in the results. Counts can be restricted to only userspace or kernel events, as well as specifying whether monitoring should be per-process or system-wide.

3.3 Off-Core and Shared Counter Resources

In addition to the counters found on CPUs, various other hardware subsystems found in modern computers also provide performance counters. This includes hardware such as hard disks, graphical processing units (GPUs), and networking equipment.

Many network switches and network interface cards (NICs) contain counters that can monitor various performance and reliability events. Possible events include checksum errors, dropped packets, and packets sent and received. Communication in OS-bypass networks such as Myrinet and Infiniband is asynchronous to the application; therefore hardware monitoring, in addition to being low overhead, may be the only way to obtain important data about communication performance.

As processor densities climb, the thermal properties and energy usage of high performance systems are becoming increasing important. Such systems contain large numbers of densely packed processors that require a great deal of electricity. Power and thermal management are becoming critical to successful resource utilization [133]. Modern systems are beginning to provide interfaces for monitoring thermal and power consumption sensors.

Unlike on-processor counters, off-processor counters and sensors usually measure events in a system-wide rather than a process- or thread-specific context. When an application has exclusive use of a machine partition (or runs on a single core of a multi-core node) it may be possible to interpret such events in the context of the application. Alternately, when running a multi-threaded application it may be possible to deconvolve the system-wide counts and develop a coarse picture of single thread response.

Section 3.6 discusses extensions to the PAPI hardware counter library to handle off-processor counters. There is currently no standard mechanism to provide such performance information in a way that is useful for application performance tuning. New methods need to be developed to collect and interpret hardware counter information for shared resources in an appropriate manner. Research in underway in the PAPI project to explore these issues.

3.4   Platform Examples

Hardware performance counters are available for most modern CPU implementations. This includes not only the popular x86 line of processors, but also VLIW systems (ia64), RISC systems (Alpha, MIPS, Power and SPARC) as well as embedded systems (ARM, SH, AVR32). Unfortunately counter implementations vary wildly across architectures, and even within chip families. Below is a brief sampling of available implementations.

3.4.1   AMD

AMD performance counter implementation has remained relatively stable from the 32-bit Athlon processors up to and including current 64-bit systems. Each core supports four counters, with the available events varying slightly across models.

Performance counter measurements become more complicated as multiple cores are included per chip package. Some measurements are package-wide as opposed to being per-core. These values include counts such as per-package cache and memory controller events. The L3 cache events in AMD Opteron quad-core processors are not monitored in four independent set of hardware counter registers but in a single set of registers not associated with a specific core (often referred to as "shadow" registers). When an L3 event is programmed into one of the four per-core counters, it gets copied by hardware to the shadow register. Thus, only the last event programmed into any core is the one used by all cores. When several cores try to share a shadow register, the results are not clearly defined. Performance counter measurement at the process or thread level relies on the assumption that counter resources can be isolated to a single thread of execution. That assumption is generally no longer true for resources shared between cores.

Newer AMD processors also support an advanced feature known as Instruction Based Sampling (IBS) that can provide more information than usually provided with regular counters. This is described in detail in Section 3.7.2.

3.4.2 Intel

Intel processor counter implementations are model specific. The Pentium 4 has a complicated implementation that is completely different from anything before or after it. Most modern processors are based on an enhanced version of an interface similar to that available on the original Pentium Pro. Despite similarities at the interface level, the Core2, Atom, and Nehalem implementations vary in the available number of counters and which events are available. The best references for these are the Intel manuals [179].

More recent multi-core Intel processors (starting with Nehalem) have "uncore" counters that are shared among all the cores of the chip. These are different than the off-core AMD measurements as they are implemented separately from the core counters and count only at a system-wide level. On some systems this is further complicated by the availability of HyperThreading which further splits the counters. Early Intel dual-core processors addressed the issue of shared resources by providing SELF and BOTH modifiers on events. This allowed each core to independently monitor the event stream and to collect counts for its activity or for all activities involving that resource. With the introduction of the Nehalem (core i7) architecture, Intel moved measurement of chip level shared resources off the cores and onto the chips.

Intel also supports an advanced instrumentation feature, PEBS, described in Section 3.7.2.

3.4.3 IBM Blue Gene

IBM's Blue Gene series of machines show another approach to performance counters and the problem of measuring activities on shared resources. Blue Gene/L uses dual-core processors and Blue Gene/P uses quad-core processors. In both cases hardware counters are implemented in the UPC, a Universal Performance Counter module that is completely external to any core. In Blue Gene/P, for example, the UPC contains 256 independent hardware counters [288]. Events on each core can be measured independently, but the core must be specified in the event identifier. This can create great difficulty for code that does not know (or care) about on which core it is running. Another difficulty is that these counters can only be programmed to measure events on either cores 0 and 1, or cores 2 and 3, but not on a mixture of all cores at once.

3.5 Operating System Interfaces

The interface to performance counters varies by operating system. We will focus on the interfaces available under Linux; other operating systems have similar functionality.

3.5.1 Perf Events

In December of 2008, a new approach to monitoring hardware performance counters was introduced to the Linux community by Ingo Molnar and Thomas Gleixner. This interface was rapidly accepted by Linus Torvalds and incorporated into the mainline Linux kernel as of version 2.6.31. This approach was originally informally called "Performance Counters for Linux" (PCL) but recently the scope of the project has been broadened and it now goes by "Perf Events." The speed with which this code was accepted into the Linux kernel was startling to many in the community, as perfctr and perfmon had both met with intense resistance to inclusion in the kernel for many years. The code base is evolving rapidly, incorporating counter grouping capabilities, virtual counters for metrics such as page faults and context switches, and inheritance capabilities for monitoring performance events in child processes. Unfortunately the code is still not as mature and feature-rich as the previous perfmon2 and perfctr projects.

Perf Events introduces a single system call as an entry point to a single counter abstraction. The call, `sys_perf_counter_open`, returns a file descriptor that provides access to a hardware event. A continuing point of contention between Perf Events and other approaches is that Perf Events incorporates all information about counters and events into the kernel code itself, adding a significant amount of descriptive code to the kernel. This is a significant departure from tools such as perfmon that put as little knowledge about specific events as possible in the kernel and provide event description information in user space libraries. Embedding such data in the kernel can cause difficulties for user-space tools to make decisions about what events can be counted simultaneously and for vendors to provide updates without resorting to patching the kernel or waiting on new kernel releases. It is likely that some combination of user-space library and kernel level infrastructure will continue to exist after Perf Events becomes part of a released Linux kernel.

3.5.2   Perfmon2

The perfmon2 project was the leading contender for inclusion in the Linux kernel before the surprise emergence of Perf Events. The development of perfmon2 has slowed now that Perf Events has been merged, and parts of the infrastructure (such as the libpfm4 library) have been adapted so that they can work with the Perf Events subsystem.

The goal of the perfmon2 project is to design a generic low-level performance monitoring interface to access all hardware performance monitoring unit (PMU) implementations and support a variety of performance tools. Although PMUs exhibit considerable variations between process architectures and even within the same processor architecture, the perfmon2 developers aimed to exploit some common characteristics. For example, all modern PMUs use a register-based interface with simple read and write operations. A PMU commonly exports a set of control registers for configuring what is to be measured. Perfmon2 developers aimed to support the diverse needs of performance tools. Some tools simply collect counts of events while others sample to collect profiles. Some collect measurements for a single thread (per-thread) while others measure the entire system (system-wide).

The interface consists of the perfmonctl system call. A system call was preferred over the device driver model because it is built in and offers greater flexibility for the type and number of parameters. The system call also makes it easier for implementations to support a per-thread monitoring mode which requires saving and restoring the PMU state on thread context switch. The entire PMU state is encapsulated by a software abstraction called the perfmon **context**.

The PMU hardware interface is abstracted by considering it to be comprised of two sets of registers: a set of control (PMC) and a set of data (PMD) registers. The interface exports all PMU registers as 64-bit data, even though many PMU implementations have fewer. The interface does not know what each register does, how many there are, or how they are associated with each other. The implementation for each platform maps the logical PMC and PMD onto the actual PMU registers. All event specific information is relegated to the user level where it can be easily encapsulated in a library.

The interface supports time-based sampling (TBS) at the user level. The sampling period is determined by a timeout. When the timeout occurs, a sample is recorded. The user can implement the timeout using existing interfaces such as setitimer(). The interface also supports event-based sampling (EBS) for which the sampling period is determined by the number of occurrences of an event rather than by time. This mode requires the PMU to be capable of generating an interrupt when a counter overflows.

3.5.3   Oprofile

Oprofile [213] is a low-overhead profiler for Linux which is capable of using hardware performance counters to profile processes, shared libraries, and the kernel (including device drivers). Oprofile is modeled after Compaq/HP's Digital Continuous Profiling Infrastructure for the Alpha processor and was first introduced in 2002 for Linux 2.4 kernels.

Oprofile does not modify the kernel, but rather uses a daemon to monitor hardware counter performance across the entire system. It works by allowing the user to specify a hardware event to be monitored and an event threshold to be counted. It relies on interrupt-on-overflow to statistically profile where events are occurring in the system image. Using debug symbol tables, it can then map addresses back to source code to expose where the system is spending large fractions of time. More recent versions can also provide call-graph profiles.

3.6   PAPI in Detail

The PAPI library [64] was originally developed to address the problem of accessing the processor hardware counters found on a diverse collection of modern microprocessors in a portable manner. The original PAPI design consists of two internal layers: a large portable layer optimized for platform independence and a smaller hardware specific layer containing platform dependent code. By compiling and statically linking the independent layer with the hardware specific layer, an instance of the PAPI library can be produced for a specific operating system and hardware architecture.

3.6.1   Extension to Off-Processor Counters

Modern complex architectures have made it important to be able to access off-processor performance monitoring hardware. The current situation with off-processor counters is similar to the situation that existed with on-processor counters before PAPI. A number of different platform-specific interfaces exist, some of which are poorly documented or not documented at all. PAPI, as of version 4.0, has been extended to encompass off-processor counters and sensors. Version 4.0 (called Component-PAPI, or PAPI-C) is a complete redesign of the PAPI library. The one-to-one relationship between the platform independent layer and a hardware specific layer has been replaced by a one-to-many coupling between the independent framework layer and a collection of hardware specific components. Using PAPI-C, it is now possible to access counters from different subsystems simultaneously, for example from processor, network, and thermal sensor components.

```
Available events and hardware information.
-------------------------------------------------------------------------
PAPI Version                : 4.0.0.2
Vendor string and code      : GenuineIntel (1)
Model string and code       : Intel Core i7 (21)
CPU Revision                : 5.000000
CPUID Info                  : Family: 6  Model: 26  Stepping: 5
CPU Megahertz               : 2926.000000
CPU Clock Megahertz         : 2926
Hdw Threads per core        : 1
Cores per Socket            : 4
NUMA Nodes                  : 2
CPU's per Node              : 4
Total CPU's                 : 8
Number Hardware Counters    : 7
Max Multiplex Counters      : 32
-------------------------------------------------------------------------
The following correspond to fields in the PAPI_event_info_t structure.

      Name        Code     Avail Deriv Description (Note)
PAPI_L1_DCM   0x80000000   Yes   No    Level 1 data cache misses
PAPI_L1_ICM   0x80000001   Yes   No    Level 1 instruction cache misses
PAPI_L2_DCM   0x80000002   Yes   Yes   Level 2 data cache misses
PAPI_L2_ICM   0x80000003   Yes   No    Level 2 instruction cache misses
PAPI_L3_DCM   0x80000004   No    No    Level 3 data cache misses
PAPI_L3_ICM   0x80000005   No    No    Level 3 instruction cache misses
PAPI_L1_TCM   0x80000006   No    No    Level 1 cache misses
PAPI_L2_TCM   0x80000007   Yes   No    Level 2 cache misses
PAPI_L3_TCM   0x80000008   No    No    Level 3 cache misses
...
PAPI_DP_OPS   0x80000068   Yes   Yes   Floating point operations
PAPI_VEC_SP   0x80000069   Yes   No    Single precision vector/SIMD instructions
PAPI_VEC_DP   0x8000006a   Yes   No    Double precision vector/SIMD instructions
-------------------------------------------------------------------------
Of 107 possible events, 35 are available, of which 9 are derived.
```

FIGURE 3.1: Sample output from the `papi_avail` command.

Example components have been implemented in the initial release of C-PAPI for ACPI temperature sensors, the Myrinet network counters, and the lm-sensors interface. Work is ongoing on Infiniband and Lustre components.

3.6.2 Countable Events

Countable events in PAPI are either preset events (defined across architectures) or native events (unique to a specific architectures). To date, preset events have only been defined for processor hardware counters; all events on off-processor components are native events.

A preset event can be defined in terms of a single event native to a given CPU, or it can be derived as a linear combination of native events. More complex derived combinations of events can be expressed in reverse Polish notation and computed at run-time by PAPI. For many platforms, the preset event definitions are provided in a comma separated values file, `papi_events.csv`, which can be modified by developers to explore novel or alternative definitions of preset events. Because not all present events are implemented on all platforms, a utility called `papi_avail` is provided to examine the list of available preset events on the underlying platform. A portion of the output for an Intel Nehalem (core i7) processor is shown in Figure 3.1.

PAPI components contain tables of native event information that allow native events to be programmed in essentially the same way as preset events. Each native event may have a number of attributes, called unit masks, that can act as filters to determine exactly what gets counted. An example of a native event name with unit masks for the Intel Nehalem architecture is `L2_DATA_RQSTS:DEMAND_M_STATE:DEMAND_I_STATE`.

Attributes can be appended in any order and combination, and are separated by colon characters. Some components such as LM-SENSORS may have hierarchically defined native events. An example of such a hierarchy is `LM_SENSORS.max1617-12c-0-18.temp2.tempw_input`. In this case, levels of the hierarchy are separated by period characters. Complete listings of the native events available on a given platform can be obtained by running a utility analogous to `papi_avail`, called `papi_native_avail`.

The PAPI low-level interface has support for advanced features such as multiplexing and sampling.

3.7   Counter Usage Modes

Hardware performance monitors are used in one of two modes:

1. Counting mode (to collect aggregate counts of event occurrences.

2. Statistical sampling mode (to collect profiling data based on counter overflows).

Both modes have their uses in performance modeling, analysis, tuning, and in feedback-directed compiler optimization. In some cases, one mode is required or preferred over the other. Platforms vary in their hardware and operating system support for the two modes. Either mode may be derived from the other. On platforms that do not support hardware interrupt on counter overflow, timer interrupts can be used to periodically check for overflow and implement statistical sampling in software. On platforms that primarily support statistical profiling, event counts can be estimated by aggregating profiling data. The degree of platform support for a particular mode can greatly affect the accuracy of that mode.

3.7.1   Counting Mode

A first step in performance analysis is to measure the aggregate performance characteristics of the application or system under study. Aggregate event counts are determined by reading hardware event counters before and after the workload is run. Events of interest include cycle and instruction counts, cache and memory access at different levels of the memory hierarchy, branch mispredictions, and cache coherence events. Event rates, such as completed instructions per cycle, cache miss rates, and branch mispredictions rates, can be calculated by dividing counts by the elapsed time.

3.7.2   Sampling Mode

Sampling is a powerful methodology that can give deeper insight into program behavior than that obtained through simple counting. Different platforms implement sampling in differing ways.

Intel PEBS: Precise Event-Based Sampling (PEBS) is an advanced sampling feature of the Intel Core-based processors in which the processor directly records samples into a designated memory region. With PEBS the format of the samples is mandated by the processor. Each sample contains the machine state of the processor at the time of counter overflow. The precision of PEBS comes from the instruction pointer recorded in each sample being at most one instruction away from where the counter actually overflowed. This instruction pointer difference (known as "skid") is minimized compared to other methods of gathering performance data. Another key advantage of PEBS is that it minimizes overhead by only involving the operating system when the PEBS buffer is full. A constraint of PEBS is that it only works with certain events, and there is no flexibility in what is recorded in each sample. For instance, in system-wide mode the process identification is not recorded.

AMD IBS: Instruction-Based Sampling (IBS) is a sampling technique available on AMD Family 10h processors [30]. Traditional performance counter sampling is not precise enough to isolate performance issues to individual instructions. IBS, however, precisely identifies instructions responsible for sampled events.

There are two types of IBS sampling. The first, IBS fetch sampling, is a statistical sampling method that counts completed fetch operations. When the number of completed fetch operations reaches the maximum fetch count (the sampling period), IBS tags the fetch operation and monitors that operation until it either completes or aborts. When a tagged fetch completes or aborts, a sampling interrupt is generated, and an IBS fetch sample is taken. An IBS fetch sample contains a timestamp, the identifier of the interrupted process, the virtual fetch address, and several event flags and values that describe what happened during the fetch operation. The second type of sampling is IBS op sampling, which selects, tags, and monitors macro-ops as issued from AMD64 instructions.

Two types of sample selection are available. Cycles-based selection counts CPU clock cycles; the op is tagged and monitored when the count reaches a threshold (the sampling period) and a valid op is available. The second option is dispatched op-based selection which counts dispatched macro-ops. When the count reaches a threshold, the next valid op is tagged and monitored. With both options, an IBS sample is generated only if the tagged op retires. Thus, IBS op event information does not measure activity due to speculative execution. The execution stages of the pipeline monitor the tagged macro-op; when the tagged macro-op retires, the processor generates a sampling interrupt and takes an IBS op sample. An IBS op sample contains a timestamp, the identifier of the interrupted process, the virtual address of the AMD64 instruction from which the op was issued, and several event flags and values that describe what happened when the macro-op executed.

3.7.3   Data Address Sampling

Data address sampling captures the operand address of a memory instruction (load or store) upon the occurrence of a user-specified threshold for events such as data cache or TLB misses. Hardware support for data address sampling is available to some extent on several platforms, including IBM POWER5 [236], Intel Itanium [179], Sun UltraSparc [182], and AMD 10h processors (e.g., Barcelona and Shanghai) [30]. Hardware data address sampling is especially useful for analyzing performance on multicore architectures, since the interaction among cores typically occurs through the sharing of data [35].

3.8   Uses of Hardware Counters

Hardware performance counters have a variety of uses, including many which probably were not envisioned by the original designers. The most common use of performance counters is for performance optimization. Performance on a multicore system can often be evaluated by means of core counter metrics.

3.8.1   Optimizing Cache Usage

If each core has its own performance monitoring unit (as for example on the Intel Core 2 and Intel Core Duo), and if it is possible to count the number of unhalted cycles on each core, then the sum of this metric over all the cores divided by the total execution time converted to cycles (assuming dynamic frequency scaling is disabled) will yield a scaling metric. The scaling metric will range between 1 for completely serialized execution and n (where $n =$ number of cores) for perfect scaling. Poor scaling due to contention for a shared cache occurs when cache sharing results in increased cache line eviction and reloading. Faster and/or bigger caches may not help with this problem, since the issue is an effective reduction of useful cache size. Most PMUs provide an event that counts last level cache (LLC) misses. In the case of strong scaling, non-contention would mean that the number of LLC misses remains constant as the number of cores is increased. In the case of weak scaling, non-contention would mean that the number of misses per core remains constant. If poor scaling is occurring due to cache contention, it may be possible to use hardware counter data to pin down the exact cause of the problem. Possible causes include:

1. True or false sharing of cache lines.

2. Parallel threads evicting each others cache lines due to capacity limitations.

True or false cache line sharing can be detected on some processors by measuring a hardware event that counts the number of times an LLC cache miss is caused by the required cache line being in another cache in a modified state (known as a "hit modified" event – for example EXT_SNOOPS.ALL_AGENTS.HITM on the Intel Core 2). Comparing the number of these events between single-thread and multi-threaded executions can reveal when such problems are occurring. False sharing is likely to occur in parallelized loops for which the iterations have been distributed to different threads. Two threads evicting each others cache lines due to competition for space would result in an increase in the total number of cache lines being evicted but not an increase in the hit-modified metric.

3.8.2 Optimizing Bus Bandwidth

A resource that is always shared by multiple cores on a chip is the front side bus (FSB). The FSB bandwidth is affected by a number of factors, including bus frequency, processor frequency, cache coherency requirements, and configuration of hardware prefetchers. The only reliable way to determine bus bandwidth on a specific platform is to measure it with a synthetic benchmark. Once the bus bandwidth has been determined, the likelihood of FSB saturation during parallel execution can be predicted from hardware counter measurements taken during a single-threaded execution. If the metric of cache lines transferred per cycle can be determined, then the expected value for a multi-threaded execution can be obtained by multiplying by the number of threads. Some platforms provide hardware counter metrics that break down the various contributions to FSB utilization.

3.8.3 Optimizing Prefetching

Another problem that can occur on multicore systems, and that may become more of an issue as the number of cores is increased, is excessive cache

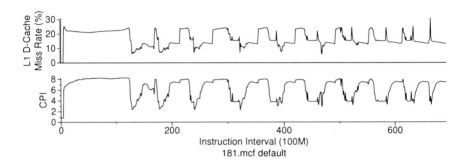

FIGURE 3.2: Performance counter measurements showing repeating program phase behavior in the `mcf` benchmark from SPEC CPU 2000.

line prefetching. While prefetching can substantially improve performance for some applications, it can degrade performance for others by aggravating bus bandwidth limits. There are different types of hardware prefetchers available (some of which can be enabled or disabled, depending on the system configuration) as well as software prefetching. Some platforms provide hardware counter events related to hardware prefetches. These counters can be used to determine the number of unread cache lines that hardware prefetchers have requested. A nonzero value for a hardware event associated with a hardware prefetcher will indicate that the prefetcher is enabled.

3.8.4 TLB Usage Optimization

Chip multi-processor (CMP) systems may have shared DTLBs or a DTLB dedicated to each core. In either case, the ideal behavior is for the number of DTLB misses per core to decrease proportionally with problem size in the case of strong scaling, and to remain constant with weak scaling. An increase in DTLB misses, which can usually be measured by a hardware counter event, may be the cause of poor scaling behavior.

3.8.5 Other Uses

Performance counters can be used in ways other than code optimization. This includes:

1. Visualization of program behavior (see Figure 3.2).

2. Computer simulator validation [113].

3. Power or temperature estimation [305].

4. Optimized operating system schedulers [132].

5. Multi-thread deterministic execution [266].

6. Feedback directed optimization [91].

3.9 Caveats of Hardware Counters

While performance counters are a powerful tool, care must be taken when using the results.

3.9.1 Accuracy

Although aggregate event counts are sometimes referred to as "exact counts," and profiling is statistical in nature, sources of error exist for both modes. As in any physical system, the act of measuring perturbs the phenomenon being measured. The counter interfaces necessarily introduce overhead in the form of extra instructions (including system calls), and the interfaces cause cache pollution that can change the cache and memory behavior of the monitored application. The cost of processing counter overflow interrupts can be a significant source of overhead in sampling-based profiling. Furthermore, a lack of hardware support for precisely identifying an event's address may result in incorrect attribution of events to instructions on modern superscalar, out-of-order processors, thereby making profiling data inaccurate.

Hardware designers use performance counters for internal testing, and they publish their use mainly as a courtesy. When asked they will admit that very little testing goes into the counters, and that caution should be exercised when using any counter (though commonly used ones such as retired instructions or cycle count can generally be counted on to be reliable).

Before using an event, it is a good idea to read the documentation for your chip and check the vendor supplied errata to make sure there are no known problems with the event. Also a few sanity checks should be run on small programs to make sure the results are what you expect. Modern chips have advanced features that can cause unexpected results, such as hardware prefetchers that can vastly change cache behavior compared to a traditional view of how caches behave.

Another issue to worry about is multiplexing. When enough counters are not available, the operating system might multiplex the counters. This adds a certain amount of inaccuracy to returned results.

3.9.2 Overhead

Even though performance counters have relatively low overhead, the overhead is enough that it can be measured. On a counter overflow an interrupt routine interrupts program flow, which can cause timing (and even cache) behavior changes that would not happen if counters were not being used.

Also, libraries such as PAPI add a certain amount of overhead each time a routine is called, which can show up in counter results.

3.9.3   Determinism

It is often expected that if the same program is run twice, the resulting performance results should be the same between runs. Even if a program is specifically set up to be deterministic, various factors can cause different counts [356]. The operating system can put parts of the program in different parts of memory run-to-run (address space randomization, a security feature of Linux that can be disabled). Some counts can vary due to operating system conditions and other programs running on the system. Branch prediction and cache rates can be affected by the behavior of other programs and by the operating system accessing memory. Especially on x86, some counters increment whenever a hardware interrupt happens, which can artificially increase counts.

These determinism problems are magnified on multi-core workloads, as the different interacting threads are scheduled in a non-deterministic fashion.

3.10   Summary

Hardware performance counters are a powerful tool that can offer insight into the underlying system behavior. As machines become more complex, interactions between applications and systems become non-intuitive and harder to understand. This is compounded by the move to large massively parallel systems. The use of performance counters and tools such as PAPI can greatly increase understanding and enable performance improvements of difficult codes.

3.11   Acknowledgment

This work was sponsored in part by the Performance Engineering Research Institute, which is supported by the SciDAC Program of the U.S. Department of Energy, and National Science Foundation grants NSF CNS 0910899 and NSF OCI 0722072.

Chapter 4

Measurement and Analysis of Parallel Program Performance Using TAU and HPCToolkit

Allen D. Malony

University of Oregon

John Mellor-Crummey

Rice University

Sameer S. Shende

University of Oregon

4.1 Introduction

Over the last several decades, platforms for high performance computing (HPC) have become increasingly complex. Today, the largest systems consist of tens of thousands of nodes. Each node is equipped with one or more multi-core microprocessors. Individual processor cores support additional levels of parallelism including pipelined execution of multiple instructions, short vector operations, and simultaneous multithreading. Microprocessor-based nodes rely on deep multi-level memory hierarchies for managing latency and improving data bandwidth to processor cores. Subsystems for interprocessor communication and parallel I/O add to the overall complexity of these platforms.

Achieving high efficiency on such HPC systems with sophisticated applications is imperative for tackling "grand challenge" problems in engineering and science. Without careful attention, data-intensive scientific applications typically achieve less than 10% of peak performance on the sort of microprocessor-based parallel systems common today [264]. As a result, there is an urgent need for effective and scalable tools that can pinpoint performance and scalability bottlenecks in sophisticated applications running on HPC platforms. A challenge for tools is that bottlenecks may arise from a myriad of causes both within and across nodes. A performance analysis tool should enable one to pinpoint performance bottlenecks and identify their underlying causes.

The performance of a parallel program on a large-scale parallel system depends primarily on how its work, load balance, communication, and I/O costs scale with the size of its input data and the number of processors used. Additionally, one must consider how well an application makes use of the resources available on the target platform. Understanding the performance of a parallel program on a particular system requires measurement and analysis of many different aspects of program behavior. Some aspects worthy of attention include work (i.e., operation counts), resource consumption (e.g., memory bandwidth, communication bandwidth, I/O bandwidth), and inefficiency (e.g., contention, delays, and insufficient parallelism).

For performance tools to analyze aspects of a parallel program's execution, they must be observable. For any given parallel system, not all aspects of interest will be measurable. For those that are, performance tools must gather metrics and correlate them to the program contexts in which they occur. It is also useful for tools to gather data about how execution behavior unfolds over time. While some aspects of program performance (e.g., load balance) can be diagnosed with simple summary information, others (e.g., serialization) require temporal or ordering information to identify them.

Performance tools must balance the need for information against the cost of obtaining it. Too little information leads to shallow and inaccurate analysis results; very detailed information is difficult to measure reliably and cumbersome to analyze. Unfortunately, there is no formal approach to determine, given a parallel performance evaluation problem, how to best observe a parallel execution to produce only the information essential for analysis. Furthermore, any parallel performance methodology ultimately is constrained by the capabilities of the mechanisms available for observing performance.

A parallel program execution can be regarded as a sequence of concurrent operations reflecting aspects of its computational activities. While the execution of instructions by processor cores is of course critical to high performance, other aspects of an execution are equally important, including activities in the memory system, interconnection network, and I/O subsystem. Operations of interest must be measured to be observed. *Events* are operations made visible to the measurement system. Not all operations can be monitored, and thus events only represent operations that are observable. In addition to encoding information about an operation, an event can be associated with the execution context in which it occurred and when it occurred in time.

A fundamental challenge for performance tools is how to gather detailed information to identify performance problems without unduly distorting an application's execution. The nature of performance measurements, the cost of collecting them, and the analyses that they support follow directly from the approach used to gather performance data. In this chapter, we discuss several different approaches for measuring the performance of parallel programs and then describe in detail two leading performance toolkits that use rather different approaches for performance measurement and analysis. The next section provides some precise definitions of terms that one must understand to appreciate the commonality and differences between the measurement approaches described in this chapter. The chapter continues with descriptions of the HPCTOOLKIT performance tools and the TAU Performance System. The section about each toolkit describes its features, outlines its capabilities, and demonstrates how it can be used to analyze the performance of large-scale applications. The chapter concludes with a brief discussion of open issues and challenges ahead for performance measurement and analysis on next generation parallel systems.

4.2   Terminology

Performance measurement relies on access to events. The terms *synchronous* and *asynchronous* are useful to distinguish the way an event is measured. A synchronous measurement of an event is one that is collected by code embedded in a program's control flow. An asynchronous measurement of an event is one that is collected by a signal handler invoked in response to an event occurrence.

Two orthogonal techniques used by performance tools are *instrumentation* and *sampling*. Performance tools for parallel systems typically augment programs with additional instructions (i.e., *instrumentation*) to monitor various aspects of program behavior. Most commonly, when people think of instrumentation, they think of using it to measure performance characteristics; this use of instrumentation is described in Section 4.3.1. However, instrumentation is also used by performance tools to observe certain significant operations, such as creation and destruction of processes and threads. Instrumentation of such operations is used by tools to ensure that necessary tool control operations, such as initialization or finalization of performance data for a thread, can be performed at appropriate times.

Monitoring all events in a program execution can be costly. To reduce the cost of monitoring, performance tools often use *sampling* to observe a fraction of program events. Most commonly, when people think of sampling in the context of performance tools, they think of *asynchronous sampling*, in which program control flow is periodically interrupted to associate an event or cost with the context in which it occurred. However, sampling need not be asynchronous. One can use sampling in conjunction with instrumentation injected into a program to make operations observable. For example, when instrumentation is encountered during program execution, the measurement code may choose to collect data only for some subset of the times it is executed. This style of synchronous sampling was used by Photon MPI [341], which used instrumentation of message passing routines, but used sampling to collect measurements for only a subset of messages transmitted. In general, this points out the important distinction between how events are made observable and what is measured about the events.

There are two basic ways to record performance measurements of a program execution: as a stream of events, known as a *trace*, or a statistical summary, known as a *profile*. Each of these approaches has strengths and weaknesses. Traces can provide detailed insight into how an execution unfolds over time. While this is a powerful capability, it comes at a cost: for large-scale parallel systems, detailed traces can be enormous. In contrast, profiles are relatively compact—their size does not grow linearly with program execution time—and they can be recorded with low overhead. As a statistical summary, a profile by definition lacks detail about how events unfold over time in an execution; for this reason, profiles cannot yield insight into a program's transient execution behavior. Despite their limitations, profiles can be surprisingly useful; for example, Section 4.4.3.3 describes how profiles can be used to pinpoint and quantify scalability bottlenecks with low overhead.

4.3 Measurement Approaches

In this section, we review two different measurement approaches used by performance tools: measurement based on instrumentation, and measurement based on asynchronous sampling.

4.3.1 Instrumentation

A straightforward approach to observing the performance of each thread in a parallel program execution is to directly instrument the program at multiple points with measurement code, known as a *probe* or *caliper*. Probes can be added to an application prior to compile time using source-level instrumentation, during compilation, or afterward using static [206] or dynamic [169] binary rewriting. The instrumentation process causes execution of probes to become part of the normal control flow of the program.

Probes can be used to measure different aspects of program performance. For instance, one can measure the time between a pair of probes inserted at procedure entry and exit to determine the duration of a procedure call. Modern microprocessors also contain hardware performance counters that can be used by probes to gather information. Using these counters, one can count events between probes that represent work (e.g., execution of instructions or floating-point operations), resource consumption (e.g., cycles or memory bus transactions), or inefficiency (e.g., cache misses or pipeline stall cycles). Typically, microprocessors support multiple counters so that several events can be monitored by probes in a single execution. Another use for probes is to directly observe semantic information of interest associated with software operations (e.g., bytes allocated, messages sent, or bytes communicated).

While probes measuring software activity such as messages or data allocation can provide exact information, using timers or hardware counters to measure activity between pairs of probes is subject to perturbation caused by execution of the probes themselves. It is important to understand that when relying on probes for measurement, only program code instrumented with probes is actually observed and measured. To avoid incomplete measurement of an execution, binary rewriting can be used to insert probes everywhere they are needed, even in libraries for which no source code is available.

Probe-based measurement is a robust approach that has provided the foundation for many parallel performance tools, including Scalasca [373], Vampir [251], mpiP [342], and IPM [307]. It is the primary technique used by the TAU performance system discussed in Section 4.5.

4.3.2 Asynchronous Sampling

An effective, unobtrusive way to monitor a program's execution is to use periodic sampling based on a recurring sample trigger. When a sample trigger fires, the operating system sends a signal to the executing program. When the program receives the signal, a signal handler installed in the program by a performance tool will record the program counter location where the signal was received and perhaps the current calling context and/or detailed information associated with the sample event (e.g., the effective address of a load or store). For parallel applications, one can set up sample triggers to monitor and signal each thread independently.

The recurring nature of a sample trigger means that typically a thread's execution will be sampled many times, producing a histogram of program contexts for which samples were received. Asynchronous sampling can measure and attribute detailed performance information at a fine grain accurately as long as (1) code segments are executed repeatedly, (2) the execution is sufficiently long to collect a large number of samples, and (3) the sampling frequency is uncorrelated with a thread's behavior. If these conditions are met, the distribution of samples will approximate the true distribution of the thread activity that the sample triggers are measuring. For modest sampling frequencies, the overhead and distortion introduced by asynchronous sampling is typically much lower than that introduced by probe-based instrumentation [143].

Sample-based measurement has proven to be a useful technique that is used by many performance measurement tools, including PerfSuite [201] and Sun's Performance Analyzer [321]. It is also the primary technique used in the HPCToolkit system discussed in Section 4.4.

There are two principal types of asynchronous sampling: *event-based sampling* and *instruction-based sampling*. The difference between these approaches lies in how samples are triggered. We briefly describe these two methods and the nature of their differences.

4.3.2.1   Event-Based Sampling (EBS)

Event-based sampling uses the occurrence of specific kinds of program events to trigger samples. Sample triggers based on different kinds of events measure different aspects of program performance. The classical approximation used by event-based sampling is that the program context in which a sample event is received is charged for all of the activity since the last sample event.

The most common type of event trigger is an interval timer. One can configure an interval timer to periodically interrupt a program execution with a specified sample period. The result of sampling an execution based on an interval timer is a quantitative record of where the program was observed to have spent time during its execution. For decades, this approach has been used by the Unix *prof* [333] and *gprof* [152] tools.

For over a decade, microprocessors have routinely provided hardware performance counters that can be used as event triggers. Using these counters, one can count events that represent work, resource consumption, or inefficiency. One can configure a counter to trigger an interrupt when a particular count threshold is reached. Typically, processors support multiple performance counters and each counter can be configured independently as a recurring event trigger; this enables multiple events to be monitored in a single execution.

A problem for event-based sampling using hardware performance counters on modern out-of-order microprocessors is that long pipelines, speculative execution, and the delay between when an event counter reaches a pre-determined threshold and when it triggers an interrupt make it difficult for performance monitoring hardware to precisely attribute an event to the instruction that caused it. As a result, information gathered using conventional event-based sampling on out-of-order processors is too imprecise to support fine-grain analysis of a program execution at the level of individual instructions.

An approach developed to address this problem on Intel processors is *precise event-based sampling* (PEBS) [315]. PEBS uses a special microassist and an associated microcode-assist service routine to precisely attribute an event sample. On today's Intel Core microarchitecture, only nine events support PEBS and PEBS can only be used with one of the programmable hardware performance counters [179]. Because of the imprecision of conventional event-based sampling, unless PEBS is used, one should not expect instruction-level accuracy in the attribution of sample events. Despite the fact that conventional event-based samples are inaccurate at the instruction level, event-based samples without instruction-level accuracy are not useless: they can provide meaningful aggregate information at the loop level and above, except for loops that execute only a trivial number of instructions each time they are encountered.

4.3.2.2   Instruction-Based Sampling (IBS)

To provide a more comprehensive solution for performance monitoring on out-of-order processors, Dean et al. [110] developed an approach for the DEC Alpha 21264 that they called *ProfileMe*, which is based on sampling instructions. Periodically, an instruction is selected for monitoring before it is dispatched. As a monitored instruction moves through the execution pipeline, special hardware records a detailed record of significant events (e.g., whether the instruction suffered misses in the instruction cache, data cache, or TLB), latencies associated with pipeline stages, the effective address of a branch target or memory operand, and whether the instruction completed successfully or aborted. As a sampled instruction completes, an interrupt is triggered and the details of the instruction's execution history are available for inspection. Support for exploiting ProfileMe's instruction-based sampling was incorporated into DEC's Dynamic Continuous Profiling Infrastructure, known as DCPI [32].

Today, AMD's Family 10h Opteron processors support two flavors of instruction-based sampling: *fetch sampling* and *op sampling* [125]. For any sampled fetch, fetch sampling will log the fetch address, whether a fetch completed or not, associated instruction cache and TLB misses, the fetch latency, and the page size of the address translation [125]. Knowing when speculative fetches abort late can provide insight into an important source of speculation overhead. Op sampling provides information about instructions that complete. Information recorded about an instruction in flight includes the latency from when the instruction is tagged until it retires, the completion to retire latency, status bits about a branch's behavior and the accuracy of branch predictions, the effective address (virtual and physical) for a load/store and details about the memory hierarchy response (e.g., cache and TLB miss; misalignment, and page size of address translation). This wealth of information can be used for detailed analysis of pipeline and memory hierarchy utilization.

Instruction-based sampling is used to diagnose processor performance issues by AMD's CodeAnalyst performance tool [25]. In the past, HPCTOOLKIT exploited instruction-based sampling measurements collected using ProfileMe on the Alpha 21264. In recent years, HPCTOOLKIT's focus has been exploiting event-based sampling because it is ubiquitously available whereas instruction-based sampling is not. At this writing, HPCTOOLKIT is being extended to capture and analyze precise performance information gathered using instruction-based sampling.

4.3.3   Contrasting Measurement Approaches

In general, the aspect of program behavior or performance being studied defines the requirements for observation. No single measurement approach is inherently superior. Asynchronous sampling and instrumentation have different observational capabilities and each has their place. Asynchronous sampling is great for capturing and attributing detailed performance information at a very fine grain with low overhead. However, only instrumentation is capable of capturing semantic information at particular points in a program's execution. Detailed information about a communication event or how many times a procedure was invoked, or measuring costs within an interval defined by a pair of events, requires instrumented measurement.

When collecting performance measurements, measurement accuracy is always a concern. Accuracy is influenced by several factors, including both overhead and granularity. Any performance measurement, whether based on instrumentation or asynchronous sampling, incurs overhead when executing measurement code. Measurement overhead not only slows a program execution, it also can perturb behavior. Measurement overhead is generally linearly proportional to the total number of events measured. For asynchronous sampling, this depends upon the sampling period. For instrumentation, overhead depends upon the execution frequency of instrumented operations (e.g., procedures). Monitoring overhead also depends upon whether measurements are recorded in a profile or a trace.

A critical difference between asynchronous sampling and instrumentation is the relationship between measurement granularity and accuracy. When instrumentation is used to measure fine-grained events (i.e., where the quantity being measured is on the order of the measurement overhead) accuracy is compromised. Unfortunately, omitting instrumentation for fine-grain events introduces distortion of another kind: if an instrumentation-based performance measurement tool attributes each unit of cost (e.g., time), omitting instrumentation for fine-grain events has the effect of redistributing their costs to other measured events. In contrast, asynchronous sampling can measure and attribute detailed performance information at a fine grain accurately as long as the conditions for accurate sampling described in Section 4.3.2 are met.

Pragmatic concerns also influence the measurement approach used by performance tools. Measurement tools based on source-level instrumentation require little in the way of operating system support; for that reason, they can be available on new platforms immediately. However, source-level instrumentation can interfere with compiler optimizations as well as fail to accurately attribute costs to libraries that exist in binary only form. In contrast, asynchronous sampling can be applied to an entire executable, even dynamically-linked shared libraries. However, asynchronous sampling requires more operating system support; in the past, this dependency has delayed deployment of sampling-based tools on leading edge parallel systems such as Cray XT and IBM Blue Gene/P [324]. No approach is without drawbacks.

4.3.4   Performance Measurement in Practice

The nature of an execution under study and the types of performance problems suspected ultimately determines how performance tools are applied in practice. The goal of a performance tool is to collect measurements that yield insight into where a parallel program suffers from scalability losses and where performance losses accumulate due to a failure to make efficient use of processors, communication, or I/O. While performance bottlenecks may be identified with asynchronous sampling, fixing their causes may require understanding in detail the pattern of interactions between processes and threads, which can only be captured with instrumentation. For this reason, asynchronous sampling and instrumentation-based measurement are complementary. The next section provides an overview of the HPCTOOLKIT performance tools, which focuses on performance measurement and analysis based on asynchronous sampling; Section 4.5 describes the TAU Performance System, which focuses on performance measurement and analysis based on instrumentation.

4.4   HPCToolkit Performance Tools

This section provides an overview of the HPCTOOLKIT performance tools [22, 284] developed at Rice University. The principal design goals for HPCTOOLKIT were that it be simple to use and provide fine-grain detail about performance bottlenecks in complex parallel applications. Both of these goals have been achieved.

HPCTOOLKIT consists of tools for measuring the performance of fully optimized executables using asynchronous sampling, analyzing application binaries to understand the structure of optimized code, correlating measurements with both static and dynamic program structure, and presenting the resulting performance data in a top-down fashion to facilitate rapid analysis. Section 4.4.1 outlines the design principles that shaped HPCTOOLKIT's development and provides an overview of some of HPCTOOLKIT's key components. Sections 4.4.2 and 4.4.3, respectivelyi, describe HPCTOOLKIT's measurement and analysis components in more detail. Section 4.4.4 describes two user interfaces for analyzing performance data collected with HPCTOOLKIT. We illustrate HPCTOOLKIT's capabilities for analyzing the performance of complex scientific applications using two codes: one that employs structured adaptive mesh refinement to model astrophysical thermonuclear flashes and another that uses a particle-in-cell method to simulate turbulent plasma in a tokamak. Section 4.4.5 offers some final thoughts and sketches our plans for enhancing HPCTOOLKIT for emerging parallel systems.

4.4.1   Design Principles

HPCTOOLKIT's design is based on a set of complementary principles that form a coherent synthesis that is greater than their constituent parts. Although we have outlined these principles in earlier work [22, 235], it is useful to review them here so that the reader can understand the motivation behind HPCTOOLKIT's approach to performance measurement and analysis.

Be language independent. Modern parallel scientific programs typically consist of an application core, along with a collection of libraries for communication, I/O, and mathematical operations. It is common for components to be written in different programming languages. To support measurement and attribution of program performance across components independent of the programming language in which they are expressed, HPCTOOLKIT works directly with application binaries.

Avoid instrumentation for measuring performance. The overhead of instrumentation can distort application performance through a variety of mechanisms [250]. In prior work [143, 327], we found that approaches that add instrumentation to every procedure to collect call graphs can slow execution by a factor of two to four across the SPEC integer benchmarks. To reduce instrumentation overhead, tools such as TAU intentionally refrain from instrumenting certain procedures [299]. However, the more this approach reduces overhead, the more it reduces precision. To avoid the high overhead common with routine-level instrumentation, as well as to provide loop and line-level performance details with low measurement overhead, HPCTOOLKIT uses asynchronous sampling as its principal method of measurement.

Avoid blind spots. Applications frequently link against libraries for which source code is not available. For instance, a compiler's implementation of OpenMP typically relies on a runtime support library of this sort. To avoid systematic error, one must measure and attribute costs for routines in all such libraries. To handle routines in binary-only libraries, HPCTOOLKIT performs several types of binary analysis.

Context is essential for understanding modern software. Parallel applications and libraries often call communication primitives in many different contexts. Similarly, C++ programs may use the same algorithm or data structure template in many different contexts. In both cases, understanding performance requires understanding it *in context*. For this reason, HPCTOOLKIT supports call path profiling to attribute costs to the full calling contexts in which they are incurred.

Any one performance measure produces a myopic view. Measuring time or only one species of event seldom diagnoses a correctable performance problem. One set of metrics may be necessary to identify a problem and others may be necessary to diagnose its causes. For this reason, HPCTOOLKIT supports collection, correlation and presentation of multiple metrics.

Derived performance metrics are essential for effective analysis. Typical metrics such as elapsed time are useful for identifying program hot spots. However, tuning a program usually requires a measure of not where resources are consumed, but where they are consumed *inefficiently*. For this purpose, derived metrics, such as total aggregate losses accumulated as a result of differences between peak and actual performance, are far more useful than raw metrics such as operation counts. HPCTOOLKIT supports computation of user-defined derived metrics and enables users to rank and sort program scopes using such metrics.

Performance analysis should be top-down. It is unreasonable to require users to wade through mountains of data to hunt for evidence of important problems. To make analysis of large programs tractable, performance tools should present measurement data in a hierarchical fashion, prioritize what appear to be important problems, and support a top-down analysis methodology that helps users quickly locate bottlenecks. HPCTOOLKIT's user interface supports hierarchical presentation of performance data according to both static and dynamic contexts, along with ranking and sorting based on metrics.

Hierarchical aggregation is vital. Instruction-level parallelism in processor cores can make it difficult or expensive for hardware counters to precisely attribute particular events to specific instructions. However, even if fine-grain attribution of events is flawed, aggregate event counts within loops or procedures are typically accurate. HPCTOOLKIT's hierarchical attribution and presentation of measurement data deftly addresses this issue; loop level information available with HPCTOOLKIT is particularly useful.

Measurement and analysis must be scalable. For performance tools to be useful on large-scale parallel systems, measurement and analysis techniques must scale to tens and even hundreds of thousands of threads. HPCTOOLKIT's sampling-based measurements are compact and the data for large-scale systems is not unmanageably large. As we describe later, HPCTOOLKIT supports a novel approach for quantifying and pinpointing scalability bottlenecks conveniently on systems independent of scale.

4.4.1.1 From Principles to Practice

From these principles, we have devised a general methodology embodied by the workflow depicted in Figure 4.1. The workflow is organized around four principal activities: *measurement* of performance metrics while an application executes; *analysis* of application binaries to recover program structure; *correlation* of performance measurements with source code structure; and *presentation* of performance metrics and associated source code.

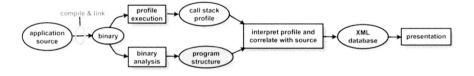

FIGURE 4.1: HPCTOOLKIT workflow.

To use HPCTOOLKIT to measure and analyze an application's performance, one first compiles and links the application for a production run, using *full* optimization. Second, one launches an application with HPCTOOLKIT's measurement tool, `hpcrun`, which uses asynchronous sampling to collect a performance profile. Third, one invokes `hpcstruct`, HPCTOOLKIT's tool to analyze an application binary and recover information about files, functions, loops, and inlined code.[1] Fourth, one uses `hpcprof` to combine information about an application's structure with performance measurements to produce a performance database. Finally, one explores a performance database with HPCTOOLKIT's `hpcviewer` graphical user interface.

At this level of detail, much of the HPCTOOLKIT workflow approximates other performance analysis systems, with the most unusual step being binary analysis. In the following sections, we outline how the methodological principles described above suggest several novel approaches to both accurate measurement (Section 4.4.2) and effective analysis (Section 4.4.3).

Our approach is *accurate*, because it assiduously avoids systematic measurement error (such as that introduced by instrumentation), and *effective*, because it associates useful performance metrics (such as parallel idleness or memory bandwidth) with important source code abstractions (such as loops) as well as dynamic calling context.

4.4.2   Measurement

This section highlights the ways in which the methodological principles from Section 4.4.1 are applied to measurement. Without accurate performance measurements for fully optimized applications, analysis is unproductive. Consequently, one of the chief concerns for HPCTOOLKIT has been designing an accurate measurement approach that exposes low-level execution details while avoiding systematic measurement errors that come with high measurement overhead. For this reason, HPCTOOLKIT uses sampling (both synchronous and asynchronous) for low overhead measurement.

Asynchronous event triggers. Different kinds of event triggers measure different aspects of program performance. HPCTOOLKIT initiates asynchronous samples using either an interval timer, hardware performance counter events, or hardware support for instruction-based sampling. Hardware performance counters enable HPCTOOLKIT to profile events such as cache misses and issue-stall cycles. Instruction-based sampling enables precise fine-grained attribution of costs such as stall cycles, which is problematic with event-based sampling.

Synchronous event triggers. Synchronous triggers are generated via direct program action. Examples of interesting events for synchronous monitoring are lock acquisition and release, memory allocation, I/O, and inter-

---

[1] For the most detailed attribution of application performance data using HPCTOOLKIT, one should ensure that the compiler includes line map information in the object code it generates. While HPCTOOLKIT does not require this information, it helps users by improving the correlation between measurements and source code. Since compilers can usually provide line map information for fully optimized code, this need not require a special build process.

process communication. For such events, one can measure quantities such as lock waiting time, bytes allocated, bytes written, or bytes communicated. To reduce the overhead of instrumentation-based synchronous measurements, HPCTOOLKIT uses sampling to record data for only a fraction of the monitored events.

Maintaining control over parallel applications. To control measurement of an executable, HPCTOOLKIT intercepts certain process control routines including those used to coordinate thread/process creation and destruction, signal handling, dynamic loading, and MPI initialization. To support measurement of unmodified, dynamically linked, optimized application binaries, HPCTOOLKIT uses the library preloading feature of modern dynamic loaders to preload a measurement library as an application is launched. With library preloading, process control routines defined by HPCTOOLKIT are called instead of their default implementations. To measure the performance of statically linked executables, HPCTOOLKIT provides a script for use at link time that arranges for process control routines to be intercepted during execution.

Call path profiling and tracing. Experience has shown that comprehensive performance analysis of modern modular software requires information about the full *calling context* in which costs are incurred. The calling context for a sample event is the set of procedure frames active on the call stack at the time the event trigger fires. The process of monitoring an execution to record the calling contexts in which event triggers fire is known as *call path profiling*. To provide insight into an application's dynamic behavior, HPCTOOLKIT also offers the option to collect *call path traces*.

When synchronous or asynchronous event triggers fire, `hpcrun` records the full calling context for each event. A calling context collected by `hpcrun` is a list of instruction pointers, one for each procedure frame active at the time the event occurred. The first instruction pointer in the list is the program address at which the event occurred. The rest of the list contains the return address for each active procedure frame. Rather than storing the call path independently for each sample event, `hpcrun` represents all of the call paths for events as a calling context tree (CCT) [31]. In a calling context tree, the path from the root of the tree to a node corresponds to a distinct call path observed during execution; a count at each node in the tree indicates the number of times that the path to that node was sampled. Since the calling context for a sample may be completely represented by a leaf node in the CCT, to record a complete trace of call path samples `hpcrun` simply records a sequence of (CCT node id, time stamp) tuples along with a CCT.

Coping with fully optimized binaries. Collecting a call path profile or trace requires capturing the calling context for each sample event. To capture the calling context for a sample event, `hpcrun` must be able to unwind the call stack at *any* point in a program's execution. Obtaining the return address for a procedure frame that does not use a frame pointer is challenging since the frame may dynamically grow (i.e., when space is reserved for the caller's registers and local variables, the frame is extended with calls to `alloca`, or arguments to called procedures are pushed) and shrink (as space for the aforementioned purposes is deallocated) as the procedure executes. To cope with this situation, `hpcrun` uses a fast, on-the-fly binary analyzer that examines a routine's machine instructions and computes how to unwind a stack frame for the procedure [327]. For each address in a routine, there must be a recipe for how to unwind. Different recipes may be needed for different intervals of addresses within the routine. Each interval ends in an instruction that changes the state of the routine's stack frame. Each recipe describes (1) where to find the current frame's return address, (2) how to recover the value of the stack pointer for the caller's frame, and (3) how to recover the value that the base pointer register had in the caller's frame. Once `hpcrun` computes unwind recipes for all intervals in a routine, it memorizes them for later reuse.

To compute unwind recipes with binary analysis, `hpcrun` must know where each routine starts and ends. When working with applications, one often encounters partially stripped libraries or executables that are missing information about function boundaries. To address this problem, `hpcrun` uses a binary analyzer to infer routine boundaries [327].

HPCTOOLKIT's use of binary analysis for call stack unwinding has proven to be very effective, even for fully optimized code. At present, HPCTOOLKIT provides binary analysis for stack unwinding on the x86_64, Power, and MIPS architectures. A detailed study of HPCTOOLKIT's x86_64 unwinder on versions of the SPEC CPU2006 benchmarks optimized with several different compilers showed that the unwinder was able to recover the calling context for all but a vanishingly small number of cases [327]. For other architectures (e.g., Itanium), HPCTOOLKIT uses the `libunwind` library [249] for unwinding.

Handling dynamic loading. Modern operating systems such as Linux enable programs to load and unload shared libraries at run time, a process known as *dynamic loading*. Dynamic loading presents the possibility that multiple functions may be mapped to the same address at different times during a program's execution. During execution, `hpcrun` ensures that all measurements are attributed to the proper routine in such cases.

4.4.3   Analysis

This section describes HPCTOOLKIT's general approach to analyzing performance measurements, correlating them with source code, and preparing them for presentation.

4.4.3.1   Correlating Performance Metrics with Optimized Code

To enable effective analysis, performance measurements of fully optimized programs must be correlated with important source code abstractions. Since measurements are made with reference to executables and shared libraries, for analysis it is necessary to map measurements back to the program source.

To correlate sample-based performance measurements with the static structure of fully optimized binaries, we need a mapping between object code and its associated source code structure.[2] HPCTOOLKIT's `hpcstruct` constructs this mapping using binary analysis; we call this process *recovering program structure*.

`hpcstruct` focuses its efforts on recovering procedures and loop nests, the most important elements of source code structure. To recover program structure, `hpcstruct` parses a load module's machine instructions, reconstructs a control flow graph, combines line map information with interval analysis on the control flow graph in a way that enables it to identify transformations to procedures such as inlining and account for transformations to loops [327].[3]

Several benefits naturally accrue from this approach. First, HPCTOOLKIT can expose the structure of and assign metrics to what is actually executed, *even if source code is unavailable*. For example, `hpcstruct`'s program structure naturally reveals transformations such as loop fusion and scalarized loops implementing Fortran 90 array notation. Similarly, it exposes calls to compiler support routines and wait loops in communication libraries of which one would otherwise be unaware. Finally, `hpcrun`'s function discovery heuristics identify procedures within stripped binaries.

HPCTOOLKIT's `hpcprof` tool is used to correlate measurements collected using `hpcrun` with program structure information gathered by `hpcstruct`. The resulting performance database relates performance metrics to dynamic program structure (call chains) as well as static program structure (load modules, procedures, loops, inlined code, and source lines). A unique strength of this tool is that it can integrate both static and dynamic context information and relate performance to call paths that include representations for loops and inlined procedures. We recently developed an MPI [238] version of `hpcprof` which scalably analyzes, correlates, and summarizes call path profiles from all threads and processes from an execution on a system with large-scale parallelism.

4.4.3.2   Computed Metrics

Identifying performance problems and opportunities for tuning may require synthetic performance metrics. To identify where an algorithm is not effectively using hardware resources, one should compute a metric that reflects *wasted* rather than consumed resources. For instance, when tuning a floating-point intensive scientific code, it is often less useful to know where the majority of the floating-point operations occur than where opportunities for executing floating-point operations were squandered, such as waiting for pipeline stalls or long-latency loads. A measure of how much floating point

---

[2]This object to source code mapping should be contrasted with the binary's line map, which (if present) is typically fundamentally line based.

[3]Without line map information, `hpcstruct` can still identify procedures and loops, but is not able to account for inlining or loop transformations.

resources are underutilized can be computed as (cycles in a scope × peak flop/cycle − actual flops in that scope) and displaying this measure for loops and procedures. Many other useful quantities such as memory bandwidth available can be computed similarly.

4.4.3.3 Identifying Scalability Bottlenecks in Parallel Programs

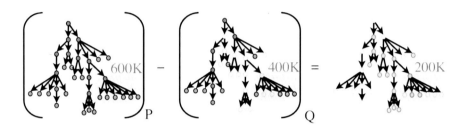

FIGURE 4.2: Differencing call path profiles to pinpoint scalability bottlenecks when weak scaling from q to p processors.

If an execution of an SPMD program is sufficiently balanced and symmetric, the behavior of the program as a whole can be characterized by examining the scaling performance of any one of the processes or threads. If behavior is sufficiently stable over time, then an entire run can be characterized by analyzing profiles that are integrated over any appropriate time interval.

Differential profiling [231] is a strategy for analyzing multiple executions of a program by combining their execution profiles mathematically. A comparison of profiles from two executions yields information about where and how much the costs in the two executions differ. To pinpoint and quantify scalability bottlenecks for parallel programs and attribute them to calling contexts, we use HPCTOOLKIT to compare call path profiles from a pair of executions using different levels of parallelism [95]. Consider two parallel executions of an application, one executed on q processors and the second executed on p processors, where $p > q$. In a weak scaling scenario, each processor in both executions computes on an identical amount of local data. If the application exhibits perfect weak scaling, then the total cost (e.g., execution time) should be the same in both executions. If every part of the application scales uniformly, then this equality should hold in *each scope* of the application.

Figure 4.2 pictorially shows an analysis of weak scaling by comparing two representative calling context trees (CCTs)[4]—one tree from a process in each execution. For instance, the difference in cost incurred in the subtrees highlighted in the CCTs for p and q processors represents parallel overhead when scaling from q to p processors. This difference in cost for the subtree can be converted into percent scalability loss by dividing its cost by the inclusive

---

[4]A CCT is a weighted tree in which each calling context is represented by a path from the root to a node and a node's weight represents the cost attributed to its associated context.

cost of the root node in the CCT for p processors and multiplying by 100. Computing inclusive scalability losses for each node in a CCT enables one to easily locate scalability bottlenecks in a top-down fashion by tracing patterns of losses down paths from the root.

In addition, we recently developed general techniques for effectively analyzing multithreaded applications using measurements based on both synchronous and asynchronous sampling [326, 328]. Using them, HPCTOOLKIT can attribute precise measures of lock contention, parallel idleness, and parallel overhead to *user-level* calling contexts—even for a multithreaded language such as Cilk [142], which uses a work-stealing run-time system.

4.4.4   Presentation

This section describes `hpcviewer` and `hpctraceview`, HPCTOOLKIT's two presentation tools. We illustrate the functionality of these tools by applying them to analyze measurements of parallel executions of FLASH [126], a code that uses adaptive mesh refinement to model astrophysical thermonuclear flashes, and the Gyrokinetic Toroidal Code (GTC) [217], a particle-in-cell (PIC) code for simulating turbulent transport in fusion plasma in devices such as the International Thermonuclear Experimental Reactor.

4.4.4.1   hpcviewer

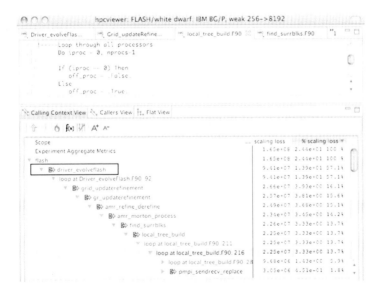

FIGURE 4.3: Using HPCTOOLKIT and differential profiling to analyze scaling losses (microseconds) for FLASH when running at 256 and 8192 cores on an IBM Blue Gene/P.

FIGURE 4.4: A Caller's (bottom-up) view of scaling loss (wallclock) for FLASH when running at 256 and 8192 cores on an IBM Blue Gene/P.

HPCTOOLKIT's `hpcviewer` user interface [23] presents performance metrics correlated to program structure (Section 4.4.3.1) and mapped to a program's source code, if available. Figure 4.3 shows a snapshot of the `hpcviewer` user interface displaying data from several parallel executions of FLASH. The top pane of the `hpcviewer` interface shows program source code. The bottom pane associates a table of performance metrics with static or dynamic program contexts. `hpcviewer` provides three different views of performance measurements collected using call path profiling.

Calling context view. This top-down view associates an execution's dynamic calling contexts with their costs. Using this view, one can readily see how much of an application's cost was incurred by a function when called from a particular context. If finer detail is of interest, one can explore how the costs incurred by a call in a particular context are divided between the callee itself and the procedures it calls. HPCTOOLKIT distinguishes calling contexts precisely by individual call sites; this means that if a procedure g contains calls to procedure f in different places, each call represents a separate calling context. Figure 4.3 shows a calling context view. This view is created by integrating static program structure (loops and inlined code) with dynamic calling contexts gathered by `hpcrun`. Analysis of the scaling study shown in the figure is discussed in the next section.

Callers view. This bottom-up view enables one to look upward along call paths. This view is particularly useful for understanding the performance of software components or procedures that are called in more than one context. For instance, a message-passing program may call communication routines in many different calling contexts. The cost of any particular call will depend upon its context. Figure 4.4 shows a caller's view of processes from two parallel runs of FLASH. Analysis of the example shown in this figure is discussed in the next section.

Flat view. This view organizes performance data according to an application's static structure. All costs incurred in any calling context by a procedure are aggregated together in the flat view. This complements the calling context view, in which the costs incurred by a particular procedure are represented separately for each call to the procedure from a different calling context.

`hpcviewer` can present an arbitrary collection of performance metrics gathered during one or more runs, or compute derived metrics expressed as formulae with existing metrics as terms. For any given scope, `hpcviewer` computes both *exclusive* and *inclusive* metric values. Exclusive metrics only reflect costs for a scope itself; inclusive metrics reflect costs for the entire subtree rooted at that scope. Within a view, a user may order program scopes by sorting them using any performance metric. `hpcviewer` supports several convenient operations to facilitate analysis: revealing a *hot path* within the hierarchy below a scope, *flattening* out one or more levels of the static hierarchy, such as to facilitate comparison of costs between loops in different procedures, and *zooming* to focus on a particular scope and its children.

Using `hpcviewer`

We illustrate the capabilities of `hpcviewer` by using it to pinpoint scaling bottlenecks in FLASH by using differential profiling to analyze two (weak scaling) simulations of a white dwarf explosion by executing 256-core and 8192-core simulations on an IBM Blue Gene/P. For the 8192-core execution, both the input and the number of cores are 32x larger than the 256-core execution. With perfect weak scaling, we would expect identical run times and process call path profiles in both configurations.

Figure 4.3 shows a portion of the residual calling context tree, annotated with two metrics: "scaling loss" and "% scaling loss." The former quantifies the scaling loss (in microseconds) while the latter expresses that loss as a percentage of total execution time (shown in scientific notation). The top-most entry in each metric column gives the aggregate metric value for the whole execution. A percentage to the right of a metric value indicates the magnitude of that particular value relative to the aggregate. Thus, for this execution, there was a scaling loss of 1.65×10^8 μs, accounting for 24.4% of the execution.[5] By sorting calling contexts according to metric values, we immediately see that the evolution phase (`Driver_evolveFlash`) of the execution (highlighted with a black rectangle) has a scaling loss that accounts for 13.9% of the total execution time on 8192 cores, which represents 57.1% of the scaling losses in the execution.

To pinpoint the source of the scalability bottleneck in FLASH's simulation, we use the "hot path" button to expand automatically the unambiguous portion of the hot call path according to this metric. Figure 4.3 shows this result. HPCTOOLKIT identifies a loop (beginning at line 213), within its full dynamic calling context – a unique capability – that is responsible for 13.7% of the scaling losses. This loop uses a ring-based all-to-all communication pattern known as a *digital orrery* to update a processor's knowledge about neighboring blocks of the adaptive mesh. Although FLASH only uses the orrery pattern to set up subsequent communication to fill guard cells, looping over all processes is inherently unscalable. Other scalability bottlenecks can be identified readily by repeating the "hot path" analysis for other subtrees in the computation where scaling losses are present. The power of HPCTOOLKIT's scalability analysis is apparent from the fact that it immediately pinpoints and quantifies the scaling loss of a key loop deep inside the application's layers of abstractions.

To quickly understand where the scaling losses for the initialization and simulation phases are aggregated, we turn to the bottom-up caller's view. The caller's view apportions the cost of a procedure (in context) to its callers. We sort the caller's view by the exclusive scaling loss metric, thus highlighting the scaling loss for each procedure in the application, exclusive of callees. Two routines in the BG/P communication library immediately emerge as responsible for the bulk of the scaling loss: `DCMF::Protocol::MultiSend::TreeAllreduceShortRecvPostMessage::advance` (advance for short) and `DCMF::BGPLockManager::globalBarrierQueryDone` (QueryDone for short). When we use `hpcviewer`'s caller's view to look up the call chain from `advance`, we find routines that represent a call to `MPI_Allreduce`. The first call, which accounts for 57% of the scaling loss (14.1% of run time), is highlighted in blue; the others, which are inconsequential, are hidden by an image overlay indicated by the thick horizontal black line. As the corresponding source code shows, this call to `MPI_Allreduce` is a global max reduce for a scalar that occurs in code managing the adaptive mesh. HPCTOOLKIT is uniquely able to pinpoint this one crucial call to `MPI_Allreduce` and distinguish it from several others that occur in the application. Next, we peer up the hot path of call chains leading to `QueryDone`. We hone in on one call to a barrier that disproportionately affects scaling. The barrier call site is within `Grid_fillGuardCells` and is visible at the bottom of Figure 4.4; it accounts for 13.6% of the scaling loss (or 3.31% of the run time). In a few simple steps, HPCTOOLKIT has

---

[5]A scaling loss of 24.4% means that FLASH is executing at 75.6% parallel efficiency on 8192 cores relative to its performance on 256 cores.

enabled us to quickly pinpoint exactly two calls that account for about 70% of FLASH's scaling loss on BG/P.

This brief study of FLASH shows how the measurement, analysis, attribution, and presentation capabilities of HPCTOOLKIT make it straightforward to pinpoint and quantify sources of scaling losses between different parallel configurations of an application.

4.4.4.2   hpctraceview

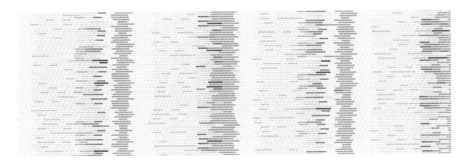

FIGURE 4.5: hpctraceview showing part of a execution trace for GTC. (See color insert.)

hpctraceview is a prototype visualization tool that was recently added to HPCTOOLKIT. hpctraceview renders space-time diagrams that show how a parallel execution unfolds over time. Figure 4.5 shows a screen snapshot from the space-time view pane of hpctraceview displaying an interval of execution for a hybrid MPI+OpenMP version of the Gyrokinetic Toroidal Code (GTC) [217] running on 64 processors. GTC is a particle-in-cell code for simulating turbulent transport in fusion plasma in devices such as the International Thermonuclear Experimental Reactor. The execution consists of 32 MPI processes with two OpenMP threads per process. The figure shows a set of time lines for threads. A thread's activity over time unfolds left to right. Time lines for different threads are stacked top to bottom. Even numbered threads (starting from 0) represent MPI processes; odd-numbered threads represent OpenMP slave threads. Although hpctraceview's space-time visualizations appear similar to those by many contemporary tools, the nature of its visualizations and the data upon which they are based is rather different.

Other performance tools, e.g., TAU and Paraver [276], render space-time diagrams based on data collected by embedded program instrumentation that *synchronously* records information about the entry and exit of program procedures, communication operations, and/or program phase markers. In contrast, hpctraceview's traces are collected using *asynchronous* sampling. Each time line in hpctraceview represents a sequence of asynchronous samples taken over the life of a thread (or process) and each colored band represents a procedure frame that was observed by an asynchronous sample. While that difference is important, it alone does not justify construction of a new engine for rendering space-time diagrams. A new viewer is necessary because hpctraceview's samples are multi-level: each sample for a thread represents the entire call stack of procedures active when the sample event occurred. hpctraceview can in fact render traces at different levels of abstraction by displaying them at different call stack depths at the user's request.

Despite the fact that hpctraceview's space-time diagrams are based on asynchronous samples, they can support analysis of transient behavior similar to their counterparts based on synchronous traces. Casual inspection of the space-time view in Figure 4.5 shows three complete repetitions of a "pattern" and part of a fourth. A closer look reveals that the first and third repetitions are somewhat different in nature than the second and fourth: the aforementioned patterns contain a band of yellow bars,[6] whereas the latter do not. These yellow bars represent a procedure that is called only in alternate iterations of the simulation. Space-time visualizations are good for spotting and understanding such time varying behavior. To relate such visual feedback back to the source program, one can use a pointing device to select a colored sample in the space-time view. Another pane in the hpctraceview GUI (not shown) will display the complete call path for a selected sample.

At present, hpctraceview is a proof-of concept prototype. At this writing, enhancements are under way to enable it to render long traces from huge numbers of threads based on out-of-core data. Not only is the underlying data displayed by hpctraceview statistical in nature, its rendering is as well. If one tries to render more samples than will fit horizontally or more threads than will fit vertically, a rule is used to determine what color should be rendered for a pixel. Paraver has shown that with appropriate rules, such low-resolution visualizations can still convey important information about transient program behavior. We expect that hpctraceview will provide similar insights based on asynchronous call stack samples.

4.4.5 Summary and Ongoing Work

The key metric for parallel performance is scalability; this is especially true as parallel systems grow beyond the petascale systems of today. Consequently, there is an acute need for application scientists to understand and address scaling bottlenecks in their codes. HPCTOOLKIT's sampling-based performance measurements make it possible to pinpoint and quantify scaling bottlenecks both within and across nodes. Measurement overhead is low, analysis is rapid, and results are actionable.

In addition to understanding scaling losses, gaining insight into node performance bottlenecks on large-scale parallel systems is a problem of growing importance. Today's parallel systems typically have between four and sixteen cores per node. IBM's forthcoming Blue Waters system will increase

---

[6]In a black and white copy, the yellow bars are those of lightest shade.

the core count per node by packing four multithreaded eight-core processors into a multi-chip module [257]. By using event- or instruction-based sampling on hardware performance counters, one can distinguish between core performance bottlenecks caused by a variety of factors including inadequate instruction-level parallelism, memory latency, memory bandwidth, and contention. One can also use sampling to understand load imbalance [326] or lock contention [328] among the threads on a multiprocessor node.

Ongoing work in the HPCTOOLKIT project is focused on making it easier to analyze performance data from huge numbers of cores in a top down fashion. HPCTOOLKIT's new MPI version of `hpcprof` [325] is capable of analyzing in parallel a set of execution profiles from many separate cores. Output of the tool provides summary statistics for all cores and a visual means for assessing similarities and differences in performance across the cores in a parallel system. HPCTOOLKIT delivers these summaries along with access to fine-grain detail in the way that it always has: mapped back to the source code with full dynamic calling context augmented with information loops and inlined code. To cope with the volume of measurement data from very large-scale parallel systems, HPCTOOLKIT's presentation tools are being enhanced to support access to out-of-core data.

Key challenges ahead for HPCTOOLKIT are to help diagnose causes of performance loss in parallel systems rather than just identifying where the losses occur, to provide deeper insight into how performance losses arise at the node level, and to offer with suggestions for how to tune your program to improve its performance. Ongoing work in this area includes a partnership with researchers at the University of Texas at Austin to build PerfExpert [70], an expert system for performance analysis, which aims to provide such guidance based on performance measurements collected by HPCTOOLKIT.

4.5   TAU Performance System

The TAU Performance System® [225, 300, 301, 339] is the product of over sixteen years of development to create a robust, flexible, portable, and integrated framework and toolset for performance instrumentation, measurement, analysis, and visualization of large-scale parallel computer systems and applications. Although the University of Oregon is home to the TAU project, its success represents the combined efforts of researchers at the University of Oregon and colleagues at the Research Centre Jülich and Los Alamos National Laboratory, especially in TAU's formative years. The following gives an overview of TAU's architecture and current suite of tools. Best practices for its use are discussed and examples drawn from work highlighting some of TAU's recent features.

4.5.1   TAU Performance System Design and Architecture

TAU is designed as a tool framework, wherein tool components and modules integrate and coordinate their operations using well-defined interfaces and data formats. The TAU framework architecture, shown in Figure 4.6, is organized into three primary layers: *instrumentation*, *measurement*, and *analysis*. Multiple modules exist within each layer that can be configured in a flexible manner by the user. The following sections discuss the layers in more detail, but first the overall design decisions that governed TAU's development are presented.

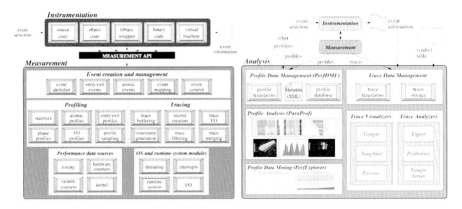

FIGURE 4.6: TAU framework architecture.

TAU is a performance systems based on probe-based observation of events in the application, library, or system code. For any performance experiment using TAU, performance events of interest must be decided and the program must be instrumented. The role of the instrumentation layer in TAU is to insert calls to the TAU measurement system at code locations where the events occur. Since performance events of interest can be found at different places in the code, TAU provides a range of instrumentation capabilities to gather events from all these locations.

TAU supports both *atomic* events and *interval* events. An atomic event denotes a single action. When it occurs, the measurement system has the opportunity to obtain the performance data associated with that action at that time. In contrast, a interval event is really a pair of events: *begin* and *end*. The measurement system uses performance data obtained from each event to determine a performance result.

TAU uses probe-based measurement to support both *profiling* and *tracing*. Profiling methods compute performance statistics at runtime based on measurements of atomic or interval events. Tracing, on the other hand, records the measurement information for each event (including when it occurred) in a file for future analysis. In profiling, the number of recorded events is fixed, whereas tracing will generate a record for every event occurrence. The TAU performance system includes tools for parallel profile analysis and targets parallel trace analysis tools from other research groups, principally Vampir [251] and Jumpshot [385].

Overall, TAU's design has proven to be robust, sound, and highly adaptable to generations of parallel systems. In recent years, we have extended the TAU performance system architecture to support kernel-level performance integration [254], performance monitoring [253, 256], and collaboration through the TAU *Portal* [340]. The extensions have been moderate, mainly in the development of new interfaces for performance data access and reporting.

4.5.2   TAU Instrumentation

From TAU's perspective, the execution of a program is regarded as a sequence of significant performance events. As events are triggered during execution, the instrumented probes engage the TAU performance measurement infrastructure to obtain the performance data. Logically, instrumentation is separated from measurement in TAU. The measurement options will determine what performance data is recorded and the type of measurement made for the events, whether profile or trace. Instrumentation is focused primarily on how the events are created and how code gets placed in the program.

4.5.2.1   Event Interface

The TAU *event interface* allows events to be defined, their visibility controlled, and their runtime data structures to be created. Each event has a *type* (*atomic* or *interval*), a *group*, and a unique *event name*. The event name is a character string and is a powerful way to encode event information. At runtime, TAU maps the event name to an efficient *event ID* for use during measurement. Events are created dynamically in TAU by providing the event interface with a unique event name. This makes it possible for runtime context information to be used in forming an event name (*context-based events*), or values of routine parameters to be used to distinguish call variants, (*parameter-based events*). TAU also supports *phase* and *sample* events.

The purpose of event control in TAU is to enable and disable a group of events at a coarse level. This allow the focus of instrumentation to be refined at runtime. All groups can be disabled and any set of groups can be selectively enabled. Similarly, all event groups can be enabled initially and then selectively disabled. It is also possible to individually enable and disable events. TAU uses this support internally to throttle high overhead events during measurement.

4.5.2.2   Instrumentation Mechanisms

Instrumentation can be introduced in a program at several levels of the program transformation process. In fact, it is important to realize that events and event information can be between levels and a complete performance view may require contribution across levels [298]. For these reasons, TAU supports several instrumentation mechanisms based on the code type and transformation level: source (manual, preprocessor, library interposition), binary/dynamic, interpreter, component, and virtual machine. There are multiple factors that affect the choice of what level to instrument, including accessibility, flexibility, portability, concern for intrusion, and functionality. It is not a question of what level is "correct" because there are trade-offs for each and different events are visible at different levels. The goal in the TAU performance system is to provide support for several instrumentation mechanisms that might be useful together.

Source Instrumentation

Instrumenting at the source level is the most portable instrumentation approach. Source-level instrumentation can be placed at any point in the program and it allows a direct association between language-and program-level semantics and performance measurements. Using cross-language bindings, TAU implements the event interface in C, C++, Fortran, Java, and Python languages, and provides a higher-level specification in SIDL [194,302] for cross-language portability and deployment in component-based programming environments [54].

Programmers can use the event API to manually annotate the source code of their program. For certain application projects, this may be the preferred way to control precisely where instrumentation is placed. Of course, manual instrumentation can be tedious and error prone. To address these issues, we have developed an automatic source instrumentation tool, `tau_instrumentor`, for C, C++, and Fortran, based on the *program database toolkit (PDT)* [218]. The TAU source instrumentor tool can place probes at routine and class method entry/exit, on basic block boundaries, in outer loops, and as part of component proxies. PDT's robust parsing and source analysis capabilities enable the TAU instrumentor to work with very large and complex source files and inserts probes at all possible points. We have translated the technology behind the TAU instrumentor into a more generic instrumentation component recently [148].

In contrast to manual instrumentation, automatic instrumentation needs direction on what performance events of interest should be instrumented in a particular performance experiment. TAU provides support for *selective instrumentation* in all automatic instrumentation schemes. An *event specification* file defines which of the possible performance events to instrument by grouping the event names in include and exclude lists. Regular expressions can be used in event name specifiers and file names can be given to restrict instrumentation focus. Selective instrumentation in TAU has proven invaluable as a means to customize performance experiments and to easily "select out" unwanted performance events, such as high frequency, small routines that may generate excessive measurement overhead.

Library wrapping is a form of source instrumentation whereby the original library routines are replaced by instrumented versions which in turn call the original routines. The problem is how to avoid modifying the library calling interface. Some libraries provide support for *interposition*, where an alternative name-shifted interface to the native library is provided and weak bindings are used for application code linking. Like other tools, TAU uses MPI's support for interposition (*PMPI* [238]) for performance instrumentation purposes. A combination of atomic and interval events are used for MPI. The atomic events allow TAU to track the size of messages in certain routines, for instance, while the interval events capture performance during routine execution. TAU provides a performance instrumented PMPI library for both MPI-1 and MPI-2. In general, automatic library wrapping, with and without interposition, is possible with TAU's instrumentation tools.

Source instrumentation can also be provided in source-to-source translation tools. TAU uses the Opari tool [245] for instrumenting OpenMP applications. Opari rewrites OpenMP directives to introduce instrumentation based on the POMP/POMP-2 [245] event model. TAU implements a POMP-compatible interface that allows OpenMP (POMP) events to be instantiated and measured.

Binary / Dynamic Instrumentation

Source instrumentation is possible only if the source code is available. To remove any dependencies on source access or compiler support, TAU can use the *Dyninst* package from the University of Wisconsin and the University of Maryland [69] to re-write the binary with instrumentation at routine boundaries. This takes place prior to code execution using the `tau_run` tool to inject the TAU measurement library as a shared object and conduct code instrumentation. TAU can also use Dyninst for online instrumentation of a running program. In each case, it is possible to use selective instrumentation for routine events.

Compiler-Based Instrumentation

Compilers from Intel, PGI, IBM, and GNU provide hooks for instrumentation of C, C++, and Fortran codes at the routine level. TAU works with all of these compilers automatically through environment parameters and compiler scripts. The compiler scripts can examine a selective instrumentation file to determine what source files to process. However, special care is needed for TAU to properly work with statically-linked and dynamically-linked shared objects. The problems arise in identifying routine names and involves the mapping of names to runtime addresses. All of the issues are correctly handled by TAU.

Interpreter-Based Instrumentation

TAU also supports the measurement of the runtime systems of both Java and Python. For these languages, TAU can interface dynamically at runtime with the virtual machine or interpreter to inject the TAU hooks into the context of the running program and capture information about events that take place. In these cases, no source code is modified by TAU and there is no need to re-compile the source code to introduce the instrumentation calls in the application. However, TAU also provides an instrumentation API for both Java and Python so that the user can set calipers around regions of interest.

4.5.2.3   Instrumentation Utilities

To deliver the richness of instrumentation TAU provides for performance observation, it helps to have utilities to reduce the impact on users. Where this is most evident is in building applications with source instrumentation. TAU provides a set of compiler scripts that can substitute for standard compilers in order to take control of the instrumentation process with little modification to application build environments. TAU comes with a PDT-based wrapper generator, `tau_wrap`, which allows library header files to be read and a new library wrapper header file created with preprocessor `DEFINE` macros to change routine names to TAU routines names. In this way, a new wrapped library is created with these routines instrumented. This is very similar to what is done by TAU for C `malloc/free` and I/O wrapping, except that `tau_wrap` can be used for any library.

4.5.3   TAU Measurement

The measurement system is the heart and soul of TAU. It has evolved over time to a highly robust, scalable infrastructure portable to all HPC platforms. The instrumentation layer defines which events will be measured and the measurement system selects which performance data metrics to observe. Performance experiments are created by selecting the key events of interest and by configuring measurement modules together to capture desired performance data [118]. TAU's measurement system provides portable timing support, integration with hardware performance counters, parallel profiling and parallel tracing, runtime monitoring, and kernel-level measurement.

4.5.3.1   Measurement System Design

As shown in Figure 4.6, the design of the measurement system is flexible and modular. It is responsible for creating and managing performance events, making measurements from available performance data sources, and recording profile and trace data for each thread in the execution.

TAU's measurement system has two core capabilities. First, the *event management* handles the registration and encoding of events as they are created. New events are represented in an *event table* by instantiating a new *event record*, recording the *event name*, and linking in storage allocated for the event performance data. The event table is used for all atomic and interval events regardless of their complexity. Event type and context information are encoded in the event names. The TAU event management system hashes and maps these names to determine if an event has already occurred or needs to be created. Events are managed for every thread of execution in the application.

Second, a runtime representation called the *event callstack* captures the nesting relationship of interval performance events. It is a runtime measurement abstraction for managing the TAU performance state for use in both profiling and tracing. In particular, the event callstack is key for managing execution context, allowing TAU to associate this context with the events being measured.

The TAU measurement system implements another novel performance observation feature called *performance mapping* [298]. The ability to associate low-level performance measurements with higher-level execution semantics is important to understanding parallel performance data with respect to the application's structure and dynamics. Performance mapping provides a mechanism whereby performance measurements, made for one instrumented event, can be associated with another (semantic) event at a different level of performance observation. TAU has implemented performance mapping as an integral part of its measurement system and uses it to implement sophisticated capabilities not found in other tools.

4.5.3.2   Parallel Profiling

Profiling characterizes the behavior of an application in terms of its aggregate performance metrics. Profiles are produced by calculating statistics for the selected measured performance data. Different statistics are kept for interval events and atomic events. For interval events, TAU computes exclusive and inclusive metrics for each event. Exclusive metrics report the performance when only the interval event is active, excluding nested events. Inclusive metrics report the exclusive performance of the interval event and all its nested events. Typically one source is measured (e.g., time), but the user may configure TAU with an option for using multiple counters and specify which are to be tracked during a single execution. For atomic events, the statistics measured include maxima, minima, mean, standard deviation, and the number of samples. When the program execution completes, a separate profile file is created for each thread of execution.

The TAU profiling system supports several profiling variants. The most basic and standard type of profiling is called *flat profiling*, which shows the exclusive and inclusive performance of each event, but provides no other performance information about events occurring when an interval is active (i.e., nested events). In contrast, TAU's *event path profiling* can capture performance data with respect to event nesting relationships. In general, an *event path* identifies a dynamic event nesting on the event stack of a certain length. An event path profile of length one is a flat profile. An event path profile of length two is often referred to as a *callgraph profile*. It provides performance data for all parents and children of an interval event. By specifying a path length in an environment variable, TAU can make a performance measurement for any event path of any length. Again, exclusive and inclusive performance information is kept in the measurement. The objective of callpath profiling is to gain more insight in how performance is distributed across the program's execution.

While callpath profiling can reveal the distribution of performance events based on nesting relationships, it is equally interesting to observe performance data relative to an execution *state*. The concept of a *phase* is common in scientific applications, reflecting how developers think about the structural, logical, and numerical aspects of a computation. A phase can also be used to interpret performance. *Phase profiling* is an approach to profiling that measures performance relative to the phases of execution [226]. TAU supports an interface to create phases (*phase events*) and to mark their entry and exit. Internally in the TAU measurement system, when a phase, P, is entered, all subsequent performance will be measured with respect to P until it exits. When phase profiles are recorded, a separate parallel profile is generated for each phase. Phases can be nested, in which case profiling follows normal scoping rules and is associated with the closest parent phase obtained by traversing up the callstack. When phase profiling is enabled, each thread of execution in an application has a default phase corresponding to the top level event. When phase profiling is not enabled, phases events acts just like interval events.

TAU also has support for recording of the current values of parallel profile measurements while the program is being executed in a form of a *parallel profile snapshot* [248]. The objective is to collect multiple parallel profile snapshots to generate a time-sequenced representation of the changing performance behavior of a program. In this manner, by analyzing a series of profile snapshot, temporal performance dynamics are revealed.

4.5.3.3   Tracing

Parallel profiling aggregates performance metrics for events, but cannot highlight the time varying aspect of the parallel execution. TAU implements robust parallel tracing support to log events in time-ordered tuples containing a time stamp, a location (e.g., node, thread), an identifier that specifies the type of event, event-specific information, and other performance-related data e.g., hardware counters). All performance events are available for tracing and a trace is created for every thread of execution. TAU will write these traces in its modern trace format as well as in VTF3 [296], OTF [191], and EPILOG [246] formats. TAU writes performance traces for post-mortem analysis, but also supports an interface for online trace access. This includes mechanisms for online and hierarchical trace merging [66, 68].

4.5.3.4   Measurement Overhead Control

TAU is a highly-engineered performance system and delivers very high measurement efficiencies and low measurement overhead. However, it is easy to naively construct an experiment that will significantly perturb the performance effect to be measured. TAU implements support to help the user manage the degree of performance instrumentation as a way to better control performance intrusion. The approach is to find performance events that have either poor measurement accuracy (i.e., they are small) or a high frequency of occurrence. Once these events are identified, the event selection mechanism described above can be used to reduce the instrumentation degree, thereby reducing performance intrusion in the next program run.

In addition, TAU offers two runtime techniques for profiling to address performance overhead. The first is *event throttling*. Here TAU regulates the active performance events by watching to see if performance intrusion is excessive. Environment variables `TAU_THROTTLE_PERCALL` and `TAU_THROTTLE_NUMCALLS` can be set to throttle when thresholds of per call overhead or number of calls are exceeded. The second is *overhead compensation*. Here TAU estimate how much time is spent in various profiling operations. TAU will then attempt to compensate for these profiling overheads while these events are being measured. This is accomplished by subtracting the estimated amount of time dedicated to profiling when calculating time spent for an event. TAU can also compensate for metrics besides time (e.g., floating-point operations). Overhead compensation requires small experiments to be run prior to execution. Its accuracy in practice will depend on how stable the estimated overhead is during the program run.

4.5.4   TAU Analysis

As the complexity of measuring parallel performance increases, the burden falls on analysis and visualization tools to interpret the performance information. If measurement is the heart and soul of the TAU performance system, the analysis tools bring TAU to life. As shown in Figure 4.6, TAU includes sophisticated tools for parallel profile analysis and leverages existing trace analysis functionality available in robust external tools, including the Vampir [251] and Expert/CUBE [313, 372] tools. This section focuses on TAU's parallel profile analysis and parallel performance data mining.

4.5.4.1   Parallel Profile Management

The TAU performance measurement system is capable of producing parallel profiles for tens to hundreds of thousands of threads of execution consisting of hundreds of events. Scalable analysis tools are required to handled this large amount of detailed performance information. Figure 4.7 shows TAU's parallel profile analysis environment. It consists of a framework for managing parallel profile data, *PerfDMF* [174], and TAU's parallel profile analysis tool, *Para-Prof* [49]. The complete environment is implemented entirely in Java.

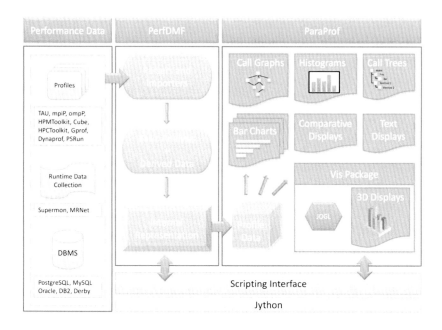

FIGURE 4.7: TAU parallel profiling environment.

PerfDMF provides a common foundation for parsing, storing, and querying parallel profiles from multiple performance experiments. It supports the importing of profile data from tools other than TAU through the use of embedded translators. These are built with PerfDMF's utilities and target a common, extensible parallel profile representation. Many profile formats are supported in addition to TAU profiles. The profile database component is the center of PerfDMF's persistent data storage and builds on robust SQL relational database engines. Once imported to the database, profile data can also be exported to a common format for interchange between tools.

To facilitate performance analysis development, the PerfDMF architecture includes a well-documented data management API to abstract query and analysis operation into a more programmatic, non-SQL form. This layer is intended to complement the SQL interface, which is directly accessible by the analysis tools, with dynamic data management and higher-level query functions. Analysis programs can utilize this API for their implementation. Access to the SQL interface is provided using the Java Database Connectivity (JDBC) API.

4.5.4.2   Parallel Profile Analysis

The *ParaProf* parallel profile analysis tool [49] included in TAU is capable of processing the richness of parallel profile information produced by the measurement system, both in terms of the profile types (flat, callpath, phase, snapshots) as well as scale. ParaProf provides users with a highly graphical tool for viewing parallel profile data with respect to different viewing scopes and presentation methods. Profile data can be input directly from a PerfDMF database and multiple profiles can be analyzed simultaneously.

To get a brief sense of what ParaProf can produce, consider the AORSA (All Orders Spectral Algorithm) [183] simulation code, one of the major applications of DOE's fusion modeling program. There are a 2-D and 3-D versions of AORSA. AORSA-2D provides a high-resolution, two-dimensional solutions for mode conversion and high harmonic fast wave heating in tokamak plasmas. AORSA-3D provides fully three-dimensional solutions of the integral wave equation for minority ion cyclotron heating in three dimensional stellarator plasmas. We ran AORSA-2D on the Cray XT5 (Jaguar) system at Oak Ridge National Laboratory to investigate its performance scaling behavior. The code was instrumented at routine boundaries and the outer loop level to evaluate the effect of multi-core CPUs on the overall code performance and to identify inefficiencies. Figure 4.8 shows a parallel profile from a 256-core execution. Here ParaProf has analyzed the full performance data and generated the three-dimensional view shown. This view is useful for highlighting performance features that reflect the relationships between events (left-side axis) and execution threads (bottom-side axis). In this case, the patterns reflect aspects of the model being simulated. Exclusive time is the performance metric (right-side axis).

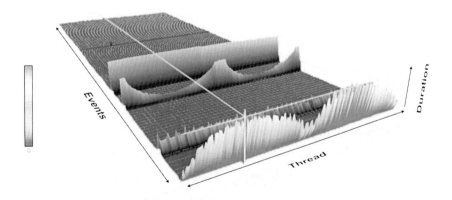

FIGURE 4.8: ParaProf displays of AORSA-2D performance on the ORNL Cray XT5 (Jaguar) system. (See color insert.)

This example shows just a part of ParaProf's capabilities. ParaProf can produce parallel profile information in the form of bargraphs, callgraphs, scalable histograms, and cumulative plots. ParaProf is also capable of integrating multiple performance profiles for the same performance experiment, but using different performance metrics for each. Phase profiles are also fully supported in ParaProf. Users can navigate easily through the phase hierarchy and compare the performance of one phase with another.

ParaProf is able to extend its analysis functionality in two ways. First, it can calculate derived statistics from arithmetic expressions of performance metrics. A simple example of this is "floating point operations per second" derived from two metrics, "floating point counts" and "time." Second, ParaProf analysis workflows can be programmed with Python, using the Jython scripting interface.

4.5.4.3   Parallel Performance Data Mining

To provide more sophisticated performance analysis capabilities, we developed support for parallel performance data mining in TAU. *PerfExplorer* [172, 173] is a framework for performance data mining motivated by our interest in automatated performance analysis and by our concern for extensible and reusable performance tool technology. PerfExplorer is built on PerfDMF and targets large-scale performance analysis for single experiments on thousands of processors and for multiple experiments from parametric studies. PerfExplorer addresses the need to manage large-scale data complexity using techniques such as clustering and dimensionality reduction, and the need to perform automated discovery of relevant data relationships using comparative and correlation analysis techniques. Such data mining operations are enabled in the PerfExplorer framework via an open, flexible interface to statistical analysis and computational packages, including WEKA [371], the R system [332], and Octave [128].

Figure 4.9 shows the architecture of TAU's PerfExplorer system. It supports process control for scripting analysis processes, persistence for recording results of intermediate analysis, provenance mechanisms for retaining analysis results and history, metadata for encoding experiment context, and reasoning/rules for capturing relationships between performance.

PerfExplorer is able to analyze multi-experiment performance data and we have used it to understand scaling performance of the S3D application on the IBM BG/P (Intrepid) system at Argonne National Laboratory. S3D [320] is a turbulent combustion simulation system developed at the Combustion Research Facility at Sandia National Laboratories. We collected S3D parallel profile data on Intrepid for jobs ranging up to 12,000 cores. Figure 4.10(a) is a plot produced by PerfExplorer showing how major S3D events scale. Figure 4.10(b) is a plot showing the correlation of these events to total execution time. Events more strongly correlated with total execution time have a positive slope. These analyses were fully automated by PerfExplorer.

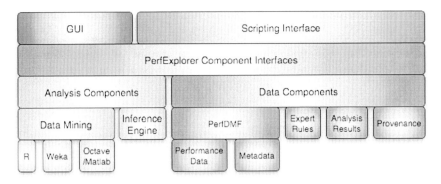

FIGURE 4.9: TAU PerfExplorer data-mining architecture.

4.5.5   Summary and Future Work

The TAU Performance System® has undergone several incarnations in pursuit of its objectives: flexibility, portability, integration, interoperability, and scalability. The outcome is a robust technology suite that has significant coverage of the performance problem solving landscape for high-end computing. TAU follows an instrumentation-based performance observation methodology. This approach allows performance events relevant to a parallel program's execution to be measured as they occur and the performance data obtained to be interpreted in the context of the computational semantics. However issues of instrumentation scope and measurement intrusion have to be addressed. Throughout TAU's lifetime, the project has pursued these issues aggressively and enhanced the technology in several ways to reduce the effects of the probe.

TAU is still evolving. We have added support for performance monitoring to TAU built on the Supermon [253] and MRNet [256] scalable monitoring infrastructures. The goal here is to enable opportunities for dynamic performance analysis by allowing global performance information to be accessed at runtime. We have extended our performance perspective to include observation of kernel operation and its effect on application performance [255]. This perspective will broaden to include parallel I/O and other sub-systems. TAU is also being extended to support performance analysis of heterogeneous parallel systems, in particular systems with GPU acceleration. In this regard, we are developing a CUDA performance measurement infrastructure [230]. Our current and long-term vision for TAU is targeted to *whole performance evaluation* for extreme scale optimization. Here we are working to integrate TAU in the system software stack and parallel application development environments being created for petascale and exascale systems.

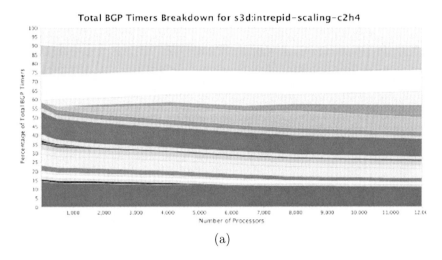

(a)

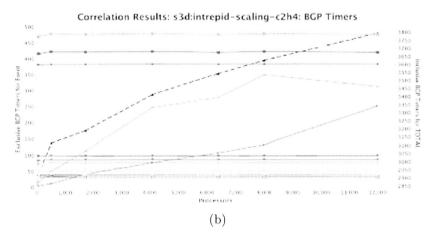

(b)

FIGURE 4.10: PerfExplorer analysis of S3D performance on IBM BG/P. (a) S3D events scale. (b) correlation of events to execution time. (See color insert.)

4.6   Summary

Over the last two decades, research in the area of performance tools has delivered important ideas, approaches, and technical solutions. HPCTOOLKIT and the TAU Performance System are representative of advances in the field. The impending emergence of extreme-scale parallel systems that incorporate heterogeneity will drive the development of next-generation tools. Performance tools for extreme-scale systems will need new measurement capabilities to support new architectural features, new strategies for data management to cope with the torrent of information from these systems, and automated parallel analysis capabilities to sift through reams of performance data effectively and efficiently. With the growing complexity of emerging systems, it will be increasingly important for next generation performance tools to increasingly focus on providing higher-level guidance for performance tuning rather than just pinpointing performance problems.

Acknowledgments

The HPCTOOLKIT project is supported by the Department of Energy's Office of Science under cooperative agreements DE-FC02-07ER25800 and DE-FC02-06ER25762. Principal contributors to HPCTOOLKIT include Laksono Adhianto, Michael Fagan, Mark Krentel, and Nathan Tallent.

The TAU project is supported by the U.S. Department of Energy, Office of Science, under contracts DE-FG03-01ER25501 and DE-FG02-03ER25561. It is also supported by the U.S. National Science Foundation, Software Development for Cyberinfrastructure (SDCI), under award NSF-SDCI-0722072. The work is further supported by U.S. Department of Defense, High-Performance Computing Modernization Program (HPCMP), Programming Environment and Training (PET) activities through Mississippi State University under the terms of Agreement No. #GSO4TO1BFC0060. The opinions expressed herein are those of the author(s) and do not necessarily reflect the views of the DoD or Mississippi State University.

This chapter reports results obtained using resources at both Argonne's Leadership Computing Facility at Argonne National Laboratory, which is supported by the Office of Science of the U.S. Department of Energy under contract DE-AC02-06CH11357; and the National Center for Computational Sciences at Oak Ridge National Laboratory, which is supported by the Office of Science of the U.S. Department of Energy under Contract No. DE-AC05-00OR22725.

Chapter 5

Trace-Based Tools

Jesus Labarta

Barcelona Supercomputing Center

5.1 Introduction

Performance analysis aims at observing the rate of progress of a computation, quantifying it in terms of some metric and inferring factors that determine the actual performance of a system. Properly identifying these factors and understanding their impact is key to effective application tuning.

As in all areas of science, observation and data acquisition is the first step towards the characterization and understanding of system behavior, and as in all areas of science, two different approaches can be followed. A first direction is to summarize the raw data stream close to where it is produced in order to minimize the amount of data that has then to be transported and subsequently processed off-line. The generic issue is how much raw data should be captured for later off-line study, and what are the algorithms and statistics that should be applied to the raw data during the acquisition process so that most useful information in the data is retained.

Performance analysis is not different. It is also characterized by a very large high production rate and restrictions in storage and post processing. The space of possible designs is a continuum between just reporting after the program run a handful of summarized values to storing all raw captured data for off-line analysis. Different positions in this spectrum have been taken by performance analysis tools.

In this chapter we will focus on approaches that capture as much data as possible including timing and ordering information at run time to post process and analyze it off-line. Trace-based tools have undergone a transformation from not being considered practical to the point of clearly demonstrating their usefulness, and today appear to be finding a solid place within the broad spectrum of performance tools for high-performance computing. Throughout the chapter we will describe general issues faced by this approach, give some details of how they have been addressed by the CEPBA-Tools environment (MPItrace, Paraver, Dimemas, . . .) and introduce some of the techniques that we consider will be relevant in the future.

The chapter is structured as follows. Section 5.2 discusses the motivations for traced based tools and Section 5.3 looks at some of the major challenges faced by this approach, whereas Section 5.4 examines data acquisition. Section 5.5 presents some analysis techniques to identify structure in traces and in Section 5.6 we look at modeling techniques. Section 5.7 looks at interoperability between different types of tools. We conclude with a view on how we perceive the future of trace-based tools and how techniques learned from them are likely to evolve.

5.2   Tracing and Its Motivation

The tracing approach is rooted in the consideration that variability and details are important elements of parallel program behavior. When looking for real insight, the search proceeds through a huge and fuzzy space where hypotheses arise as the analysis proceeds. Very precise measurements of unforeseen metrics may be needed to validate or reject the hypotheses.

As we face the multicore and exascale era, we expect high levels of asynchrony and dynamic scheduling to be supported by the programming models and run time systems in order to overlap latencies and accelerate the progress of computation on the critical path. Programs will be less SPMD-like or lockstep-like than today and run time systems will have more responsibility in optimizing the mapping between the application's demands and the architecture's resources. Such dynamic environments will result in great variability in the duration and execution order of tasks. Different runs of the same program will probably have very different internal sequences of events and just providing average statistics will miss the detailed interactions between application demands and resource availability that determine the system performance. In a world constrained by energy consumption, the efficient usage of computational resources will be even more important than in the past. We believe that this situation will make necessary the generation of detailed snapshots (traces) of a single run, as well as off-line searching for insight into important performance behavior at the microscopic level.

Trace based visualization tools [252, 275, 385] are browsers to navigate the raw data and analyze it in the search for information. These tools are able to present the information as timelines displaying the evolution of the metrics and activity in both the process number and time dimension. They can also be used to compute and present statistics for the range of processes and time selected though the GUI.

Kojak [246] presents an interesting point in the above mentioned spectrum. It is a system that presents profile data but whose internal implementation is based on traces. The profiles are generated by off-line processing of the traces through a program that can apply complex algorithms and obtain more elaborated metrics than what on-line processing would allow. In particular, having access to all the events incurred by all processes, it is possible to compute metrics that depend on the time relationship between events in two or more processes. An example of such metrics is *late senders* which quantifies the blocking delay suffered by processes waiting to receive data when the matching send takes place significantly after the receive is issued. Other metrics such as *Wait at N x N* are also defined for collective operations. All of these metrics are computed and emitted into a profile file which can then be interactively browser by an advanced profile presentation tool (CUBE [147]).

The spectrum of approaches is thus essentially differentiated by how late one computes metrics in the analysis process and to what extent the analyst is able to define new ones. Ideally the later and more flexible the mechanism, the better equipped the analyst is to iterate the hypothesis-validation loop as many times as possible and to face unexpected issues. However, the more flexible and powerful the tool, the more expertise (architecture, programming models, operating system, etc.) is required from the analyst.

Paraver [204] is a flexible trace visualizer developed at the Barcelona Supercomputer Center (BSC). It can display timelines identifying the region of the code (i.e., functions, MPI calls, etc.) being executed along the time axis, a selected set of hardware counter-derived metrics, and advanced communication-related metrics such as transfer bandwidth. A second type of display is based on a mechanism to compute tables from timelines. This mechanism can be used to compute and display both profiles and histograms. A large number of metrics and statistics are programmable through the GUI. When a timeline, profile or histogram showing a relevant metric is generated by an analyst, it can be saved into a configuration file that can later be loaded by nonexpert users on the same or different traces. The values reported by both timelines and tables can also be saved in textual form to be further processed with other data analysis tools such as Excel and Matlab.

Figure 5.1 shows an example timeline of a GPUSs [34] run on a node with 4 GPUs. The algorithm is a matrix multiply operation, and the individual tasks are implemented with CUBLAS. The timeline on the top shows in yellow the transfer of data between the host memory and the device. Brown corresponds to actual execution of a task. Detailed bandwidth for each such transfer is shown in the middle timeline for the same time interval. In the timelines we can see how transfers do not overlap with computation. We also observe that transfers to/from all GPUs tend to cluster and happen at the same time. This will tend to overload the communication link which has a limited bandwidth compared to the computational power of the devices. The effect can be observed in Figure 5.2 which shows the histogram (computed by Paraver and saved in a file read by excel) of effective bandwidths when the program is run with one, two and four GPUs. We see how the effective bandwidth of the transfers is in the order of 2.2 Gbyte/s when only one GPU is used, but significantly less and with broader distribution when more GPUs are used.

The simultaneous attempt to use the PCI bus to transfer data to/form all GPUs result in lower average bandwidth and higher variability as the number of GPUs increases. The effect does have some kind of self synchronizing behavior and the analysis suggests slightly modifying the run time to counteract this effect. By limiting the number of concurrent transfers it is possible to skew the activity of the different GPUs avoiding contention and achieving better overlap in the utilization of the resources (PCI and GPU).

Similar instrumentation in other middleware developed at BSC (CellSs [50], MPI+SMPSs [228], GridSs [37], Load balance [146]) lets us understand the detailed microscopic issues that end up having a significant impact in the macroscopic application behavior. We believe that run times for new programming models and system software developments at many different levels will have to be instrumented to capture snapshots of the activity of the application as well as details of their internal operation in order to identify potential improvements in the mechanisms and policies they apply.

5.3 Challenges

One major concern being faced by trace based tools is scalability, in at least two sense. On one hand, storing and handling the potentially huge amounts of data generated by a computation running on thousands of processors is a substantial a challenge. But visualization of these very large datasets is

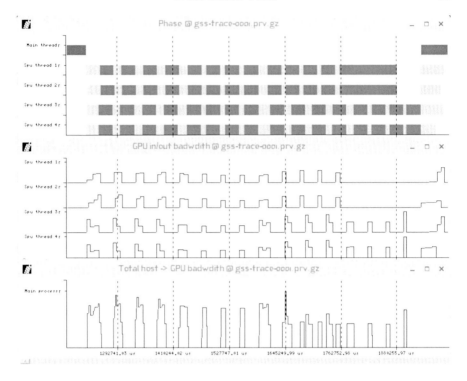

FIGURE 5.1: Trace of a GPUSs MxM on a mode with 4 GPUs.

also a daunting challenge — indeed, it is often regarded as being beyond the capabilities of current displays. Let us address these two issues.

5.3.1 Scalability of Data Handling

Instrumentation of parallel MPI programs typically involves injecting "probes" at each MPI call. Additionally, one can instrument user-level routines, as identified by a previous performance profile of the application, by inserting probes at the entry and exist of each routine. Each of these probes collects and emits trace information about the specific call and its parameters, such as message size and communication partner, and possibly a few hardware counter values, such as number of floating-point operations, instructions and cache misses, since the previous probe. Depending on the application, such instrumentation, may result in data production rates that may approach one Mbyte/s per core, although it is often closer to a few tens of Kbyte/s. Table 5.1 shows some data production rates for real applications when using the MPItrace package and instrumenting MPI calls. Similar values have been reported for other tools such as Scalasca.

This process often generates large amounts of raw data. Merely creating

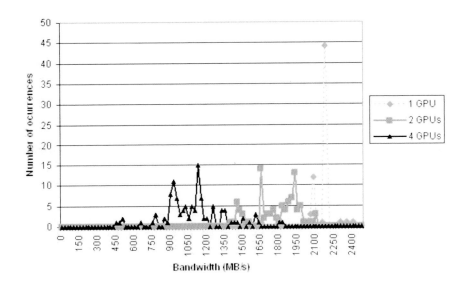

FIGURE 5.2: Histogram of bandwidth for different number of GPUs.

the files, performing the output at generation time and reading them at the post-processing stage often is in itself a significant challenge, even in an era when very large online storage facilities are available. Two approaches can be employed to address this issue: (a) use scalable data handling mechanisms, and (b) implement data filtering mechanisms.

Two tools that are representative of the first approach are VampirServer and Scalasca. VampirServer [67] is a redesign of the previous Vampir [252] visualization tool with special emphasis on being able to handle very large traces. VampirServer displays Open Trace Format (OTF) [191] traces. One way in which this trace format addresses scalability issues is that a single logical trace can be split onto several files, thus allowing for concurrent storage

TABLE 5.1: Data production rates for different applications and tracing setups

Application	Processes	Run time (minutes)	Full trace size (GB)	Per proc. Rate (MB/s)
SPECFEM3D	64	58	3	0.01
GROMACS	64	8	20	0.65
MILC	245	5.5	5.5	0.68
ZEUS-MP	256	10	22	0.14
LESLIE3D	512	56	82	0.05
GADGET	2048	3.5	37	0.09

and processing. VampirServer follows a client-server approach, wherein the visualization client can query several servers that process in parallel different parts of the trace and reply with their respective data to visualize. OTF also includes mechanisms to store partial summarized data or pointers, permitting fast access to specific parts of the trace. These data handling mechanisms permit VampirServer to handle traces up to several tens of GBbytes. Other scalable trace formats, such as the "Slog" format, also follow this hierarchical and partitioned approach to structure the raw data.

Scalasca [373] is a performance analysis system that presents profile data generated from traces. Scalasca builds on the previous experience with Kojak in addressing the scalability issues. The statistics generation program is itself an MPI parallel program that replays the trace and computes the statistics just immediately after the instrumented application run. New versions of the system attempt to economize the file access overhead of storing and reading back traces by setting up memory regions persistent across applications, so that the analyzer running immediately after the instrumented application will find the trace data in memory. Since it is a parallel program itself, it requires a large amount of memory, communication and compute power in order to compute the required metrics in a reasonable time.

Research has also been performed in the area of trace compression. For examlpe, with Paraver (and for archival purposes), we observe compression rates of 1:8 with standard lossless compression packages. Systems such as [261] or [192] are lossless in terms of preserving communication patterns but lossy in terms of retaining the actual computation time between communications. These techniques are based on the predictably repetitive structure of most applications [138] in terms of the communications sequence; the computation bursts between MPI calls tend to be exhibit higher variability. Some researchers [283] have proposed different discretizations of this time in order to reduce the trace size and still capture, with good accuracy, the global behavior of the application.

The second principle approach to address the large size of trace datasets is to examine the application structure and selectively focus on the minimal amount of raw data that captures the information relevant for the analysis. We will describe some of these techniques in the following paragraphs. Such techniques are being developed to make the whole process automatic, avoiding the need for human intervention and thus speeding up the analysis process. We will describe them in Section 5.5.

The simplest approaches require the analyst to specify what part of a run or what type of information to capture. This is done through an application performance interface (API) to start and stop the tracing at specific points in the source code. The typical practice is to skip the initialization phase and only trace a small number of iterations of the main computation phase. This approach requires access to the source code and knowledge of its general structure. Although usually not very difficult to identify by hand, it is often a cumbersome process and constitutes a non-negligible part of the analysis. The

TABLE 5.2: Trace size (Mbyte) for GADGET when using software counters and different computation burst thresholds

Processes	75 ms	150 ms	300 ms	600 ms
512				2.4
1024			4.9	3.7
2048	17.5	10.6		6.2
4096	47.7		4.9	3.7

approach may still result in large traces if the number of cores is very large, the interactions between them are frequent or the durations of iterations are long. Furthermore, it is not useful in the case we need to analyze application variability during long runs.

MPItrace implements additional mechanisms proposed in [203] to address some of the above limitations. The first idea is not to restrict the time interval, but instead to emit detailed information just for some relevant events. Based on our analysis experience, we consider as relevant events only those that characterize the long computation bursts (long serial computation between two MPI calls). Hardware counter events are only emitted at the start and end of these computation burst. This is a way of saving storage overhead for hardware counters, since they are not emitted for short computation bursts. It is often the case that a small percentage — often less than 10% — of the computation bursts represent more than 90% of the elapsed time. It is even possible not to emit any MPI events, and given that the emitted computation bursts are time-stamped, still nonetheless draw conclusions on the behavior of the communication activity of the program. This approach is able to drastically reduce the trace size. For example, a trace of 3000 seconds of a Linpack run on 10K cores only generates 527 Mbyte of data.

An intermediate approach [203] we call *software counters* is to emit summarized data for otherwise discarded events at the start of long computation bursts. Examples are a count of the number of MPI calls and an aggregate of the number of bytes transferred in the interval since the end of the previous long computation burst. Some example trace sizes for runs of the GADGET astrophysics code at different core counts and burst duration thresholds are given in Table 5.2. We can see how the sizes are drastically reduced over the full sizes reported in table 5.1 and still contain very useful information to explain the scalability behavior of the application. We will show in Section 5.3.2 how filtered traces and software counters do support detailed analyses and displays.

Paraver lets the user navigate with acceptable response times with traces up to a few hundred Mbyte in size. This does not mean that larger traces can not be analyzed. Paraver is also a trace manipulation tool. In the case the trace file is too large, Paraver offers the possibility of filtering it, generating a new trace file with a subset of the raw data in the original one. The approach is equivalent to the *software counters* functionality of MPItrace but implemented off-line. The user can specify in an xml file the type of information that should be kept and what should be dropped or aggregated using software counter techniques. A typical practice is to generate a filtered file with only the long computation bursts and possibly collective MPI calls. This file can be loaded with Paraver and the analyst can identify the structure of the application (i.e., repetitive pattern, perturbed regions,...). The GUI can be used to specify in this view a time interval of interest, and Paraver will extract a cut of the original trace with full detail for that interval.

These manipulation capabilities support the hierarchical analysis of very large traces. By operating with the full original trace file, it is possible to iteratively extract different subsets of the original data as the analysis proceeds. This often results in faster response time than resubmitting the instrumented application. It also has also the advantage that all the different partial traces that may be examined during an analysis correspond to the same run. This ensures that different observations can be precisely correlated. In practice, after the initial analysis process has been done, the original trace can usually be discarded. Typically, only the filtered trace covering the whole application run and a few representative cuts with full detail are kept for further analysis or archiving.

Other trace manipulation utilities can be used to merge different traces and to shift timestamps of different processes (see Section 5.4).

5.3.2 Scalability of Visualization

Some believe that in order to visually represent metrics, one needs at least one pixel per object. Displaying performance views for tens of thousands of processors is impossible under this hypothesis. Although often not perceived as problematic, the same issue arises in the time dimension, when it is necessary to display intervals several orders of magnitude larger than the average time between events. This is an issue not only for trace visualizers, but also for profile presentation systems that map metrics onto a physical system structure, as for example CUBE [147].

The visualization resolution issue has not been as acute in other arenas, such as data visualization tools or gaming. A lot of work has been done in data visualization, and effective rendering approaches and techniques have been developed to handle different levels of detail.

The real underlying issue is not what a screen can display, but what can the human eye perceive from the screen. The real objective is to maximize the information transfer between the visualization tool engine, which has a huge amount of data, and the human vision system, which essentially performs a low pass filtering of the image. Being able to identify information, outliers and the like is the real objective, rather than generating attractive and highly precise color plots.

In [203], we described how simple nonlinear rendering techniques can be applied in trace visualization tools. In order to compute the actual value (color) displayed for each pixel, a certain function has to be applied to all the contributions to that pixel. One possibility is to use the average of all individual values, which may be an option for views displaying continuous valued functions, but which may be totally meaningless for views displaying categorical data. A particularly interesting nonlinear rendering function is to report for every pixel the maximum of all values that would map to it. For windows with categorical data this will bias the view towards values encoded as high numbers, but the total view will still provide useful information to the analyst. For windows representing numerical data, this approach easily highlights regions of the process-time space with outliers. Even if the actual outlier occurs in one out of thousands of processes and lasts for a very short time, it will be possible to focus on it with zooming facilities and analyze it in detail. Even random rendering is often substantial effective in conveying information to the analyst. In this approach, one of the many values mapping to a given pixel is selected at random and displayed. An interesting practice in this case is to redraw the same view several times, using thus different random sequences. The similarity/differences between different images ends up showing interesting information about the underlying structure of the application.

Figure 5.3 shows the instructions per cycle (IPC) within DGEMM computations in the Linpack trace of 10K cores mentioned in Section 5.3.1. The rendering varies from green for an IPC of 2.85 to dark for an IPC of 2.95. Figure 5.4 shows the L1 miss ratio (misses per thousand instructions) which is clearly correlated to the IPC pattern. The views do show structure in the behavior of the application and exemplify how the combination of scalable tracing and display techniques support extremely precise (below 5% in this example) measurements even with very large processor counts.

Figure 5.5 is an example from a similar run with 4096 cores. The view represents the duration of the communication phases for a subset of 300 processes, during five iterations of the outer loop. Computation time appears in white, while communication time is shaded. The first iteration shows the typical skewed behavior in the linearized broadcast implementations in Linpack. Outliers that take significantly more time are shown shaded. The outliers in the second iteration are due to contention in the network. The outliers in the following iterations are caused by the propagation, to other processes, of the delays induced by the previous contending communications. This example shows how it is possible to obtain a large amount of information from a trace even if the communication records have been filtered.

The other example mentioned in Section 5.3.1 was a GADGET run on 2048 processors. Figure 5.6 shows several synchronized views of one iteration (10.5 seconds) of the core computational step of this trace. Each of them represents a different metric identified on the top left corner of each timeline. For all of them the shading corresponds from light to dark, corresponding to the maximum value in the timeline. Regions in white correspond to a zero value of the metric.

The view on top shows the duration of computation bursts longer than 75

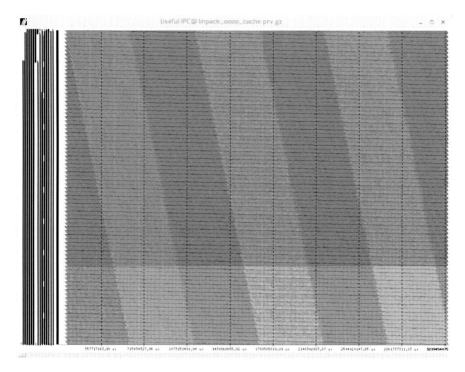

FIGURE 5.3: Evolution of DGEMM IPC over 3000 seconds of a 10K cores linpack run.

ms. We can see how during the last third of the iteration only 1024 processors perform significant computation, resulting in major performance degradation, whose cause should be further studied. All other views report information on the complementary portion of the timeline, which we will refer as communication phases, each of them representing a sequence of MPI calls separated by short computation bursts. The communication phases at the beginning of the iteration show the expected SPMD structure, where all process are either computing or communicating. Because of the load imbalance issue just identified, the last communication phase for the final 1024 processors is much longer than for the first 1024. This situation is not so typical but shows how the method used to capture data enables the identification of potentially unexpected structures.

The second view shows the percentage of time in MPI within each such phase. The values are pretty close to 100% almost everywhere, except in the light stripe just before entering the imbalanced region, where the value is 40%. The view below shows that in this communication phase there are a large number (43) of collective calls while in all other regions the number of collective calls is significantly lower. The fourth view also shows that the actual number of bytes sent by those collectives is not very high (418 Kbyte in total,

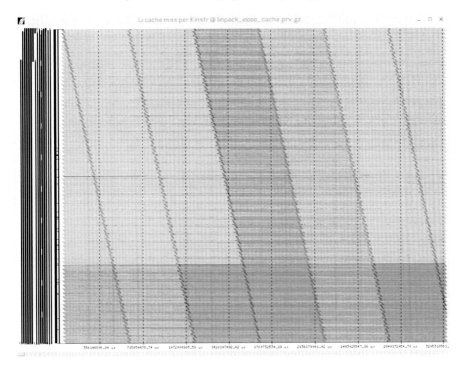

FIGURE 5.4: Evolution of DGEMM L1 miss ratio over 3000 seconds of a 10K cores Linpack run.

thus only close to 10 Kbyte on average). For example, the next communication phase has fewer collective operations but larger amounts of data sent. The fact that 60% of the time is spent in MPI in this communication phase and that it contains a tight sequence of collectives raises the concern that load imbalance and global synchronization semantics might limit the scalability of this phase. Further analysis and understanding of the detailed behavior of this phase would be desirable. The last three timelines show respectively the number of point to point calls, number of bytes transfered and effective bandwidth achieved within each communication phase. We can see that the two last communication phases is where most of the point-to-point data is transferred. From the timeline we can measure that within each such phase, each of the first 1024 processes sends a total of 13.4 Mbyte, and the effective bandwidth is 20 Mbyte/s. This value is not very high, and thus we have identified a third issue and region that deserves further analysis.

   We have shown how analysis of the software counter traces is able to identify regions in the time dimension where further analysis is needed. The trace file can also include line numbers for the first MPI call in a communication phase, thus easily pointing to the code region to examine. Even more importantly, we also have initial quantification of the performance problem and

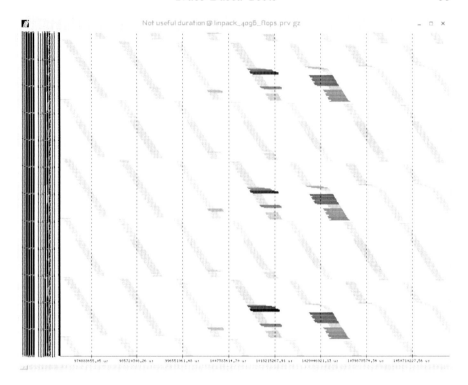

FIGURE 5.5: Duration of communication phases for 5 iterations of a Linpack run on 4096 processors.

indications of what we have to look for in a more detailed analysis (load imbalance for some regions of code, granularity and global synchronization in other regions, inefficient use of bandwidth in others, etc.). A similar analysis can be used to perform a first level of comparison between runs with different processors and check whether all computation bursts or communication phases shrink proportionally as we increase the number of processors.

5.4 Data Acquisition

Two general methods of activating the monitoring package probes are instrumentation and sampling. In the first case, probes are hooked into specific points of the application code (source or binary). This can be done through source code preprocessors [245, 246, 300] before compilation, through hooks provided by the compiler (ie. -finstrument-functions option in the gcc compiler), by binary rewriting techniques [212, 221, 316] before execution, at load

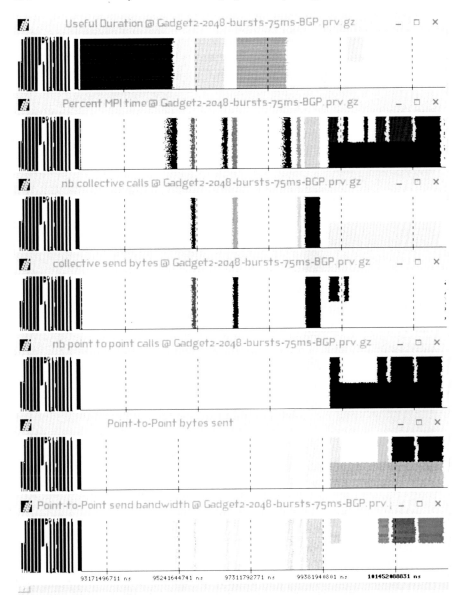

FIGURE 5.6: Several views for a trace with software counters of a GADGET run on 2048 BG/P cores.

time through dynamic linking mechanisms (such as the PMPI interface in MPI) or by totally dynamic instrumentation of the binaries in execution [69]. In the sampling case, the probes are fired when a given time interval elapses or a certain hardware counter overlaps (ie. using PAPI [65]).

When instrumenting an application, the monitoring system's activation is totally correlated to the activity of the program in execution. When sampling, the actual firing of the probe is not directly correlated to a specific point in the program, although there may be some correlation to the program general activity (i.e., if sampling on the overflow of a cache miss counter, samples will probably concentrate in parts of the code experiencing more cache misses). This is an essential difference between the two methods that influences the way statistics are then obtained from the data. In the first case, profiles or detailed statistics can be accurately computed form the captured data. In the second, statistical techniques have to be used in order to obtain the profile information. A second difference is that instrumentation does not guarantee an acquisition frequency. Instrumentation can be very coarse, in which case no data is acquired for very long intervals, which results in very poor detail in the behavior of the application. In other cases, instrumentation may lead to a large overhead, especially if the instrumented points in the code are exercised very frequently by the program. On the contrary, the sampling approach provides a simple mechanism to constrain the potential overhead by just tuning the frequency (or counter overflow value).

The trade-off between instrumentation and sampling is then in terms of precision and overhead. An interesting research direction is to further exploit the differences and complementarities between them as we will describe in Section 5.5. The objective is to instrument only major routines and to derive very precise information of the internal behavior of such routines by correlating the sampled data to the instrumented data.

An important issue in the case of acquiring data for traces and subsequent off-line analysis regards the timestamping process. When different data are captured by a probe (i.e., call arguments, hardware counters, etc.), one might be tempted to get a timestamp for each of these values, for example reading the clock and the event value for every hardware counter. To support precise off-line analysis, all data captured by each probe should have a single timestamp, representing the occurrence of the event that fired the probe. Especially when reading hardware counters, atomicity in the capture is more important than the total duration of the probe. Ideally, we would like to have a hardware mechanism where a single instruction would force a simultaneous copy of the active counters into buffer registers. The probe could then read these frozen copies without possible perturbations such as interrupts modifying the captured values. Of course perturbations would affect the running counters and thus the value for the next probe, but the important thing is that the atomic mechanism would avoid the situation where some of the counters read by one probe would include the effect of the perturbation while the effect for other counters would be accounted in the next probe. In the absence of such mechanism, it is important to read the hardware counters in a tight back-to-back sequence and, if possible, to disable interrupts as well.

Properly synchronizing clocks between processors may be an issue for some trace based analyses [373]. Timestamps indicating that a message has arrived to its destination before being sent may occur if the clocks at the send and receive node are not synchronized. This may be a big problem for tools that directly extract statistics out of the raw trace. Large clusters often lack global clock synchronization in hardware and thus some approaches have to be used by the tracing libraries to synchronize the timestamping clock across nodes. Often this is done by computing the offsets between local clocks at the MPI_init. In some cases [46] very elaborated methods are used to try and minimize the offset and even take into account skews in local clock frequencies.

Besides the clock synchronization issue, nuisances can appear in the timing process for two more reasons: the nonatomic process of firing a probe and reading the timestamp and other information to be emitted to the trace; and the discrete precision of the timer itself. In most present-day systems, timestamp resolutions are only a few nanoseconds.

The philosophy in Paraver is that one has to learn to live with this issue, try to keep it constrained, but not aim at totally eliminating it. MPItrace uses the simplest approach of introducing an implicit global synchronization at the end of the MPI_init call. In our experience this provides a sufficiently precise offset computation for most practical situations. Many of the timelines used in analyses do not depend on such offsets, and Paraver can display traces with backwards communications without problem. Some configuration files provided with the distribution do quantify the amount of backwards communication and show timelines where the user can see how the effect is distributed over time and processors.

Shifting the timestamps of some processes is a technique that has been demonstrated to be very useful in the case the original trace shows synchronization anomalies. We provide utilities to perform the shifting but also to compute, out of the original trace, the proper shift for each process. We often use this functionality to perfectly align the exit of a given collective. If applied to collectives that have global synchronization semantics, this technique typically results in views that expose the behavior of the application (i.e., imbalance before the collective call even at very fine granularities) much more clearly than the original trace. It is also an appropriate method to eliminate backwards communication operations around the specific collective and also to enable the generation of clean cuts of a trace starting just after the collective. Another way to eliminate backwards communication operations is to rely on the possibility of quantifying the effect. This is done through a Paraver configuration file that reports the aggregated backwards communication time experienced by each process. This number is used to shift the process and iterate trying to minimize the aggregated backwards time of all processes. Using Paramedir (a non graphical version of Paraver that emits the statistics to a file) the process can be automatized by a simple script. The gradient optimization process is not guaranteed to arrive to a null value of backwards time, but very good results are obtained in most practical cases.

5.5   Techniques to Identify Structure

Techniques to identify structure are of great importance when facing the analysis of large amounts of data. Indeed, numerous other fields of science, ranging from biology to astrophysics, are now facing similar challenges. Unfortunately the performance analysis community lags behind in the application of those techniques to reduce the raw data in the search for valuable information. This certainly applies to trace-based environments, but it is far more general and the same techniques should be used in online analysis as well.

Structural characterization at different levels can be of high relevance in the analysis of applications. Three such examples are discussed in this section. The first one uses signal processing techniques to detect the periodic behavior of an application. A second approach uses clustering techniques to identify similarities between the behavior of different processors and code regions in the program. Finally we look at ways to achieve extremely detailed information on the evolution of metrics with time without requiring very frequent data acquisition

The structural information provided by these techniques is useful to ease the analysis process, both to drive the basic mechanisms described in Section 5.3 and to point out precise areas of relevance.

5.5.1   Signal Processing

Signal processing theory has been widely applied in many areas of science and engineering, but not yet very widely in the performance analysis area. Opportunities nevertheless arise if we envisage the timeline or profile plots generated by many tools as two- or three-dimensional images susceptible of being processed by image analysis techniques. Similarly, the performance data generated by one process along the time axis can be considered as a signal to which the existing body of theory and practice can be applied.

Detecting the periodicity of a program's behavior is one such opportunity. Work in [81] used spectral analysis techniques to identify the hierarchical structure of the periodic behavior of applications, but mainly to automatically drive the extraction of a representative trace much shorter than the original one obtained with a blind instrumentation of the full application. Different signals derived form the trace can be used for the analysis. Examples are the instantaneous number of active processes doing useful computation, the complementary function representing the number of processes in MPI, the number of processes invoking a collective call, the average instantaneous IPC, etc. It is interesting to note how the signal itself need not have a physical meaning, so long as it exposes the periodic structure of the program activity. In particular, a signal that results in very good detection of the periodicity is the instantaneous sum over all processes of the duration of their computation phases.

This process can be combined with mathematical morphology techniques to identify regions of high perturbation (by coalescing clustered burst in a signal that represents the instantaneous number of processes flushing trace data to disk) and thus skip them in the algorithm to detect periodicity. The wavelet transform combined with mathematical morphology can be used to detect regions with different structure [80]. This allows, for example, the automatic identification of initialization and termination phases in a trace, so that they can be disregarded as we typically are interested in analyzing the core computation phase.

Other uses can be identified, for example to monitor the progress rate of the program, to compress performance metrics using the wavelet transform [145] or to detect acquisition intervals that may lead to better estimates of global metrics [82].

5.5.2   Clustering

Clustering algorithms group data points in an n-dimensional space by their similarity. This is typically measured in terms of Euclidean or Manhattan distance. A wide variety of clustering algorithms have been proposed trying to detect concentration of data points in such a space. Some classes of algorithms search for spherically shaped clusters (for example k-mean-based algorithms) or just densely populated regions but still irregular in structure (density-based algorithms).

The most frequent use of clustering techniques in parallel program performance analysis [130, 260] characterize each process as a single point in a multidimensional space, where each dimension represents a given metric computed for the whole duration of the run (i.e., percentage of time in a given routine, total number of floating point instructions, total time in MPI, ...). The k-mean scheme is typically used here, and the analysis typically aims at identifying differentiated groups of processes such that one representative of each group can be used for a detailed analysis.

In [150] we have focused at a much finer granularity. In this case, each point represents just one computation burst between exit from an MPI call and the entry to the next one. Each such point is characterized by its duration and hardware counter metrics. The number of points is much larger than in the previous approaches and the objective is to detect internal structure within the processes. The clustering algorithm is based on density (DBSCAN), since the observed distribution of points in the space shows clouds of points with different principal component directions and shape (not necessarily spherical). For example, one application may have clusters where the same number of instructions is executed with IPCs varying in a relatively broad continuum range on different processors, while for other regions of the same program the IPC may actually be correlated with the number of instructions.

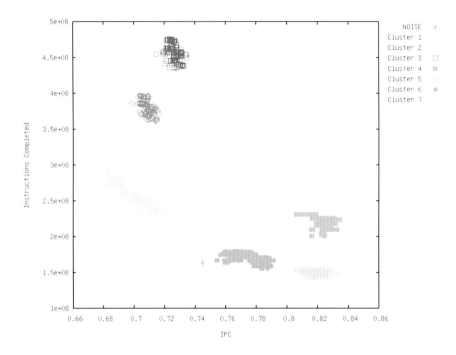

FIGURE 5.7: Clustered scatter plot of instructions vs. IPC for computation bursts in a 128 processors run of WRF. (See color insert.)

While the clustering algorithm is an interesting issue in itself, the really important point is how to validate and use the information. Two types of outputs from the clustering utility are relevant at this point: scatter plots and labeled traces. A scatter plot like the one showed in Figure 5.7 shows a two-dimensional projection of the n-dimensional space along two metrics relevant from the analysis point of view. Instructions and IPC are two typically used dimensions, since the distribution of the points in this space rapidly gives insight into possible load imbalance. Variability within a cluster in the number of instructions does point to computational load imbalance and directly suggests that the partitioning of computations should be improved. Variability in the IPC points to potential differences across processors, often related to different locality behavior. In finite element codes for example this typically suggests the need to improve the numbering algorithms used to achieve better and more uniform locality across domains. As each cluster is also referred to the section of the source code to which it corresponds, the analyst can directly point to where the tuning should be done.

A scatter plot alone may nevertheless provide not enough information to the analyst, since the points are not labeled with timestamp and processor. Thus doubt may arise whether its apparent structure actually corresponds to structure in the processor and time dimension or not. A given cluster may

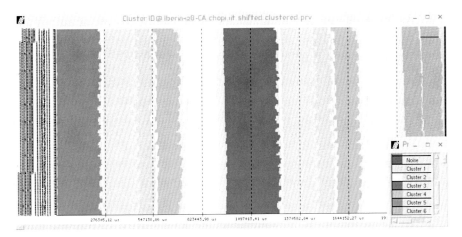

FIGURE 5.8: Clustered timeline of a 128 processor run of WRF. (See color insert.)

contain points from all processes or just a subset, may be regularly distributed in time or not, etc. Timelines showing the space/temporal distribution of the identified clusters would let the analyst answer those questions. The clustering utility feeds its results back into the original trace, introducing records that label the start and end of the clusters. Figure 5.8 shows the time distribution of the clusters with the same shading scheme as the scatter plot in Figure 5.7. Paraver can also be used to extract much more detail such as histograms of specific metrics for each individual cluster. Both the scatter plot and the visualization of the clustered trace are complementary presentation schemes that provide high-level insight into the behavior of the major regions of the parallel program.

Clustering algorithms are controlled by parameters selected by the analyst. Examples of such parameters are the desired number of clusters and a distance threshold differentiating what is considered to be similar and what is not. Different values of the parameters produce different granularities in the structure, and it may be useful to look at several of them. Selecting the proper value may be a difficult task, probably requiring several trials. For this reason, some tools [172] compute the clustering results for various values of the control parameter and let the analyst decide which is the most appropriate. In practice, this way of checking the quality of the clustering (known as the expert criterion) is a very important one. In our case, given that most programs today follow an SPMD structure, one would expect all processes to be doing the same type of computation at the same time. Looking at the timeline display of the clusters (Figure 5.8), an analyst can check whether the identified clusters do show such SPMD structure and determine whether it is appropriate to refine or coarsen the clusterization parameter.

Ideally one would like to have mechanisms for formally evaluating the quality of the clustering, but that is difficult in the general case. In [150], we present an approach that leverages algorithms from protein alignment developed in the life sciences area to compute the *spmdiness* of the clustering results. A duality can be made between a DNA fragment as a sequence of nucleotides and a process as a sequence of clusters. In our case, a good SPMD behavior would be one where the sequence of clusters for all processes can be properly aligned, indicating that they all perform the same activity at the same time. In the alignment process, some holes can be inserted, as it may be the case that different processes perform, for example, more communication operations (i.e., they are allocated domains with more neighbors) and thus may have a different number of computation bursts in the exchange phases. These computation bursts will typically be small compared to the main computation bursts between communication phases. The alignment algorithm receives as input weights that favor the alignment of long clusters. Finally, a numerical metric between zero and one is derived that quantifies the SPMD-ness of the clustering. This quantitative criteria is used to obtain an estimate of the quality of the clustering before even looking at the results. Our future work will use this metric in search algorithms to automatically identify good parameters for the clustering algorithm.

5.5.3   Mixing Instrumentation and Sampling

As mentioned in Section 5.4, instrumentation and sampling are two techniques for data acquisition with different properties regarding the precision they provide and the overhead they introduce. A challenging objective is to obtain, with very low overhead, extremely precise descriptions of the time evolution of different performance metrics such as IPC and MFlop/s rates. In [297] we showed how this can be done for programs with clearly repetitive structures, by merging instrumentation and sampling. The key underlying idea is that if the process is ergodic, the instrumentation hooks will provide the reference to which the measurements of the sampling probes may be referred. For example we can instrument the beginning of individual iterations or the entry to subroutines. Samples that happen to fire after a certain percentage of the duration of the iteration or routine will provide data on how many instructions, flops, or cycles occurred since the beginning. By collecting samples over many iterations, we can build a detailed evolution of such counts along the entire duration of a representative iteration.

MPITrace uses PAPI to set exceptions on hardware counter overflows such that the sampled data can be acquired and the appropriate records emitted to the same buffers as the instrumentation records. We typically use the cycles counter, and in order to keep the overhead low, the overflow value is set such that a sample is acquired every few milliseconds. A post processing of the trace performs the folding process, handling issues such as elimination of outliers, and the proper fitting of the projected data set by a polynomial model. A first output of this utility is a gnuplot file to display the evolution of the instantaneous rate of the different hardware counter metrics such as MFlop/s or MI/s (millions of instructions per second) along the duration of a routine or iteration of the program. The post-processing tool also emits a new Paraver trace with synthetic events at fine grain intervals, such that the full Paraver analysis power can be applied to the extrapolated trace.

Figure 5.9 shows an example of the reconstruction of the instantaneous MI/s for one iteration of two process in a parallel run of the NAS BT class B benchmark on 16 processors. The two functions at the top of the figure show the instantaneous performance in millions of instructions per second (MI/s) when outside MPI and zero when inside MPI. The figure in the middle shows for the same two processes and time scale the MPI call view. White in the timeline represents that the process is doing useful computation while shading corresponds to time inside MPI_waitall or MPI_wait. The last view shows the major user level function being executed in one iteration of the program. The successive colors represent routines copy_faces, compute_rhs, x_solve and y_solve. While the application was executed, instrumentation samples where obtained at the entry and exit of user level functions and MPI calls. The very fine level of detail presented in the top view is obtained through the projection onto this reference iteration of the random samples scattered along many iterations of the actual run. This view does show internal structure inside the different user function routines. For example, there are four phases or iteration in routine compute_rhs. Routines x_solve y_solve and z_solve also start with four phases of good MI/s values and terminate with a region of poor MI/s.

5.6   Models

5.6.1   Detailed Time Reconstruction

Conventional profiling or timeline analysis of a computational run can identify metrics that may be above standard values and point to a possible problem. Nevertheless they often lack the ability to quantify the potential impact of fixing that problem, or even determining that resolving the bottleneck will significantly improve global performance. Trace-based systems can help answer these questions. Tracing packages typically emit not only timestamped events, but also sufficient information to characterize the data exchanges and dependencies between different processes. This level of detail offers the potential for analyses well beyond merely computing metrics or statistics for a specific program run. Ordering information captures a lot of information about characteristics intrinsic to an application that can be used not only to predict its potential behavior on different target machines but also to estimate

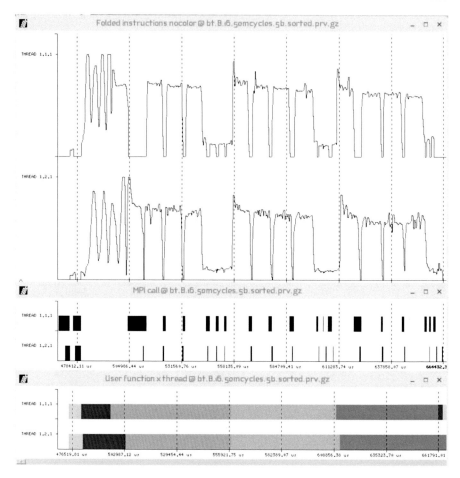

FIGURE 5.9: Instantaneous MI/s evolution for two processes of a 16-processor run of the NAS BT benchmark.

the impact of possible optimizations and rank them in terms of expected productivity. Traces can be fed to simulators that reconstruct the behavior of the program on systems with different architectural characteristics.

Dimemas [204] is one such simulator for message passing programs running on clusters of SMP nodes. Dimemas is fed with traces that can be obtained from Paraver through a utility program. The essential difference in the trace formats is that while the Paraver trace specifies the actual time when each event (computation or communication) happens, the Dimemas trace records just the sequence of computation demands (time of computation bursts) and communication demands (partner, size, and synchronization semantics). The Dimemas simulator is able to rebuild the actual timing of a hypothetical run based on the application demands and system characteristics.

Often processor or system architects use very elaborate simulators in an attempt to capture every detail of a real system. Such facilities typically result in very long simulations and do not clearly isolate the impact of individual factors. The system model in Dimemas tries to be both simple and functional. A simple model allows fast simulations and the usage of the tool in an interactive environment. The key objective at the functional level is to be able to capture factors that influence the behavior of a parallel machine but model them in an abstract way. For example, Dimemas as an MPI replay engine takes into account the blocking/nonblocking semantics as well as the buffered or rendezvous synchronization and the typical limit that standard mode calls apply to perform eager or synchronous transfers. These are key factors that influence the performance of an MPI application. At the architectural level, latency and bandwidth are two important factors, but endpoint or network contention can also have a strong impact on application performance. We model endpoint contention by a parameter describing the number of links (or adapters) between the node and the network. We model network contention by a parameter, which we call number of buses (B), limiting the number of concurrent transfers allowed on the network. This is an extremely rough approximation of the behavior of the real networks. But the important thing here is that this parameter is a model, something not necessary directly linked to a specific feature (number of levels, topology, switch design, etc.) of the actual interconnect, but nonetheless is still able to reproduce the observed behaviors. In fact, the value of number of buses that optimizes the Dimemas model's fit to the real execution is one way of characterizing the level of contention that a real network has.

This type of simulator can be used to predict the performance on machines not yet available or to configure a system in procurement processes. We believe them to be even more valuable in understanding an application and providing expectations of gains to be achieved when different optimizations are applied. For example, two typical suggestions when a program does not scale well are trying to group messages to avoid overhead and trying to use nonblocking calls to overlap communication and computation. Each of these transformations may require a significant programming effort. Thus it would be interesting to have in advance some raw estimate of the potential gain such effort would produce. Performing simulations with ideal parameters may provide useful insight on the expected impact. For example, by setting the overhead of MPI calls to zero, one can estimate an upper bound of the gain for grouping messages. Performing a simulation with infinite network bandwidth would give an estimate of the potential gain of transfers where overlapped with computation. In practice, in often happens that these two estimates indicate that the potential gains for the above two tuning techniques are less than what developers expect, indicating that there are other problems that should be addressed first.

Performing parameter sweep simulations of one trace on a wide range of parameter values is a useful way of detecting intervals where the impact of the specific parameter does not significantly affect the performance of the application, and when the impact starts to be relevant. Figure 5.10 shows an example of such a parametric sweep for two versions (ARW and NMM) of the WRF code when run on 128, 256, and 512 processors. The figures show how the ARW version is more sensitive to bandwidth that the NMM version, especially when the bandwidth is below 256 Mbyte/s. At the 256 Mbyte/s level, reducing the network connectivity (effective bisection bandwidth) can have a deleterious impact on the performance of the ARW version with 256 and 512 processors. Both versions are very insensitive to the MPI overhead (latency, in Dimemas terms). Because Dimemas can generate Paraver traces of the reconstructed behavior, an analyst can visually determine if the impact is the same over the entire program or if there are regions in the timeline where some parameter, for example bandwidth, has a bigger impact that in others.

A common problem in many applications is endpoint contention. A frequently used technique, in codes where each process has several neighbors, is to send the messages by rank order of the destination. The result is that low-ranked processes are overloaded with incoming messages, serializing all their incoming transfers. This also delays senders that get blocked while some of their neighbors are ready to receive. A good practice would be to schedule the communications such that sequences of pairwise exchanges maximize the number of different destinations targeted at each step. The question is how to detect this situation in a program and to quantify its potential severity. One solution is to perform a Dimemas simulation with no network contention and just one output adapter but infinite input adapters per node. In order to focus on the communication impact, very low bandwidth should be used in these simulations (or equivalently one should simulate infinitely fast CPUs). Figure 5.11 shows a timeline of the number of concurrent incoming messages to the different nodes under these simulation conditions for a region of the PEPC code, showing in shaded tone the points in time where a lot of data transfers would be contending for an input adapter.

The above examples show that although no machine can be built with those parameters, the results of the predictions do reflect intrinsic characteristics of the applications, something that we consider more valuable for a developer than the actual performance on a specific machine.

TaskSim is a simulator at the node level for multicore processors being developed at BSC. Starting from Paraver traces of CellSuperscalar (CellSs [50]), for programs running in a current Cell processor, the simulator can predict the behavior of the program in possible future architectures with many more cores, different memory controllers or interleaving of data across DIMMs.

5.6.2   Global Application Characterization

Given that trace-based tools occupy one extreme of the spectrum of performance tools, some ask what the other extreme should look like. Are profilers

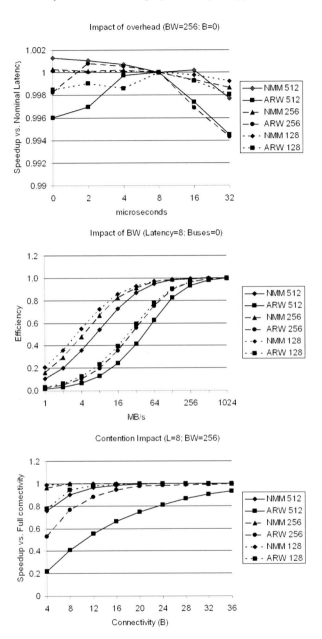

FIGURE 5.10: Impact of overhead, bandwidth and contention on WRF (versions ARM, NMM).

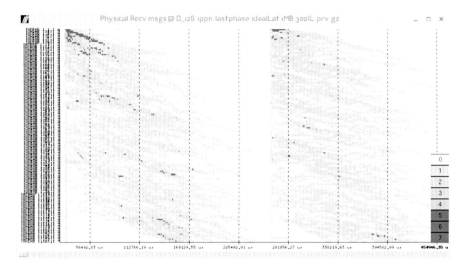

FIGURE 5.11: Incoming transfers of a 256 processes simulation of PEPC with infinite input links. (See color insert.)

today presenting a simple meaningful and abstract view of the performance of our applications? Are the statistics reported by profiler or trace visualizers neat models that clearly focus on the major causes of inefficiency at a very coarse level? Ideally users prefer reports from users to be very concise, understandable and useful, giving clear and correct hints on which directions to focus code optimization.

In [80] we propose a very general model to describe the scalability of an application in terms of general factors or concepts, without requiring deep system knowledge or examining fine details of the application. The efficiency model is multiplicative with factors quantified with a value from zero to one. A high value of the factor means that it is not limiting the efficiency (or scalability) of the application. A low value implies that this factor is responsible for a big degradation in the performance of the application.

The very first model would consider four factors: load balance, communication efficiency, raw core efficiency, and computational complexity. The parallel efficiency of a single run could be described as the product of the first three factors:

$$\eta = LB \times CommEff \times IPCeff$$

where load balance (LB) and communication efficiency ($CommEff$) are computed from the Gant diagram in Figure 5.12, which displays one row per process, ranked from highest to lowest computational load. The dark shaded area represents the total computation time of each process (T_i), and the lighter shaded areas represent the time spent in MPI by each process.

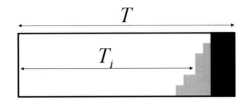

FIGURE 5.12: Gant diagram to derive efficiency model.

The different factors in the above formula can be obtained form profile data as follows.

$$eff_i = \frac{T_i}{T} \tag{5.1}$$

$$CommEff = \max(eff_i) \tag{5.2}$$

$$LB = \frac{\sum(eff_i)/P}{\max(eff_i)} \tag{5.3}$$

The $IPCeff$ factor would represent the raw core efficiency during the useful computation (shaded dark in Figure 5.12) and should be normalized to the peak value of the processor. If we are interested only in the parallel efficiency, only the first two terms should be used. If we are interested in using the model to explain the scalability between different runs an additional term should be added to the product representing the computational complexity (in terms of total number of useful instructions) of the algorithm.

The model is based on the aggregated time in each state (useful computation and MPI) and is not able to differentiate whether the possible degradation within the $CommEff$ term is actually due to transfer delays or to synchronization. It is in this point where a Dimemas based simulation can help differentiate these two factors as presented in Figure 5.13 and described by the following model:

$$\eta = LB \times CommEff \times microLB \times Transfer * IPCeff \tag{5.4}$$

The idea is that a Dimemas simulation with an ideal interconnect (zero overhead in MPI calls and infinite bandwidth) would still have to preserve the synchronizations required by the application and might delay some processes (yellow in Figure 5.13), but still compute the ideal time (T_{ideal}) bellow which the application cannot go.

The term $microLB$ may actually represent serialization (i.e., pipeline) or microscopic load imbalance. This situation arises when individual intervals between synchronizing communication phases (typically collectives) are actually imbalanced but the process that arrives late is different every phase, and when aggregating over all phases the total computation time ends up being similar across processes. Of the two indistinguishable causes, it is the most frequent one in real applications. We call this term the micro load balance factor.

$$microLB = \frac{\max(T_i)}{T_{ideal}} \qquad (5.5)$$

The gap between T_{ideal} and the original application time (T) is assigned to the actual data transfer component of MPI calls:

$$Transfer = \frac{T_{ideal}}{T} \qquad (5.6)$$

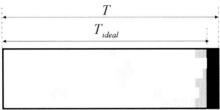

FIGURE 5.13: Model including microscopic load balance and serialization.

Figure 5.14 shows the result of such a model when applied to runs of the GROMACS molecular dynamics code at different processor counts. We can see the importance of the different factors and how they vary in a complex code. Sometimes certain factors evolve in directions that compensate others, indicating that the final application performance may be the result of intricate interactions between them.

These models have the very nice property of explaining performance based on very few factors (just three or four numbers), easily understandable by non-expert parallel programmers, and pointing in the right direction to focus optimization efforts. The first model can actually be reported by profiling tools based on instrumentation (and also in an approximate way by those based on sampling) as they already have the information to compute it. The second model requires a trace-based simulation, but is more able to identify the actual cause of possible inefficiency, separating the synchronization and transfer issues. This is important, because the two issues require very different types of tuning.

5.6.3 Projection of Metrics

The above modeling focuses on the parallel behavior of the application, but a detailed description of the behavior of the sequential computation bursts is also needed in order to properly understand where to focus the tuning efforts and to determine the potential improvements that can be achieved.

Diving down into instruction or address traces and instruction level sim-

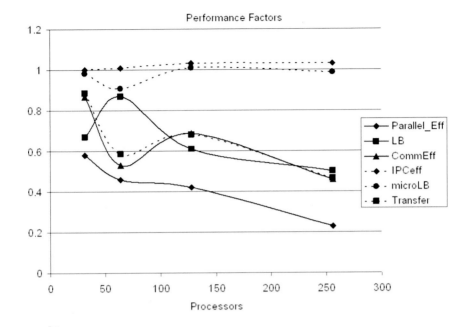

FIGURE 5.14: Performance factors for GROMACS runs at different processor counts.

ulators is a tedious and time consuming practice, and so is very seldom attempted by program developers. At the granularity that performance analysis tools can afford, hardware counters are the basic mechanism to capture information about the internal operation of the sequential computation bursts. They are a very precise and accurate monitoring tool. There are nevertheless two major difficulties in employing these measurements to understand the underlying bottlenecks of a program.

The first difficulty has to do with the actual semantics of the available counters in current processors. Hardware counters are an extremely precise mechanism to monitor the aggregate number of events at the microarchitecture level in a given time interval. They were initially introduced by hardware designers mainly for debugging purposes and, as such, the actual events counted are specific to a microarchitecture and are typically not well understood by the majority of application programmers. In heavily out-of-order machines, with lots of speculative execution and where the microarchitecture often has to reissue certain activities several times, the relationship of some of these counts to the sequential algorithm as written by the user is pretty tenuous. The PAPI [65] software tool tries to provide a standardized set of hardware counters meaningful to an analyst. Even in this case, the actual impact of certain hardware counters values on performance is not trivial. For example, it is clear that a large number of cache misses is, in general, not good, but performance degradation is not merely proportional to this number. The actual degradation depends on the actual distribution in time of the misses (i.e., the cost per miss is lower if several misses are issued in a tight burst than if they are uniformly distributed across the execution), the data layout and contention level. Different machines may have hardware counters that provide hints about such cost (direct measurement in some cases, indirect impact on certain micro architecture events in others) but properly understanding these effects is far too complicated for a normal user. From our point of view, hardware developers should provide cycles per instruction (CPI) stack models that describe the behavior of a computation burst in terms of the percentage of time lost due to specific factors, such as memory hierarchy stalls, dependences in the arithmetic computations, branch miss speculation, etc. Even approximate, abstract models of this type are often extremely useful for an analyst. An example of one such model is given in [237].

The second difficulty is the limited number of such counters that can be obtained for each individual computation burst. The issue is exemplified also by the previously mentioned CPI Stack model, which requires a very large number of hardware counts in order to properly assign costs to the different factors. This also arises in profile- or trace-based tools. The alternative is either to perform several runs to obtain different counters at each run or to use a statistical multiplexing approach. Both approaches presume that the different runs or intervals will behave the same. These approaches give averaged values for the whole application, without identifying and separating the possibly very different behavior in the application phases. The result is that a lot of the very high-fidelity data supported by the underlying mechanism is lost.

In [220] we used the clustering techniques mentioned in Section 5.5 to precisely characterize the different computation bursts with a single run of the program. The idea is to periodically change the set of hardware counters used, while keeping a few counters common across all sets. We typically use cycles and graduated instructions as the common counters across groups. These counters are then used to cluster all the computation bursts in the trace. Different points in one cluster will thus provide precise counts for different hardware events and thus a global characterization of a cluster can be made. This data can be used not only to provide a very rich set of metrics for the cluster, but also to feed the CPI Stack model mentioned above. Table 5.3 shows an example of the outcome of this process.

5.7 Interoperability

As can be seen just by examining the various chapters of this book, there are a wide variety of performance analysis tools in usage today. Each type

TABLE 5.3: Example set of metrics obtainable from a single run

Category	Metric	Cluster #1	Cluster #2	Cluster #3	Cluster #4
Performance	% of Time	46.30	31.15	15.31	4.86
Performance	IPC	0.64	0.73	0.80	0.98
Performance	MIPs	1445	1645	1809	2215
Performance	Mem. BW (Mbyte/s)	39.6	2	7	15.3
Performance	L2 load BW (Mbyte/s)	1252	172	325	897
Program	Computation intensity	0.74	0.65	0.63	0.75
Program	Memory Mix (%)	31.6	36.0	37.3	33.5
Program	Floating point Mix (%)	12.4	18.2	18.1	18.6
Program	% of FMAs	97.5	39.8	65.8	67.5
Architecture	% instr. completed	39.3	36.8	36.9	43.4
Architecture	L1 misses/Kinstr	10.6	1.1	1.4	3.1
Architecture	L2 misses/Kinstr	0.21	0.01	0.03	0.05
Architecture	Memory B/F	0.15	0.01	0.03	0.08
Architecture	nb. loads/stores per TLB miss	176864	665477	143044	301401
Architecture	Total Mbytes from Memory	124.0	11.5	10.6	0.3

of tool has different characteristics appropriate for different needs. Interoperability between these tools is thus an issue of increasing importance, so that different tools can leverage the capabilities offered by other tools, and so that analysts can combine them in the most appropriate way to address a particular application.

In the following subsections we describe some examples of interoperability between tools.

5.7.1   Profile and Trace Visualizers

Profile browsers may be the fastest approach to get a general view of the application behavior and navigate through averaged statistics. Trace-based tools are extremely useful for detailed analyses.

It is typical for profile tools to provide links to program code editors, so that metrics can be correlated to the source code to which they correspond. As shown on the top of Figure 5.15, this connection can take place in both directions (selecting a section of source code highlights its statistics or vice versa). Another, less common, requirement is to provide links from the profile to a trace visualizer. Figure 5.16 shows the graphical user interface (GUI) for the integration between Peekperf, a profile visualizer from IBM, and Paraver. A user examining profile data might think that more detailed statistics, histograms or time distributions may help understand the profile numbers. If so, it is possible for the profile GUI to select a user function or MPI call and instruct Paraver to produce pop-up views, such as showing the sequence of invocations of the specific routine or the histogram of its duration.

Such facilities can be implemented as follows. The actual profile data in the format required by peekperf (.viz xml file) is generated by a set of utilities

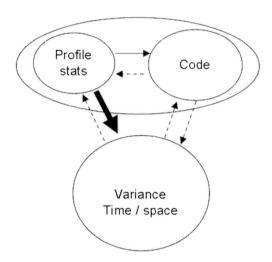

FIGURE 5.15: Interaction between profile, source code, and trace visualizers.

Label	Cycles	IPC	L2 Data Cache	NBursts	NInstr	
~ copy_faces.f						
~ copy_faces						
MPI_Isend	6.26294e+06	0.05	5720	202	304195	2
MPI_Waitall	1.12337e+09	0.18	19(			
~ error.f						
~ error_norm						
MPI_Allreduce	1.48907e+06	0.81	12(			
~ rhs_norm						
MPI_Allreduce	785505	0.68	121			
~ GlobalMetrics						
mpbt	2.44205e+10	0.04	5.3			
MPI_Allreduce	37416	0.39	4			

FIGURE 5.16: peekperf / Paraver integration GUI.

from a Paraver trace. Besides the profile data, these utilities inject into the .viz file information that will let the profile visualizer drive a Paraver version that accepts commands form another application.

The approach implemented in Kojak takes note of where in the timeline some of the metrics (e.g., late sender) have a significant contribution and is able to direct Vampir or Paraver to produce a pop-up view zoomed into that specific area.

5.7.2 Visualizers and Simulators

We have already mentioned that Paraver traces can be converted to the Dimemas format, and the result of the simulation is another trace that can be compared to the original one. Although the two different tools are quite different, this tight integration between them provides many analysis capabilities.

The first study supported by this environment is to check whether the prediction for the nominal parameters of the architecture matches the real behavior. This matching is often quite accurate. If a mismatch is detected, the environment has the potential to search for the actual cause by detailed analyses of both the original and Dimemas generated trace, but also by performing simulations with different parameters. One possible cause of mismatches are preemptions within the MPI calls in the actual run. By examining the cycles per microsecond view of the original run, Paraver can validate or negate this hypothesis. Another possible cause is contention within the MPI calls in the actual application run. Varying the parameters of the model such as contention and comparing to the real elapsed time is one way to estimate the actual level of network contention, something that cannot be measured solely from the original trace.

An interesting example of mismatch between the Dimemas simulation and the real run occurred very early in the validation process of Dimemas. We tried to determine the range of overhead and bandwidth parameters that would properly reconstruct (within 5% of their real duration) the behavior of several Pallas MPI benchmarks (now known as the Intel MPI Benchmarks) on a shared memory machine. Unfortunately the ranges of proper overhead and bandwidth figures did not overlap for the different benchmarks until the connection between the node and the network was modeled by a single half duplex link. Certainly this model did not match the low-level hardware, capable of simultaneously performing bidirectional transfers. What initially looked rather strange in the end was totally reasonable, as in a shared memory implementation the actual data transfer from user to system or destination buffers is done by program, and thus only one transfer at a time takes place.

The network model of Dimemas is quite abstract, capturing only the major conceptual factors that impact application performance. This makes it possible to rapidly run many parametric simulations with the objective of understanding the general characteristics of applications and their sensitiveness to these factors. Dimemas is thus not aimed at extremely detailed evaluation of the internals of interconnection network designs. In order to perform such architectural studies, we have implemented an application programmer interface (API) to a detailed network simulator by IBM [239] called Venus. Dimemas provides the trace parsing, MPI reply and Paraver trace generation functionalities, while the detailed data transfers through the network are simulated by Venus. These simulations can be extremely precise but are always rather slow and thus cannot be effectively applied to huge traces. It is actually desirable to use the capability of Paraver to manipulate and cut traces isolating the basic communication phases.

5.7.3   Interoperation between Visualizers

Even within a given area (i.e., profile visualizers, instrumentation and trace visualizers), different tools may have particular specializations and may provide specific mechanisms to share data. This can be handled by means of standardized interchange formats or by implementing translators. Such interoperation capabilities are increasingly available. For example, interoperation capabilities between environments such as Kojak, Vampir, and Paraver are either already in place or under development.

5.8   The Future

In this chapter we tried to show how traces have been used to understand the behavior of programs. The general approach has always been to blindly obtain a trace and then patiently browse it with a flexible visualizer in order to squeeze the information in it. The process has led to the identification of techniques to help improve the scalability of the trace-based approach itself and to better understand the type of automatic analyses that provide more abstract and useful information than the raw data. Thus the real underlying progress in better understanding how to process the raw data is to maximize the ratio of the information to the amount of data delivered to the analyst.

Many of the techniques described in this chapter are ready to be integrated in an online environment. The objective is to let the online analysis system to detect the iterative structure of the application, and directly emit a clustered trace of just one (or a few) periods as well as scatter plots, detailed CPIStack models and a wide range of statistics for each of the clusters.

The online application of basic techniques has required additional research on appropriate mechanisms to reduce the clustering cost at scale. The current approach is based on selecting a subset of all computation bursts for the actual clusterization plus classification techniques to determine the cluster each burst belongs to. Of the several potentail approaches to select the data to be used in the actual clusterization, the most promising one is to select all bursts from some processes and a random sample of individual bursts from other processes. The clustering process is periodically applied to the data being stored in circular buffers at the application nodes. Additionally, a mechanism to compare the similarity of successive clustering results is used to detect that the application is in a stationary regime, so that the result for it can be emitted.

The scalable infrastructure on which we are applying these techniques is MRNET [286]. The intermediate nodes in the tree structure can be used, for example, to filter data as it proceeds towards the root node where the clustering is applied. On the way back from the root note, the classification phase can be performed and the result merged with the raw data for the selected region of the circular buffer.

In some sense, this is quite similar to approaches used in other areas of observational science, where a huge amount of raw data can be generated, but only summarized data is actually collected for further processing and archiving. If some interesting situation is observed by such continuous monitoring, the sensors still have a significant amount of buffered raw data that can then be kept for detailed analysis.

Although great challenges lie ahead of us in the multicore and exascale systems, we believe that performance tools under development will be able to handle them by integrating several different approaches to performance analysis. It is our belief that trace-based tools will always be an important component of such integrated environments, assisting in the detailed analyses of the difficult cases, addressing variance issues and guiding the process of identification of relevant metrics and techniques to be used in automatic environments, both online and offline. Of course, there is still a need to improve current techniques and propose new ones, but we believe that the evolution in the recent past allows us to be optimistic about the future.

Acknowledgments

This chapter summarizes the views of the author, but is based on work by many other people in the BSC tools team (especially Judit Gimenez, Harald Servat, Juan Gonzalez, German Llort, Eloy Martinez, Pedro Gonzalez, and Xavi Pegenaute) and others who left the team (Vincent Pillet, Sergi Girona, Luis Gregoris, Xavi Aguilar, and others). This work is partially funded by the IBM, through the IBM-BSC Mareincognito project, the USAF grant FA8655-09-1-3075 and the Spanish Ministry of Education under grant TIN2007-60625.

Chapter 6

Large-Scale Numerical Simulations on High-End Computational Platforms

Leonid Oliker

CRD, Lawrence Berkeley National Laboratory

Jonathan Carter

NERSC, Lawrence Berkeley National Laboratory

Vincent Beckner

CRD, Lawrence Berkeley National Laboratory

John Bell

CRD, Lawrence Berkeley National Laboratory

Harvey Wasserman

NERSC, Lawrence Berkeley National Laboratory

Mark Adams

APAM Department, Columbia University

Stéphane Ethier

Princeton Plasma Physics Laboratory, Princeton University

Erik Schnetter

CCT, Louisiana State University

6.1   Introduction

After a decade where high-end computing was dominated by the rapid pace of improvements to CPU frequencies, the performance of next-generation supercomputers is increasingly differentiated by varying interconnect designs and levels of integration. Understanding the tradeoffs of these system designs is a key step towards making effective petascale computing a reality. In this work, we conduct an extensive performance evaluation of five key scientific application areas: plasma micro-turbulence, quantum chromodynamics, micro-finite-element solid mechanics, supernovae, and general relativistic astrophysics that use a variety of advanced computation methods, including adaptive mesh refinement, lattice topologies, particle in cell, and unstructured finite elements. Scalability results and analysis are presented on three current high-end HPC systems, the IBM Blue Gene/P at Argonne National Laboratory, the Cray XT4 and the Berkeley Laboratory's NERSC Center, and an Intel Xeon cluster at Lawrence Livermore National Laboratory.[1] In this chapter, we present each code as a section, where we describe the application, the parallelization strategies, and the primary results on each of the three platforms. Then we follow with a collective analysis of the codes performance and make concluding remarks.

6.2   HPC Platforms and Evaluated Applications

In this section we outline the architecture of the systems used in this chapter. We have chosen three three architectures which are often deployed

---

[1] The Hyperion cluster used was configured with 45-nanometer Intel Xeon processor 5400 series, which are quad-core "Harpertown nodes."

at high-performance computing installations, the Cray XT, IBM BG/P, and an Intel/Infiniband cluster. As the CPU and node architectures are fully described in Chapter 13, we confine ourselves here to a description of the interconnects. In selecting the systems to be benchmarked, we have attempted to cover a wide range of systems having different interconnects. The Cray XT is designed with tightly integrated node and interconnect fabric. Cray have opted to design a custom network ASIC and messaging protocol and couple this with a commodity AMD processor. In contrast, the Intel/IB cluster is assembled from off the shelf high-performance networking components and Intel server processors. The final system, BG/P, is custom designed for processor, node and interconnect with power efficiency as one of the primary goals. Together these represent the most common design trade-offs in the high performance computing arena. Table 6.1 shows the size and topology of the three system interconnects.

TABLE 6.1: Highlights of architectures and file systems for examined platforms. All bandwidth figures are in Gbyte/s. "MPI Bandwidth" uses unidirection MPI benchmarks to exercise the fabric between nodes with 0.5 Mbyte messages.

System Name	Processor Architecture	Interconnect	Total Nodes	MPI Bandwidth
Franklin	Opteron	Seastar2+/3D Torus	9660	1.67
Intrepid	BG/P	3D Torus/Fat Tree	40960	1.27
Hyperion	Xeon	Infiniband 4×DDR	576	0.37

Franklin: Cray XT4: Franklin, a 9660 node Cray XT4 supercomputer, is located at Lawrence Berkeley National Laboratory (LBNL). Each XT4 node contains a quad-core 2.3 GHz AMD Opteron processor, which is tightly integrated to the XT4 interconnect via a Cray SeaStar2+ ASIC through a HyperTransport 2 interface capable of capable of 6.4 Gbyte/s. All the SeaStar routing chips are interconnected in a 3D torus topology with each link is capable of 7.6 Gbyte/s peak bidirectional bandwidth, where each node has a direct link to its six nearest neighbors. Typical MPI latencies will range from $4--8\mu$s, depending on the size of the system and the job placement.

Intrepid: IBM BG/P: Intrepid is a BG/P system located at Argonne National Labs (ANL) with 40 *racks* of 1024 nodes each. Each BG/P node has four PowerPC 450 CPUs (0.85 GHz) and 2 Gbyte of memory. BG/P implements three high-performance networks: a 3D torus with a peak bandwidth of 0.4 Gbyte/s per link (6 links per node) for point to point messaging; a collectives network for broadcast and reductions with 0.85 Gbyte/s per link (3 links per node); and a network for a low-latency global barrier. Typical MPI latencies will range from 3-10μs, depending on the size of the system and the job placement.

Hyperion: Intel Xeon Cluster: The Hyperion cluster, located at Lawrence Livermore National Laboratory (LLNL), is composed of four scalable units, each consisting of 134 dual-socket nodes utilizing 2.5 GHz quad-core Intel Harpertown processors. The nodes within a scalable unit are fully connected via a 4× IB DDR network with an peak bidirectional bandwidth of 2.0 Gbyte/s. The scalable units are connected together via spine switches providing full bisection bandwidth between scalable units. Typical MPI latencies will range from $2 - -5\mu s$, depending on the size of the system and the job placement.

The applications chosen as benchmarks come from diverse scientific domains and employ quite different numerical algorithms and are summarized in Table 6.2.

TABLE 6.2: Overview of scientific applications examined in our study.

Name	Code Lines	Discipline	Computational Structure
GTC	15,000	Magnetic Fusion	Particle/Grid
OLYMPUS	30,000	Solid Mechanics	Unstructured FE
CARPET	500,000	Relativistic Astrophysics	AMR/Grid
CASTRO	300,000	Compressible Astrophysics	AMR/Grid
MILC	60,000	Quantum Chromodynamics	Lattice

6.3 GTC: Turbulent Transport in Magnetic Fusion

6.3.1 Gyrokinetic Toroidal Code

The Gyrokinetic Toroidal Code (GTC) is a 3D particle-in-cell (PIC) code developed to study the turbulent transport properties of tokamak fusion devices from first principles [216, 217]. The current production version of GTC scales extremely well with the number of particles on the largest systems available. It achieves this by using multiple levels of parallelism: a 1D domain decomposition in the toroidal dimension (long way around the torus geometry), a multi-process particle distribution within each one of these toroidal domains, and a loop-level multitasking implemented with OpenMP directives. The 1D domain decomposition and particle distribution are implemented with MPI using two different communicators: a toroidal communicator to move particles from one domain to another, and an intra-domain communicator to gather the contribution of all the particles located in the same domain. Communication involving all the processes is kept to a minimum. In the PIC method, a grid-based field is used to calculate the interaction between the charged particles instead of evaluating the N^2 direct binary Coulomb interactions. This field is evaluated by solving the gyrokinetic Poisson equation [210] using the particles' charge density accumulated on the grid. The basic steps of the PIC method are: *(i)* Accumulate the charge density on the grid from the contribution of each particle to its nearest grid points. *(ii)* Solve the Poisson equation to evaluate the field. *(iii)* Gather the value of the field at the position of the particles. *(iv)* Advance the particles by one time step using the equations of motion ("push" step). The most time-consuming steps are the charge accumulation and particle "push," which account for about 80% to 85% of the time as long as the number of particles per cell per process is about two or greater.

In the original GTC version described above, the local grid within a toroidal domain is replicated on each MPI process within that domain and the particles are randomly distributed to cover that whole domain. The grid work, which comprises of the field solve and field smoothing, is performed redundantly on each of these MPI processes in the domain. Only the particle-related work is fully divided between the processes. This has not been an issue until recently due to the fact that the grid work is small when using a large number of particles per cell. However, when simulating very large fusion devices, such as the international experiment ITER [181], a much larger grid must be used to fully resolve the microturbulence physics, and all the replicated copies of that grid on the processes within a toroidal domain make for a proportionally large memory footprint. With only a small amount of memory left on the system's nodes, only a modest amount of particles per cell per process can fit. This problem is particularly severe on the IBM Blue Gene system where the amount of memory per core is small. Eventually, the grid work starts dominating the calculation even if a very large number of processor cores is used.

The solution to our non-scalable grid work problem was to add another level of domain decomposition to the existing toroidal decomposition. Although one might think that a fully 3D domain decomposition is the ideal choice, the dynamics of magnetically confined charged particles in tokamaks tells us otherwise. The particle motion is very fast in both the toroidal and poloidal (short way around the torus) directions, but is fairly slow in the radial direction. In the toroidal direction, the domains are large enough that only 10% of the particles on average leave their domain per time step in spite of their high velocities. Poloidal domains would end up being much smaller, leading to a high level of communication due to a larger percentage of particles moving in and out of the domains at each step. Furthermore, the poloidal grid points are not aligned with each other in the radial direction, which makes the delineation of the domains a difficult task. The radial grid, on the other hand, has the advantage of being regularly spaced and easy to split into several domains. The slow average velocity of the particles in that direction insures that only a small percentage of them will move in and out of the domains per time step, which is what we observe.

One disadvantage, however, is that the radial width of the domains needs to decrease with the radius in order to keep a uniform number of particles in each domain since the particles are uniformly distributed across the whole volume. This essentially means that each domain will have the same volume but a different number of grid points. For a small grid having a large number of radial domains, it is possible that a domain will fall between two radial grid points. Another disadvantage is that the domains require a fairly large number of ghost cells, from three to eight on each side, depending on the maximum velocity of the particles. This is due to the fact that our particles are not point particles but rather charged "rings," where the radius of the ring corresponds to the Larmor radius of the particle in the magnetic field. We actually follow the guiding center of that ring as it moves about the plasma, and the radius of the ring changes according to the local value of the magnetic field. A particle with a guiding center sitting close to the boundary of its radial domain can have its ring extend several grid points outside of that boundary. We need to take that into account for the charge deposition step since we pick four points on that ring and split the charge between them [210]. As for the field solve for the grid quantities, it is now fully parallel and implemented with the Portable, Extensible Toolkit for Scientific Computation (PETSc) [272].

Overall, the implementation of the radial decomposition in GTC resulted in a dramatic increase in scalability for the grid work and decrease in the memory footprint of each MPI process. We are now capable of carrying out an ITER-size simulation of 130 million grid points and 13 billion particles using 32,768 cores on the BG/P system, with as little as 512 Mbytes per core. This would not have been possible with the old algorithm due to the replication of the local poloidal grid (2 million points).

6.4   GTC Performance

Figure 6.1 shows a weak scaling study of the new version of GTC on Franklin, Intrepid, and Hyperion. In contrast with previous scaling studies that were carried out with the production version of GTC and where the computational grid was kept fixed [264, 265], the new radial domain decomposition in GTC allows us to perform a true weak scaling study where both the grid resolution and the particle number are increased proportionally to the number of processor cores. In this study, the 128-core benchmark uses 0.52 million grid points and 52 million particles while the 131,072-core case uses 525 million grid points and 52 billion particles. This spans three orders of magnitude in computational problem size and a range of fusion toroidal devices from a small laboratory experiment of 0.17 m in minor radius to an ITER-size device of 2.7 m, and to even twice that number for the 131,072-core test parameters. A doubling of the minor radius of the torus increases its vol-

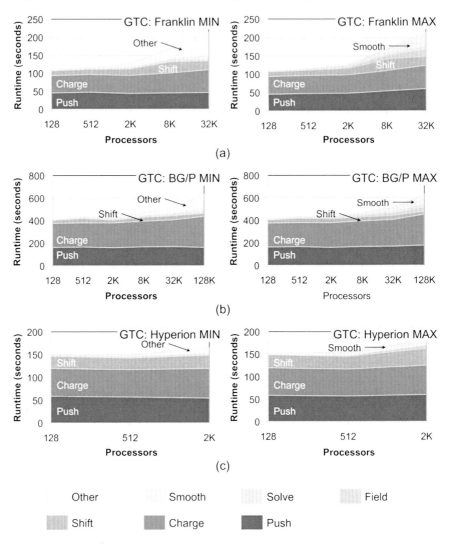

FIGURE 6.1: GTC weak scaling, showing minimum and maximum times, on the (a) Franklin XT4, (b) Intrepid BG/P, and (c) Hyperion Xeon systems. The total number of particles and grid points are increased proportionally to the number of cores, describing a fusion device the size of ITER at 32,768 cores.

ume by eight if the aspect ratio is kept fixed. The Franklin Cray XT4 numbers stop at the ITER-size case on 32,768 cores due to the number of processors available on the system although the same number of cores can easily handle the largest case since the amount of memory per core is much larger than on BG/P. The concurrency on Hyperion stops at 2048, again due to the limited number of cores on this system. It is worth mentioning that we did not use the shared-memory OpenMP parallelism in this study although it is available in GTC.

The results of the weak scaling study are presented as area plots of the wall clock times for the main steps in the time-advanced loop as the number of cores increases from 128 to 32,768 in the case of the XT4, from 128 to 131,072 for the BG/P, and from 128 to 2048 for Hyperion. The main steps of the time loop are: accumulating the particles' charge density on the grid ("charge" step, memory scatter), solving the Poisson equation on the grid ("solve" step), smoothing the potential ("smooth" step), evaluating the electric field on the grid ("field" step), advancing the particles by one time step ("push" phase including field gather), and finally, moving the particles between processes ("shift"). The first thing we notice is that the XT4 is faster than the other two systems for the same number of cores. It is about 30% faster than Hyperion up to the maximum of 2048 cores available on that system. Compared to BG/P, Franklin is four times faster at low core count but that gap decreases to 2.4 times faster at 32,768 cores. This clearly indicates that the new version of GTC scales better on BG/P than Franklin, a conclusion that can be readily inferred visually from the area plots. The scaling on BG/P is impressive and shows the good balance between the processor speed and the network speed. Both the "charge" and "push" steps have excellent scaling on all three systems as can be seen from the nearly constant width of their respective areas on the plots although the "charge" step starts to increase at large processor count. The "shift" step also has very good scaling but the "smooth" and "field" steps account for the largest degradation in the scaling at high processor counts. They also account for the largest differences between the minimum and maximum times spent by the MPI tasks in the main loop as can be seen by comparing the left (minimum times) and right (maximum times) plots for each system. These two steps hardly show up on the plots for the minimum times while they grow steadily on the plots for the maximum times. They make up for most of the unaccounted time on the minimum time plots, which shows up as "Other." This indicates a growing load imbalance as the number of processor-cores increases. We note that the "push" "charge" and "shift" steps involve almost exclusively particle-related work while "smooth" and "field" involve only grid-related work.

One might conclude that heavy communication is responsible for most of the load imbalance but we think otherwise due to the fact that grid work seems to be the most affected. We believe that the imbalance is due to a large disparity in the number of grid points handled by the different processes at high core count. It is virtually impossible to have the same number of particles and grid points on each core due to the toroidal geometry of the computational volume and the radially decomposed domains. Since we require a uniform density of grid points on the cross-sectional planes, this translates to a constant arc length (and also radial length) separating adjacent grid points, resulting in less points on the radial surfaces near the center of the circular plane compared to the ones near the outer boundary. Furthermore, the four-point average method used for the charge accumulation requires six to eight radial ghost surfaces on each side of the radial zones to accommodate particles with a large Larmor radius. For large device sizes, this leads to large differences in the total number of grid points that the processes near the outer boundary have to handle compared to the processes near the center. Since the particle work accounts for 80%–90% of the computational work, as shown by the sum of the "push" "charge" and "shift" steps in the area plots, it is more important to have the same number of particles in each radial domain rather than the same number of grid points. The domain decomposition in its current implementation thus targets a constant average number of particles during the simulation rather than a constant number of grid points since both cannot be achieved simultaneously. It should be said, however, that this decomposition has allowed GTC to simulate dramatically larger fusion devices on BG/P and that the scaling still remains impressive.

The most communication intensive routine in GTC is the "shift" step, which moves the particles between the processes according to their new locations after the time advance step. By looking at the plots of wall clock times for the three systems we immediately see that BG/P has the smallest ratio of time spent in "shift" compared to the total loop time. This translates to the best compute to communication ratio, which is to be expected since BG/P has the slowest processor of the three systems. Hyperion, on the other hand, delivered the highest ratio of time spent in "shift," indicating a network performance not as well balanced to its processor speed than the other two systems. In terms of raw communication performance, the time spent in "shift" on the XT4 is about half of that on the BG/P at low core count. At high processor count, the times are about the same. It is worth noting that on 131,072 cores on BG/P, process placement was used to optimize the communications while this was not yet attempted on Franklin at 32,768 cores.

6.5 OLYMPUS: Unstructured FEM in Solid Mechanics

Olympus is a finite element solid mechanics application that is used for, among other applications, micro-FE bone analysis [20, 45, 219]. Olympus is a general purpose, parallel, unstructured finite element program and is used to simulate bone mechanics via micro-FE methods. These methods generate finite element meshes from micro-CT data that are composed of voxel elements and accommodate the complex micro architecture of bone in much the same way as pixels are used in a digital image. Olympus is composed of a parallelizing finite element module *Athena*. Athena uses the parallel graph

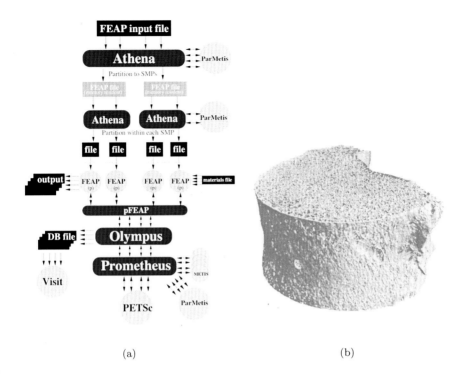

(a)   (b)

FIGURE 6.2: Olympus code architecture (a), vertebra input CT data (b).

partitioner ParMetis [186] to construct a sub-problem on each processor for
a serial finite element code *FEAP* [131]. Olympus uses a parallel finite element object *pFEAP*, which is a thin parallelizing layer for FEAP, that primarily maps vector and matrix quantities between the local FEAP grid and
the global Olympus grid. Olympus uses the parallel algebraic multigrid linear
solver *Prometheus*, which is built on the parallel numerical library *PETSc* [43].
Olympus controls the solution process including an inexact Newton solver and
manages the database output (SILO from LLNL) to be read by a visualization
application (eg, VISIT from LLNL). Figure 6.2 shows a schematic representation of this system.

The finite element input file is read, in parallel, by Athena which uses
ParMetis to partition the finite element graph. Athena generates a complete
finite element problem on each processor from this partitioning. The processor
sub-problems are designed so that each processor computes all rows of the
stiffness matrix and entries of the residual vector associated with vertices
that have been partitioned to the processors. This eliminates the need for
communication in the finite element operator evaluation at the expense of
a small amount of redundant computational work. Given this sub-problem

and a small global text file with the material properties, FEAP runs on each processor much as it would in serial mode; in fact FEAP itself has no parallel constructs but only interfaces with Olympus through the pFEAP layer.

Explicit message passing (MPI) is used for performance and portability and all parts of the algorithm have been parallelized for scalability. Hierarchical platforms, such as multi and many core architectures, are the intended target for this project. Faster communication within a node is implicitly exploited by first partitioning the problem onto the nodes, and then recursively calling Athena to construct the subdomain problem for each core. This approach implicitly takes advantage of any increase in communication performance within the node, though the numerical kernels (in PETSc) are pure MPI, and provides for good data locality.

6.5.1 Prometheus: Parallel Algebraic Multigrid Solver

The largest portion of the simulation time for these micro-FE bone problems is spent in the unstructured algebraic multigrid linear solver *Prometheus*. Prometheus is equipped with three multigrid algorithms, has both additive and (true) parallel multiplicative smoother preconditioners (general block and nodal block versions) [19], and several smoothers for the additive preconditioners (Chebyshev polynomials are used in this study [21]).

Prometheus has been optimized for ultra-scalability, including two important features for complex problems using many processors [20]: (1) Prometheus repartitions the coarse grids to maintain load balance and (2) the number of active processors, on the coarsest grids, is reduced to keep a minimum of a few hundred equations per processor. Reducing the number of active processors on coarse grids is all but necessary for complex problems when using tens or hundreds of thousands of cores, and is even useful on a few hundred cores. Repartitioning the coarse grids is important for highly heterogeneous topologies because of severe load imbalances that can result from different coarsening rates in different domains of the problem. Repartitioning is also important on the coarsest grids when the number of processors are reduced because, in general, the processor domains become fragmented which results in large amounts of communication in the solver.

6.5.2 Olympus Performance

This section investigates the performance of our project with geometrically nonlinear micro-FE bone analyses. The runtime cost of our analyses can be segregated into five primary sections: *(i)* Athena parallel FE mesh partitioner; *(ii)* linear solver *mesh* setup (construction of the coarse grids); *(iii)* FEAP element residual, tangent, and stress calculations; *(iv)* solver *matrix* setup (coarse grid operator construction); and finally *(v)* the solve for the solution.

This study uses a preconditioned conjugate gradient solver—preconditioned with one multigrid V-cycle. The smoothed aggregation multigrid method is used with a second order Chebyshev smoother [21]. Four meshes of this vertebral body are analyzed with discretization scales (h) ranging from 150 microns to 30 microns. Pertinent mesh quantities are shown in Table 6.3. The bone tissue is modeled as a finite deformation (neo-Hookean) isotropic elastic material with a Poisson's ratio of 0.3. All elements are geometrically identical trilinear hexahedra (eight-node bricks). The models are analyzed with displacement control and one solve (i.e., just the first step of the Newton solver). A scaled

TABLE 6.3: Scaled vertebral body mesh quantities. Number of cores used on BG/P is four times greater than shown in this table.

h (μm)	150	120	40	30
# degrees of freedom (10^6)	7.4	13.5	237	537
# of elements (10^6)	1.08	2.12	57	135
# of compute cores used	48	184	1696	3840
# of solver iterations (Franklin)	27	38	26	23

speedup study is conducted on each architecture with the number of cores listed in Table 6.3. Note, the iteration counts vary by as much as 10% in this data due to the non-deterministic nature of the solver, and the fact that the physical domain of each test case is not exactly the same due to noise in the CT data and inexact meshing. In particular, the second test problem (120 μm) always requires a few more iterations to converge.

This weak scaling study is designed to keep approximately the same average number of equations per node/core. Note, a source of growth in the solution time on larger problems is an increase in the number of flops per iteration per fine grid node. This is due to an increase in the average number non-zeros per row which in turn is due to the large ratio of volume to surface area in the vertebral body problems. These vertebral bodies have a large amount of surface area and, thus, the low resolution mesh (150 μm) has a large ratio of surface nodes to interior nodes. As the mesh is refined the ratio of interior nodes to surface nodes increases, resulting in, on average, more non-zeros per row—from 50 on the smallest version to 68 on the largest. Additionally, the complexity of the construction of the coarse grids in smoothed aggregation has the tendency to increase in fully 3D problems. As this problem is refined, the meshes become more fully 3D in nature—resulting in grids with more non-zeros per row and higher complexities—this would not be noticeable with a fully 3D problem like a cube. Thus, the flop rates are a better guide as to the scalability of the code and scalability of each machine.

Figure 6.3(a) shows the total flop rate of the solve phase and the matrix setup phase (the two parts of the algorithm that are not amortized in a full Newton nonlinear solve) on the Cray XT4 Franklin at NERSC. This shows very good parallel efficiency from 48 to just under 4K cores and much higher flop rates for the matrix setup. The matrix setup uses small dense matrix

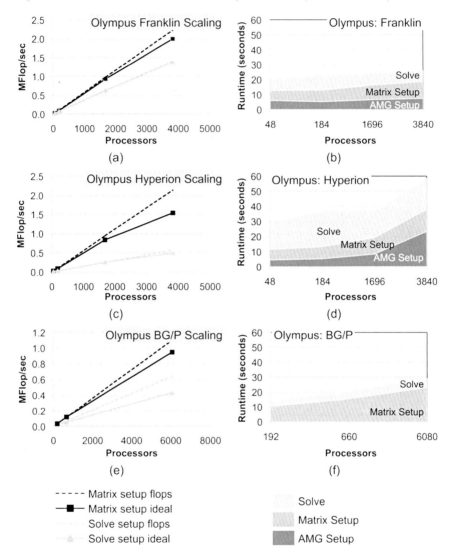

FIGURE 6.3: Olympus weak scaling, flop rates (left), run times (right), for the Cray XT4 (Franklin), Xeon cluster (Hyperion) and IBM BG/P (Intrepid).

matrix multiplies (BLAS3) in its kernel which generally run faster than the dense matrix vector multiplies (BLAS2) in the kernel of the solve phase. Figure 6.3(b) shows the times for the major components of a run with one linear solver on the Cray XT4. This data shows, as expected, good scalability (ie, constant times as problem size and processor counts are scaled up).

Figure 6.3(c) shows the flop rates for the solve phase and the matrix setup phase on the Xeon cluster Hyperion at LLNL. Again, we see much higher flop rates for the matrix setup phase and the solve phase is scaling very well up to about 4K processors, but we do observe some degradation the performance of the matrix setup on larger processor counts. This is due to the somewhat complex communication patterns required for the matrix setup as the fine grid matrix (source) and the course grid matrix (product) are partitioned separately and load balancing becomes more challenging because the work per processor is not strictly proportional to the number of vertices on that processor. Figure 6.3(d) shows the times for the major components of one linear solver on Hyperion. This data shows that matrix setup phase is running fast relative to the solve phase (relatively faster than on the Cray) and the AMG setup phase is slowing down on the larger processor counts. Overall the Hyperion data shows that communication fabric is not as good as the Cray's given that tasks with complex communication requirements are not scaling as well on Hyperion as on the XT4.

Figure 6.3(e) shows the flop rates for the solve phase and the matrix setup phase on the IBM BG/P Intrepid at ANL. Figure 6.3(f) shows run times for the solve phase and the matrix setup phase on the BG/P. Note, the scaling study on the BG/P uses four times as many cores as the other two tests, due to lack of memory. Additionally, we were not able to run the largest test case due to lack of memory. These are preliminary results in that we have not addressed some serious performance problems that we are observing on the IBM. First we have observed fairly large load imbalance of about 30%. This imbalance is partially due to the larger amount of parallelism demanded by the BG/P (i.e., we have four times as many MPI processes as on the other systems). We do observe good efficiency in the solve phase but we see significant growth in the matrix setup phase. We speculate that this is due to load imbalance issues. Thus, we are seeing good scalability in the actual solve phase of the solver but we are having problems with load balance and performance on the setup phases of the code. This will be the subject of future investigation.

6.6   Carpet: Higher-Order AMR in Relativistic Astrophysics

6.6.1   The Cactus Software Framework

Cactus [72, 151] is an open-source, modular, and portable programming environment for collaborative HPC computing. It was designed and written specifically to enable scientists and engineers to develop and perform the large-scale simulations needed for modern scientific discovery across a broad range of disciplines. Cactus is used by a wide and growing range of applications, prominently including relativistic astrophysics, but also including quantum gravity, chemical engineering, lattice Boltzmann Methods, econometrics, computational fluid dynamics, and coastal and climate modeling [33, 74, 116, 189, 193, 224, 285, 323]. The influence and success of Cactus in high performance computing was recognized with the IEEE Sidney Fernbach prize, which was awarded to Edward Seidel at Supercomputing 2006.

Among the needs of the Cactus user community have been ease of use, portability, support of large and geographically diverse collaborations, and the ability to handle enormous computing resources, visualization, file I/O, and data management. Cactus must also support the inclusion of legacy code, as well as a range of programming languages. Some of the key strengths of Cactus have been its portability and high performance, which led to it being chosen by Intel to be one of the first scientific applications deployed on the IA64 platform, and Cactus's widespread use for benchmarking computational architectures. For example, a Cactus application was recently benchmarked on the IBM Blue Gene/P system at ANL and scaled well up to 131,072 processors.

Cactus is a so-called "tightly coupled" framework; Cactus applications are single executables that are intented to execute within one supercomputing system. Cactus components (called *thorns*) delegate their memory management, parallelism, and I/O to a specialized *driver* component. This architecture enables highly efficient component coupling with virtually no overhead.

The framework itself (called *flesh*) does not implement any significant functionality on its own. It rather offers a set of well-designed APIs which are implemented by other components. The specific set of components which are used to implement these can be decided at run time. For example, these APIs provide coordinate systems, generic interpolation, reduction operations, hyperslabbing, and various I/O methods, and more.

6.6.2 Computational Infrastructure: Mesh Refinement with Carpet

Carpet [9, 291] is an adaptive mesh refinement (AMR) driver for the Cactus framework. Carpet is a driver for Cactus, providing adaptive mesh refinement, memory management for grid functions, efficient parallelization, and I/O. Carpet provides spatial discretization based on highly efficient block-structured, logically Cartesian grids. It employs a spatial domain decomposition using a hybrid MPI/OpenMP parallelism. Time integration is performed via the recursive Berger-Oliger AMR scheme [52], including subcycling in time. In addition to mesh refinement, Carpet supports multi-patch systems [115, 290, 392] where the domain is covered by multiple, possibly overlapping, distorted, but logically Cartesian grid blocks.

Cactus parallelizes its data structures on distributed memory architectures via spatial domain decomposition, with *ghost zones* added to each MPI process' part of the grid hierarchy. Synchronization is performed automatically, based on declarations for each routine specifying variables it modifies, instead of via explicit calls to communication routines.

Higher order methods require a substantial amount of ghost zones (in our case, three ghost zones for fourth order accurate, upwinding differencing stencils), leading to a significant memory overhead for each MPI process. This can be counteracted by using OpenMP within a multi-core node, which is especially attractive on modern systems with eight or more cores per node; all performance critical parts of Cactus support OpenMP. However, non-uniform memory access (NUMA) systems require care in laying out data structures in memory to achieve good OpenMP performance, and we are therefore using a combination of MPI processes and OpenMP threads on such systems.

We have recently used *Kranc* [6, 177, 208] to generate a new Einstein solver *McLachlan* [62, 233]. Kranc is a *Mathematica*-based code generation system that starts from continuum equations in *Mathematica* notation, and auto-matically generates full Cactus thorns after discretizing the equations. This approach shows a large potential, not only for reducing errors in complex systems of equations, but also for reducing the time to implement new discretization methods such as higher-order finite differencing or curvilinear coordinate systems. It furthermore enables automatic code optimizations at a very high level, using domain-specific knowledge about the system of equations and the discretization method that is not available to the compiler, such as cache or memory hierarchy optimizations or multi-core parallelization. Such optimizations are planned for the future.

Section 6.7 below describes *CASTRO*, an AMR infrastructure using a similar algorithm. We also briefly compare both infrastructures there.

6.6.3   Carpet Benchmark

The Cactus–Carpet benchmark solves the Einstein equations, i.e., the field equations of General Relativity, which describe gravity near compact objects such as neutron stars or black holes. Far away from compact objects, the Einstein equations describe gravitational waves, which are expected to be detected by ground-based and space-based detectors such as LIGO [7], GEO600 [4], and LISA [8] in the coming years. This detection will have groundbreaking effects on our understanding, and open a new window onto the universe [334].

We use the BSSN formulation of the Einstein equations as described e.g. in [26, 27], which is a set of 25 coupled second-order non-linear partial differential equations. In this benchmark, these equations are discretized using higher order finite differences on a block-structured mesh refinement grid hierarchy. Time integration uses Runge-Kutta type explicit methods. We further assume that there is no matter or electromagnetic radiation present, and use radiative (absorbing) outer boundary conditions. We choose Minkowski (flat spacetime) initial conditions, which has no effect on the run time of the simulations. We use fourth order accurate spatial differencing, Berger-Oliger style mesh refinement [52] with subcycling in time, with higher order order Lagrangian interpolation on the mesh refinement boundaries. We pre-define a fixed mesh refinement hierarchy with nine levels, each containing the same number of grid

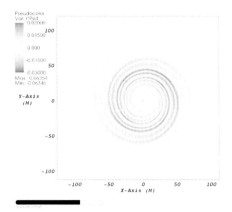 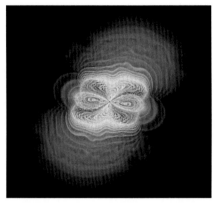

FIGURE 6.4: *Left:* Gravitational radiation emitted during in binary black hole merger, as indicated by the rescaled Weyl scalar $r \cdot \Psi_4$. This simulation was performed with the Cactus-Carpet infrastructure with nine levels of AMR tracking the inspiralling black holes. The black holes are too small to be visible in this figure. [Image credit to Christian Reisswig at the Albert Einstein Institute.] *Right:* Volume rendering of the gravitational radiation during a binary black hole merger, represented by the real part of Weyl scalar $r \cdot \psi_4$. (See color insert.) [Image credit to Werner Benger at Louisiana State University.]

points. This is a *weak scaling* benchmark where the number of grid points increases with the used number of cores. Figure 6.4 shows gravitational waves from simulated binary black hole systems.

The salient features of this benchmark are thus: Explicit time integration, finite differencing with mesh refinement, many variables (about 1 Gbyte/core), complex calculations for each grid point (about 5000 flops). This benchmark does *not* use any libraries in its kernel loop, such as BLAS, LAPACK, or PETSc, since no efficient high-level libraries exist for stencil-based codes. However, we use automatic code generation [6,177,199,208] which allows some code optimizations and tuning at build time [29].

Our benchmark implementation uses the Cactus software framework [72, 151], the Carpet adaptive mesh refinement infrastructure [9, 290, 291], the Einstein Toolkit [3, 289], and the McLachlan Einstein code [233]. We describe this benchmark in detail in [329].

6.6.4 Carpet Performance

We examined the Carpet weak scaling benchmark described above on several systems with different architectures. Figure 6.5 shows results for the Cray XT4 Franklin, a Blue Gene/P, and Hyperion in terms of runtime and parallel

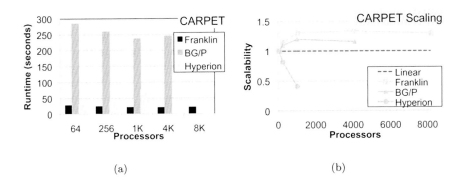

FIGURE 6.5: Carpet benchmark with nine levels of mesh refinement showing (a) weak scaling performance on three examined platforms and (b) scalability relative to concurrency. For ideal scaling the run time should be independent of the number of processors. Carpet shows good scalability except on Hyperion.

scalability. Overall, Carpet shows excellent weak scalability, achieving super-linear performance on Franklin and Intrepid. On Hyperion scalability breaks down at around 1024 processors, the source of the anomaly is under investigation, as Carpet has demonstrated high scalability up to 12288 processors on other systems with similar architecture, including the Xeon/Infiniband Queen Bee [11] and the AMD/Infiniband Ranger [12] clusters. We were unfortunately not able to obtain interactive access to Hyperion for the Cactus team, and were thus not able to examine this problem in detail — future efforts will focus on identifying the system-specific performance constraints.

We note that we needed to modify the benchmark parameters to run on the BG/P system, since there is not enough memory per core for the standard settings. Instead of assigning AMR components with 25^3 grid points to each core, we were able to use only 13^3 grid points per component. This reduces the per-core computational load by a factor of about eight, but reduces the memory consumption only by a factor of about four due to the additional inter-processor ghost zones required. Consequently, we expect this smaller version of the benchmark to show less performance and be less scalable.

In some cases (such as Franklin), scalability even improves for very large number of cores. We assume that it is due to the AMR interface conditions in our problem setup. As the number of cores increases, we increase the total problem size to test weak scaling. Increasing the size of the domain by a factor of N increases the number of evolved grid points by N^3, but increases the number of AMR interface points only by a factor of N^2. As the problem size increases, the importance of the AMR interface points (and the associated computation and communication overhead) thus decreases.

Number of Stars	Level1 Grids	Level2 Grids
one	512 (L0) 1728 (L1)	512 (L0) 512 (L1) 1728 (L2)
two	1024 3456	1024 1024 3456
four	2048 6912	2048 2048 6912
eight	4096 13824	4096 4096 13824

(a)   (b)

FIGURE 6.6: CASTRO (a) Close-up of a single star in the CASTRO test problem, shown here in shades of grey on the slice planes, are contours of density. An isosurface of the velocity magnitude is shown, and the vectors represent the radially-inward-pointing self-gravity. (b) Number of grids for each CASTRO test problem.

6.7 CASTRO: Compressible Astrophysics

CASTRO is a finite volume evolution code for compressible flow in Eulerian coordinates and includes self-gravity and reaction networks. CASTRO incorporates hierarchical block-structured adaptive mesh refinement and supports 3D Cartesian, 2D Cartesian and cylindrical, and 1D Cartesian and spherical coordinates. It is currently used primarily in astrophysical applications, specifically for problems such as the time evolution of Type Ia and core collapse supernovae.

The hydrodynamics in CASTRO is based on the unsplit methodology introduced in [96]. The code has options for the piecewise linear method in [96] and the unsplit piecewise parabolic method (PPM) in [240]. The unsplit PPM has the option to use the less restrictive limiters introduced in [97]. All of the hydrodynamics options are designed to work with a general convex equation of state.

CASTRO supports two different methods for including Newtonian self-gravitational forces. One approach uses a monopole approximation to compute a radial gravity consistent with the mass distribution. The second approach is based on solving the Poisson equation,

$$-\Delta\phi = 4\pi G\rho,$$

for the gravitational field, ϕ. The Poisson equation is discretized using standard finite difference approximations and the resulting linear system is solved using geometric multigrid techniques, specifically V-cycles and red-black Gauss-Seidel relaxation. A third approach in which gravity is externally specified is also available.

Our approach to adaptive refinement in CASTRO uses a nested hierarchy of logically-rectangular grids with simultaneous refinement of the grids in both space and time. The integration algorithm on the grid hierarchy is a recursive procedure in which coarse grids are advanced in time, fine grids are advanced multiple steps to reach the same time as the coarse grids, and the data at different levels are then synchronized. During the regridding step, increasingly finer grids are recursively embedded in coarse grids until the solution is sufficiently resolved. An error estimation procedure based on user-specified criteria evaluates where additional refinement is needed and grid generation procedures dynamically create or remove rectangular fine grid patches as resolution requirements change.

For pure hydrodynamic problems, synchronization between levels requires only a "reflux" operation in which coarse cells adjacent to the fine grid are modified to reflect the difference between the original coarse-grid flux and the integrated flux from the fine grid. (See [48, 51]). For processes that involve implicit discretizations, the synchronization process is more complex. The basic synchronization paradigm for implicit processes is discussed in [28]. In particular, the synchronization step for the full self-gravity algorithm is similar to the algorithm introduced by [242].

CASTRO uses a general interface to equations of state and thermonuclear reaction networks, that allows us to easily add more extensive networks to follow detailed nucleosynthesis.

The parallelization strategy for CASTRO is to distribute grids to processors. This provides a natural coarse-grained approach to distributing the computational work. When AMR is used a dynamic load balancing technique is needed to adjust the load. We use both a heuristic knapsack algorithm and a space-filling curve algorithm for load balancing. Criteria based on the ratio of the number of grids at a level to the number of processors dynamically switches between these strategies.

CASTRO is written in C++ and Fortran-90. The time stepping control, memory allocation, file I/O, and dynamic regridding operations occur primarily in C++. Operations on single arrays of data, as well as the entire multigrid solver framework, exist in Fortran-90.

6.7.1 CASTRO and Carpet

The AMR algorithms employed by CASTRO and Carpet (see Section 6.6.1 above) share certain features, and we outline the commonalities and differences below.

Both CASTRO and Carpet use nested hierarchies of logically rectangular grids, refining the grids in both space and time. The integration algorithm on the grid hierarchy is a recursive procedure in which coarse grids are first advanced in time, fine grids are then advanced multiple steps to reach the same time as the coarse grids, and the data at different levels are then synchronized. This AMR methodology was introduced by Berger-Oliger (1984) [52] for hyperbolic problems.

Both CASTRO and Carpet support different coordinate systems; CASTRO can be used with Cartesian, cylindrical, or spherical coordinates while Carpet can handle multiple patches with arbitrary coordinate systems [290].

The basic AMR algorithms in both CASTRO and Carpet is independent of the spatial discretization and the time integration methods which are employed. This cleanly separates the formulation of physics from the computational infrastructure which provides the mechanics of gridding and refinement.

The differences between CASTRO and Carpet are mostly historic in origin, coming from the different applications areas that they are or were targeting. CASTRO originates in the hydrodynamics community, whereas Carpet's background is in solving the Einstein equations. This leads to differences in the supported feature set, while the underlying AMR algorithm is very similar. Quite likely both could be extended to support the feature set offered by the respective other infrastructure.

For example, CASTRO offers a generic regridding algorithm based on user-specified criteria (e.g. to track shocks), refluxing to match fluxes on coarse and fine grids, vertex- and cell-centred quantities. These features are well suited to solve flux-conservative hydrodynamics formulations. CASTRO also contains a full Poisson solver for Newtonian gravity.

Carpet, on the other hand, supports features required by the Einstein equations, which have a wave-type nature where conservation is not relevant. Various formulations of the Einstein equations contain second spatial derivatives, requiring a special treatment of refinement boundaries (see [291]). Also, Carpet does not contain a Poisson solver since gravity is already described by the Einstein equations, and no extra elliptic equation needs to be solved.

CASTRO and Carpet are also very similar in terms of their implementation. Both are parallelised using MPI, and support for multi-core architectures via OpenMP has been or is being added.

6.7.2 CASTRO Performance

The test problem for the scaling study was that of one or more self-gravitating stars in a 3D domain with outflow boundary conditions on all

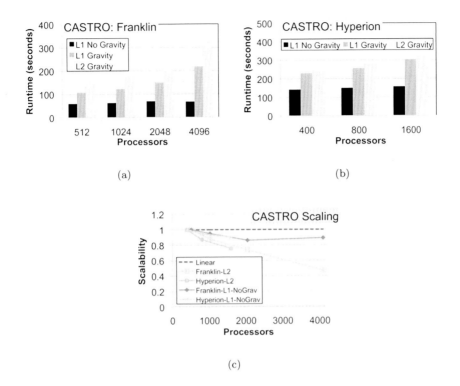

FIGURE 6.7: CASTRO performance behavior with and without the gravity solver using one (L1) and two (L2) levels of adaptivity on (a) Franklin and (b) Hyperion. (c) Shows scalability of the various configurations.

sides. We used a stellar equation of state as implemented in [336] and initialized the simulation by interpolating data from a 1D model file onto the 3D Cartesian mesh. The model file was generated by a 1D stellar evolution code, Kepler [355]. An initial velocity perturbation was then superimposed onto the otherwise quiescent star — a visualization is shown in Figure 6.6(a).

The test problem was constructed to create a weak scaling study, with a single star in the smallest runs, and two, four, and eight stars for the larger runs, with the same grid layout duplicated for each star. Runs on Hyperion were made with one, two, and four stars, using 400, 800, and 1600 processors respectively. Hyperion was in the early testing stage when these runs were made and 1600 processors was the largest available pool at the time, so the smaller runs were set to 400 and 800 processors. Runs on Franklin were made with one, two, four, and eight stars, using 512, 1024, 2048, and 4096 processors respectively. The number of grids for each problem is shown in Table 6.6(b).

Figure 6.7 shows the performance and scaling behavior of CASTRO on

Franklin and Hyperion. Results were not collected on Intrepid because the current implementation of CASTRO is optimized for large grids and requires more memory per core than is available on that machine. This will be the subject of future investigation. The runs labeled with "no gravity" do not include a gravity solver, and we note that these scale very well from 512 to 4096 processors. Here *Level1* refers to a simulation with a base level and one level of refinement, and *Level2* refers to a base level and two levels of refinement. Refinement here is by a factor of two between each level. For the *Level2* calculations, the *Level0* base grid size was set to half of the *Level1* calculation's *Level0* base size in each direction to maintian the same effective calculation resolution. Maximum grid size was 64 cells in each direction.

Observe that the runs with "gravity" use the Poisson solver and show less ideal scaling behavior; this is due to the increasing cost of communication in the multigrid solver for solving the Poisson equation. However, the *Level1* and *Level2* calculations show similar scaling, despite that fact that the*Level2* calculation has to do more communication than the *Level1* calculation in order to syncronize the extra level and do more work to keep track of the overhead data required by the extra level. The only exception to this is the highest concurrency Franklin benchmark, where the *Level2* calculation scales noticeably less well than the *Level1*.

Comparing the two architectures, Franklin shows the same or better scaling than Hyperion for the one, two, and four star benchmarks despite using more processors in each case. For the *Level2* calculations with "gravity," even at the lowest concurrency, Franklin is roughly 1.6 times faster than Hyperion, while a performance prediction based on peak flops would be about 1.2.

In a real production calculation, the number and sizes of grids will vary during the run as the underlying data evolves. This changes the calculation's overall memory requirements, communication patterns, and sizes of communicated data, and will therefore effect the overall performance of the entire application Future work will include investigations on optimal gridding, effective utilization of shared memory multicore nodes, communication locality, reduction of AMR metadata overhead requirements, and the introduction of a hybrid MPI/OpenMP calculation model.

6.8 MILC: Quantum Chromodynamics

The MIMD Lattice Computation (MILC) collaboration has developed a set of codes written in the C language that are used to study quantum chromodynamics (QCD), the theory of the strong interactions of subatomic physics. "Strong interactions" are responsible for binding quarks into protons and neutrons and holding them all together in the atomic nucleus. These codes are designed for use on MIMD (multiple instruction multiple data) parallel platforms, using the MPI library. A particular version of the MILC suite, enabling simulations with conventional dynamical Kogut-Susskind quarks is studied in this chapter. The MILC code has been optimized to achieve high efficiency on cache-based superscalar processors. Both ANSI standard C and assembler-based codes for several architectures are provided in the source distribution.[2]

In QCD simulations, space and time are discretized on sites and links of a regular hypercube lattice in four-dimensional space time. Each link between nearest neighbors in this lattice is associated with a 3-dimensional SU(3) complex matrix for a given field [200]. The simulations involve integrating an equation of motion for hundreds or thousands of time steps that requires inverting a large, sparse matrix at each step of the integration. The sparse matrix problem is solved using a conjugate gradient (CG) method, but because the linear system is nearly singular many CG iterations are required for convergence. Within a processor the four-dimensional nature of the problem requires gathers from widely separated locations in memory. The matrix in the linear system being solved contains sets of complex 3-dimensional "link" matrices, one per 4-D lattice link but only links between odd sites and even sites are non-zero. The inversion by CG requires repeated three-dimensional complex matrix-vector multiplications, which reduces to a dot product of three pairs of three-dimensional complex vectors. The code separates the real and imaginary parts, producing six dot product pairs of six-dimensional real vectors. Each such dot product consists of five multiply-add operations and one multiply [53]. The primary parallel programing model for MILC is a 4-D domain decomposition with each MPI process assigned an equal number of sublattices of contiguous sites. In a four-dimensional problem each site has eight nearest neighbors.

Because the MILC benchmarks are intended to illustrate performance achieved during the "steady-state" portion of an extremely long Lattice Gauge simulation, each benchmark actually consists of two runs: a short run with a few steps, a large step size, and a loose convergence criterion to let the lattice evolve from a totally ordered state, and a longer run starting with this "primed" lattice, with increased accuracy for the CG solve and a CG iteration count that is a more representative of real runs. Only the time for the latter portion is used.

6.8.1   MILC Performance

We conduct two weak-scaling experiments: a small lattice per core and a larger lattice per core, representing two extremes of how MILC might be used in a production configuration. Benchmark timing results and scalability characteristics are shown in Figures 6.8(a–b) and Figures 6.8(c–d) for small and large problems, respectively. Not surprisingly, scalability is generally bet-

---

[2]See http://www.physics.utah.edu/~detar/milc/index.html for a further description of MILC.

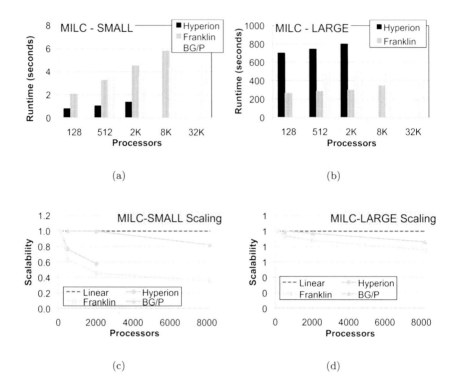

FIGURE 6.8: MILC performance and scalability using (a–b) small and (c–d) large problem configurations.

ter for the larger lattice than for the smaller. This is particularly true for the Franklin XT4 system, for which scalability of the smaller problem is severely limited. Although the overall computation/communication is very low, one of the main sources of performance degradation is the CG solve. MILC was instrumented to measure the time in five major parts of the code, and on 1024 cores the CG portion is consuming about two-thirds of the runtime. Micro-kernel benchmarking of the MPI_Allreduce operation on three systems (not shown) shows that the XT4's SeaStar interconnect is considerably slower on this operation at larger core counts. Note, however, that these microkernel data were obtained on a largely dedicated Hyperion system and a dedicated partition of BG/P but in a multi-user environment on the XT4, for which job scheduling is basically random within the interconnect's torus; therefore, the XT4 data likely include the effects of interconnect contention. For the smaller MILC lattice scalability suffers due to the insufficient computation, required to hide this increased Allreduce cost on the XT4.

In contrast, scalability doesn't differ very much between the two lattice sizes on the BG/P system, and indeed, scalability of the BG/P system is generally best of all the systems considered here. Furthermore, it is known that carefully mapping the 4-D MILC decomposition to the BG/P torus network can sometimes improve performance; however, this was not done in these studies and will be the subject of future investigations.

The Hyperion cluster shows similar scaling characteristics to Franklin for the large case, and somewhat better scaling than Franklin for the small.

Turning our attention to absolute performance, for the large case we see that Franklin significantly outperforms the other platforms, with the BG/P system providing the second best performance. For the small case, the order is essentially reversed, with Hyperion having the best performance across the concurrencies studied. MILC is a memory bandwidth-intensive application. Nominally, the component vector and corresponding matrix operations consume 1.45 bytes input/flop and 0.36 bytes output/flop; in practice, we measure for the entire code a computational intensity of about 1.3 and 1.5, respectively, for the two lattice sizes on the Cray XT4. On multicore systems this memory bandwidth dependence leads to significant contention effects within the socket. The large case shows this effect most strongly, with the Intel Harpertown-based Hyperion system lagging behind the other two architectures despite having a higher peak floating-point performance.

6.9   Summary and Conclusions

Computational science is at the dawn of petascale computing capability, with the potential to achieve simulation scale and numerical fidelity at hitherto unattainable levels. However, increasing concerns over power efficiency and the economies of designing and building these large systems are accelerating recent trends towards architectural diversity through new interest in customization and tighter system integration on the one hand and incorporating commodity components on the other. Understanding the tradeoffs of these differing design paradigms, in the context of high-end numerical simulations, is a key step towards making effective petascale computing a reality.

In this study, we examined the behavior of a number of key large-scale scientific computations. To maximize the utility for the HPC community, performance was evaluated on the full applications, with real input data and at the scale desired by application scientists in the corresponding domain; these types of investigations require the participation of computational scientists from highly disparate backgrounds. Performance results and analysis were presented on three leading HPC platforms: Franklin XT4, Hyperion Xeon cluster, and Intrepid BG/P, representing some of the most common design trade-offs in the high performance computing arena.

Figure 6.9 presents a summary of the results for largest comparable con-

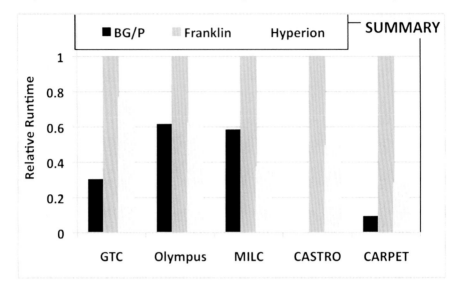

FIGURE 6.9: Results summary for the largest comparable concurrencies of the five evaluated codes on three leading HPC platforms, showing relative runtime performance normalized to fastest system. Note that due to BG/P's restricted memory capacity: Olympus uses 4x the number of BG/P cores, Carpet uses reduced BG/P problem domains, and CASTRO was unable to conduct comparable BG/P experiments.

currencies of the five evaluated codes, showing relative runtime performance normalized to fastest system. Observe that the tightly integrated Cray XT system achieves the highest performance, consistently outperforming the Xeon cluster — assembled from commodity components. Comparing to the BG/P is more difficult, as two of the benchmarks use different numbers of processors or different weak scaling parameters. For GTC and MILC, the two directly comparable benchmarks — in one case the Xeon platform outperforms the BG/P platform, but in the other the situation is reversed. For MILC, the low memory bandwidth of the Xeon Clovertown and the relatively poor performance of collective communication of IB as compared to the custom interconnect of BG/P means that BG/P comes out ahead. For the higher computational intensity GTC, the Intel/IB cluster dominates. In the case of Olympus, we can compare across architectures based on the idea that the differences in number of processors used is representative of how a simulation would be run in practice. Here BG/P and the Xeon cluster have comparable performance, with the XT ahead. For Carpet, the smaller domain sizes used for the BG/P runs makes accurate comparison impossible, but one can certainly say that absolute performance is very poor compared with both the XT and Xeon cluster.

However, as the comparison in Figure 6.9 is at relatively small concurrencies due to the size of the Intel/IB cluster, this is only part of the picture. At higher concurrencies the scalability of BG/P exceeds that of the XT for GTC and MILC, and substantially similar scaling is seen for Olympus. However, for Carpet, XT scalability is better than on BG/P, although it should be noted that both scale superlinearly. While both BG/P and XT have custom interconnects, BG/P isolates portions of the interconnect (called partitions) for particular jobs much more effectively than on the XT, where nodes in a job can be scattered across the torus intermixed with other jobs that are competing for link bandwidth. This is one of the likely reasons for the poorer scalability of some applications on the XT.

Overall, these extensive performance evaluations are an important step toward effectively conducting simulations at the petascale level and beyond, by providing computational scientists and system designers with critical information on how well numerical methods perform across state-of-the-art parallel systems. Future work will explore a wider set of computational methods, with a focus on irregular and unstructured algorithms, while investigating a broader set of HEC platforms, including the latest generation of multi-core technologies.

6.10   Acknowledgments

We thank Bob Walkup and Jun Doi of IBM for the optimized version of MILC for the BG/P system and Steven Gottlieb of Indiana University for many helpful discussions related to MILC benchmarking. We also kindly thank Brent Gorda of LLNL for access to the Hyperion system. This work was supported by the Advanced Scientific Computing Research Office in the DOE Office of Science under contract number DE-AC02-05CH11231. Dr. Ethier is supported by the U. S. Department of Energy Office of Fusion Energy Sciences under contract number DE-AC02-09CH11466. This work was also supported by the NSF awards 0701566 *XiRel* and 0721915 *Alpaca*, by the LONI *numrel* allocation, and by the NSF TeraGrid allocations TG-MCA02N014 and TG-ASC090007. This research used resources of NERSC at LBNL and ALCF at ANL which are supported by the Office of Science of the DOE under Contract No. DE-AC02-05CH11231 and DE-AC02-06CH11357 respectively.

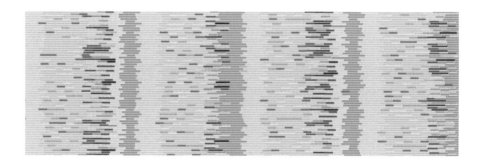

FIGURE 4.5: hpctraceview showing part of an execution trace for GTC..

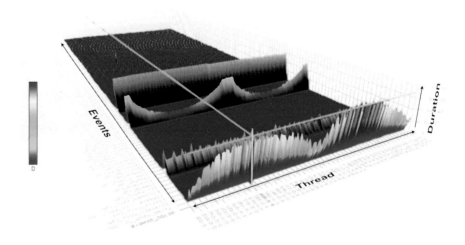

FIGURE 4.8: ParaProf displays of AORSA-2D performance on the ORNL Cray XT5 (Jaguar) system.

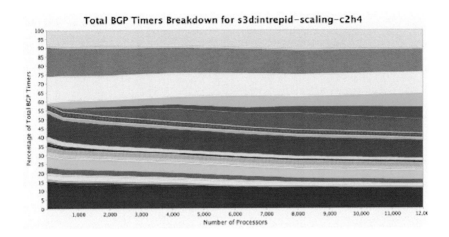

FIGURE 4.10a: PerfExplorer analysis of S3D performance on IBM BG/P. (a) S3D events scale.

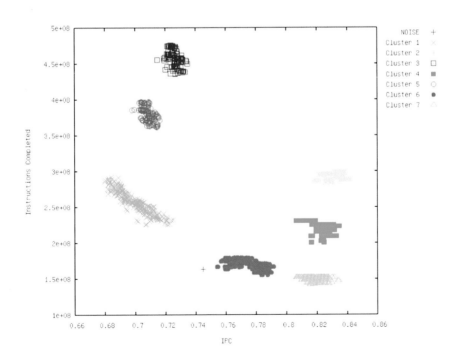

FIGURE 5.7: Clustered scatter plot of instructions vs. IPC for computation bursts in a 128 processors run of WRF.

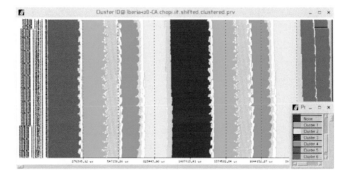

FIGURE 5.8:   Clustered timeline of a 128 processor run of WRF.

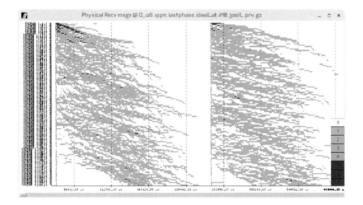

FIGURE 5.11:   Incoming transfers of a 256 processes simulation of PEPC with infinite input links.

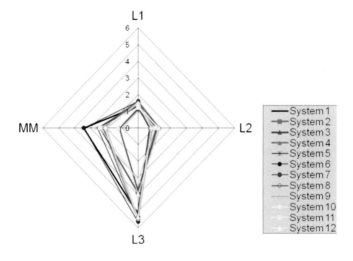

FIGURE 7.3:   Memory profiles of systems out to 2012 (anonymous).

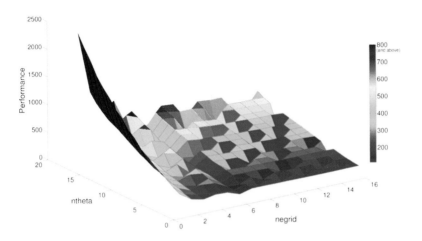

GS2 Performance as a function of two tunable parameters when the third parameter is fixed

FIGURE 10.1: GS2 performance plot with two tunable parameters. Figure from [322] [©2005 IEEE] Reprinted with Permission.

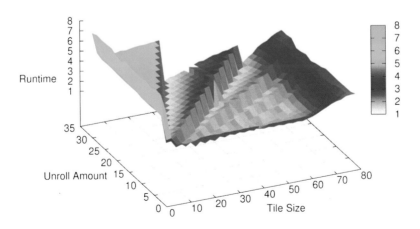

Parameter Interaction (Tiling and Unrolling for MM, N=800)

FIGURE 10.3: Parameter Search Space for Tiling and Unrolling. Figure from [337] [©2005 IEEE] Reprinted with permission.

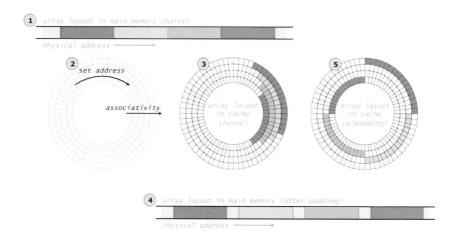

FIGURE 13.5: Conceptualization of array padding in both the physical (linear) and cache (periodic) address spaces.

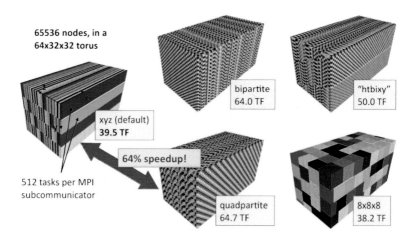

FIGURE 14.4: Performance on different node mappings.

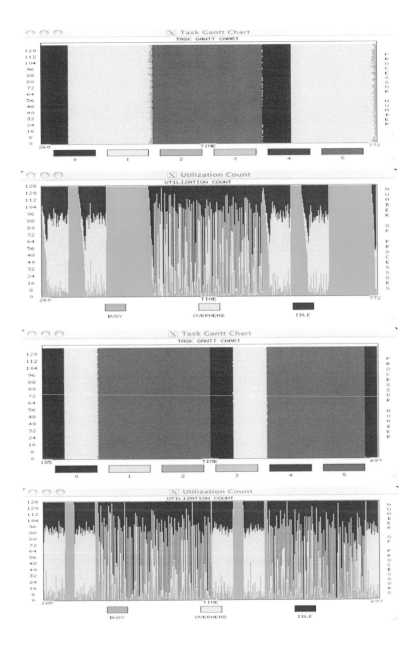

FIGURE 15.8: POP task graph and utilization count histogram on Cray X1: without and with vectorization. Horizontal axis is time. (From [384].)

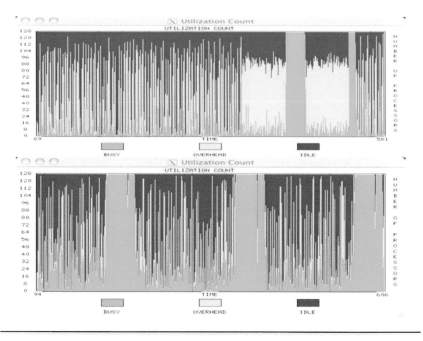

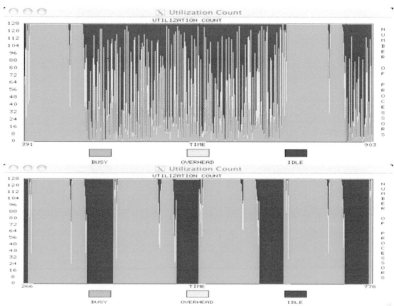

FIGURE 15.9: POP utilization count histograms on Cray X1: without and with MPI optimization (top) and Co-Array Fortran optimizations (bottom). Horizontal axis is time. (From [384].)

Chapter 7

Performance Modeling: The Convolution Approach

David H Bailey

Lawrence Berkeley National Laboratory

Allan Snavely

University of California, San Diego

Laura Carrington

San Diego Supercomputer Center

7.1 Introduction

The goal of performance modeling is to gain understanding of a computer systems performance on various applications, by means of measurement and analysis, and then to encapsulate these characteristics in a compact formula. The resulting model can be used to gain greater understanding of the performance phenomena involved and to project performance to other system/application combinations. The focus here, as in other chapters in this volume, is on large-scale scientific computation, although many of the techniques we describe apply equally well to single-processor systems and to business-type applications.

The next three chapters describe three approaches to performance modeling: (a) the *convolution* approach (this chapter), (b) the *ApexMap* approach (Chapter 8), and (c) the *roofline* approach (Chapter 9). A number of important performance modeling activities are also being done by other groups, for example at Los Alamos National Laboratory [167].

The performance profile of a given system/application combination depends on numerous factors, including: (1) system size; (2) system architecture; (3) processor speed; (4) multi-level cache latency and bandwidth; (5) inter-processor network latency and bandwidth; (6) system software efficiency; (7) type of application; (8) algorithms used; (9) programming language used; (10) problem size; (11) amount of I/O; and others. Indeed, a comprehensive model must incorporate most if not all of the above factors. Because of the difficulty in producing a truly comprehensive model, present-day performance modeling researchers generally limit the scope of their models to a single system and application, allowing only the system size and job size to vary. Nonetheless, as we shall see below, some recent efforts appear to be effective over a broader range of system/application choices.

Performance models can be used to improve architecture design, inform procurement, and guide application tuning. Unfortunately, the process of producing performance models historically has been rather expensive, requiring large amounts of computer time and highly expert human effort. This has severely limited the number of high-end applications that can be modeled and studied. Someone has observed that, due to the difficulty of developing performance models for new applications, as well as the increasing complexity of new systems, our supercomputers have become better at predicting and explaining natural phenomena (such as the weather) than at predicting and explaining the performance of themselves or other computers.

7.2 Applications of Performance Modeling

Performance modeling can be used in numerous ways. Here is a brief summary of these usages, both present-day and future possibilities:

- *Runtime estimation.* The most common application for a performance model is to enable a scientist to estimate the runtime of a job when the input parameters for the job are changed, or when a different number of processors is used in a parallel computer system. One can also estimate the largest size of system that can be used to run a given problem before the parallel efficiency drops to an unacceptable area.

- *System design.* Performance models are frequently employed by computer vendors in their design of future systems. Typically engineers construct a performance model for one or two key applications, and then compare future technology options based on performance model projections. Once performance modeling techniques are better developed, it may be possible to target many more applications and technology options in the design process. As an example of such "what-if" investigations, model parameters can be used to predict how performance rates would change with a larger or more highly associative cache. In a similar way, the performance impact of various network designs can be explored. We can even imagine that vendors could provide a variety of system customizations, depending on the nature of the users anticipated applications.

- *System tuning.* One example of using performance modeling for system tuning is given in [78]. Here a performance model was used to diagnose and rectify a misconfigured message passing interface (MPI) channel buffer, which yielded a doubling of network performance for programs sending short messages. Along this line, Adolfy Hoisie of LANL recalls that when a recent system was installed, its performance fell below model predictions by almost a factor of two. However, further analysis uncovered some system difficulties, which, when rectified, improved performance to almost the same level the model predicted [167]. When observed performance of a system falls short of that predicted by a performance model, it may be the system that is wrong not the model!

- *Application tuning.* If a memory performance model is combined with application parameters, one can predict how cache hit-rates would change if a different cache-blocking factor were used in the application. Once the optimal cache blocking has been identified, then the code can be permanently changed. Simple performance models can even be incorporated into an application code, permitting on-the-fly selection of different program options.

- *Algorithm choice.* Performance models, by providing performance expectations based on the fundamental computational characteristics of algorithms, can also enable algorithmic choice before going to the trouble to implement all the possible choices. For example, in some recent work one of the present authors employed a performance model to estimate the benefit of employing an "inspector" scheme to reorder data-structures before being accessed by a sparse-matrix solver, as part of software being developed by the SciDAC Terascale Optimal PDE Simulations (TOPS) project. It turned out that the overhead of these "inspector" schemes is more than repaid provided the sparse-matrices are large and/or highly randomized.

- *System procurement.* Arguably the most compelling application of performance modeling, but one that heretofore has not been used much, is to simplify the selection process of a new computing facility for a university or laboratory. At the present time, most large system procurements involve a comparative test of several systems, using a set of application benchmarks chosen to be typical of the expected usage. In one case that the authors are aware of, 25 separate application benchmarks were specified, and numerous other system-level benchmark tests were required as well. Preparing a set of performance benchmarks for a large laboratory acquisition is a labor-intensive process, typically involving several highly skilled staff members. Analyzing and comparing the benchmark results also requires additional effort. These steps involved are summarized in the recent HECRTF report [295].

What is often overlooked in this regard is that each of the prospective vendors must also expend a comparable (or even greater) effort to implement and tune the bench-marks on their systems. Partly due to the high personnel costs of benchmark work, computer vendors often can afford only a minimal effort to implement the bench-marks, leaving little or no resources to tune or customize the implementations for a given system, even though such tuning and/or customization would greatly benefit the customer. In any event, vendors must factor the cost of implementing and/or tuning benchmarks into the price that they must charge to the customer if successful. These costs are further multiplied because for every successful proposal, they must prepare several unsuccessful proposals.

Once a reasonably easy-to-use performance modeling facility is available, it may be possible to greatly reduce, if not eliminate, the benchmark tests that are specified in a procurement, replacing them by a measurement of certain performance model parameters for the target systems and applications. These parameters can then be used by the computer center staff to project performance rates for numerous system options. It may well be that a given center will decide not to rely completely on performance model results. But if even part of the normal application suite can be replaced, this will save considerable resources on both sides.

7.3   Basic Methodology

The performance modeling framework presented in this chapter is based upon application signatures, machine profiles and convolutions. An *application signature* is a detailed but compact representation of the fundamental operations performed by an application, independent of the target system. A *machine profile* is a representation of the capability of a system to carry out fundamental operations, independent of the particular application. A *convolution* is a means to rapidly combine application signatures with machine profiles in order to predict performance. In a nutshell, our methodology is to:

- Summarize the requirements of applications in ways that are not too expensive in terms of time/space required to gather them but still contain sufficient detail to enable modeling.

- Obtain the application signatures automatically.

- Generalize the signatures to represent how the application would stress arbitrary (including future) machines.

- Extrapolate the signatures to larger problem sizes than what can be actually run at the present time.

With regards to application signatures, note that the source code of an application can be considered a high-level description, or application signature, of its computational resource requirements. However, depending on the language it may not be very compact (Matlab is compact, while Fortran is not). Also, determining the resource requirements the application from the source code may not be very easy (especially if the target machine does not exist!). Hence we need cheaper, faster, more flexible ways to obtain representations suitable for performance modeling work. A minimal goal is to combine the results of several compilation, execution, performance data analysis cycles into a signature, so these steps do not have to be repeated each time a new performance question is asked.

A dynamic instruction trace, such as a record of each memory address accessed (using a tool such as Dyninst [69] can also be considered to be an application signature. But it is not compact and address traces alone can run to several Gbytes even for short-running applications and it is not machine independent.

A general approach that we have developed to analyze applications, which has resulted in considerable space reduction and a measure of machine independence, is the following: (1) statically analyze, then instrument and trace an application on some set of existing machines; (2) summarize, on-the-fly, the operations performed by the application; (3) tally operations indexed to the source code structures that generated them; and (4) perform a merge operation on the summaries from each machine [78,79,311,312]. From this data, one can obtain information on memory access patterns (namely, summaries of the stride and range of memory accesses generated by individual memory operations) and communications patterns (namely, summaries of sizes and type of communications performed).

The specific scheme to acquire an application signature is as follows: (1) conduct a series of experiments tracing a program, using the techniques described above; (2) analyze the trace by pattern detection to identify recurring sequences of messages and loads/store operations; and (3) select the most important sequences of patterns. With regards to (3), infrequent paths through the program are ignored, and sequences that map to insignificant performance contributions are dropped.

As a simple example, the performance behavior of CG (the Conjugate Gradient benchmark from the NAS Parallel Benchmarks [38]), which is more 1000 lines long, can be represented from a performance standpoint by one random memory access pattern. This is because 99% of execution is spent in the following loop:

```
do k = rowstr(j), rowstr(j+1)-1
sum = sum + a(k)*p(colidx(k))
enddo
```

This loop has two floating-point operations, two stride-1 memory access patterns, and one random memory access pattern (the indirect index of p). On almost all of today's deep memory hierarchy machines the performance cost of the random memory access pattern dominates the other patterns and the floating-point work. As a practical matter, all that is required to predict the performance of CG on a machine is the size of the problem (which level of the memory hierarchy it fits in) and the rate at which the machine can do random loads from that level of the memory. Thus a random memory access pattern succinctly represents the most important demand that CG puts on any machine.

Obviously, many full applications spend a significant amount of time in more than one loop or function, and so the several patterns must be combined and weighted. Simple addition is often not the right combining operator for these patterns, because different types of work may be involved (say memory accesses and communication). Also, our framework considers the impact of different compilers or different compiler flags in producing better code (so trace results are not machine independent). Finally, we develop models that include scaling and not just ones that work with a single problem size. For this, we use statistical methods applied to series of traces of different input sizes and/or CPU counts to derive a scaling model.

The second component of this performance modeling approach is to represent the resource capabilities of current and proposed machines, with emphasis on memory and communications capabilities, in an application-independent form suitable for parameterized modeling. In particular, we use low-level benchmarks to gather machine profiles, which are high-level representations of the rates at which machines can carry out basic operations (such as memory loads and stores and message passing), including the capabilities of memory units at each level of the memory hierarchy and the ability of machines to overlap memory operations with other kinds of operations (e.g., floating-point or communications operations). We then extend machine profiles to account for reduction in capability due to sharing (for example, to express how much the memory subsystems or communication fabrics capability is diminished by sharing these with competing processors). Finally, we extrapolate to larger systems from validated machine profiles of similar but smaller systems.

To enable time tractable modeling we employ a range of simulation techniques [78, 270] to combine applications signatures with machine profiles:

- Convolution methods for mapping application signatures to machine pro-files to enable time tractable statistical simulation.

- Techniques for modeling interactions between different memory access patterns within the same loop. For example, if a loop is 50% stride-1 and 50% random stride, we determine whether the performance is some composable function of these two separate performance rates.

- Techniques for modeling the effect of competition between different applications (or task parallel programs) for shared resources. For example, if program A is thrashing L3 cache with a large working set and a random memory access pattern, we determine how that impacts the performance of program B with a stride-1 access pattern and a small working set that would otherwise fits in L3.

- Techniques for defining "performance similarity" in a meaningful way. For example, we determine whether loops that "look" the same in terms of application signatures and memory access patterns actually perform the same. If so, we define a set of loops that span the performance space.

In one sense, cycle-accurate simulation is the performance modeling baseline. Given enough time, and enough details about a machine, we can always explain and predict performance by stepping through the code instruction by instruction. However, simulation at this detail is exceedingly expensive. So we have developed fast-to-evaluate machine models for current and proposed machines, which closely approximate cycle-accurate predictions by accounting for fewer details.

Our convolution method allows for relatively rapid development of performance models (full application models take one or two months now). Performance predictions are very fast to evaluate once the models are constructed (a few minutes per prediction). The results are fairly accurate. Figure 7.1 qualitatively shows the accuracy results across a set of machines and problem sizes and CPU counts for POP, the Parallel Ocean Program.

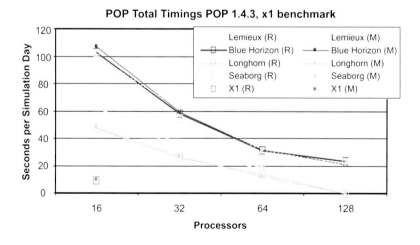

FIGURE 7.1: Results for Parallel Ocean Program (POP). (R) is real runtime (M) is modeled (predicted) runtime.

We have carried out similar exercise for several sizes and inputs of POP problems. And we have also modeled several applications from the DOD High-Performance Computing Modernization Office work-load, including AVUS, a computational fluid dynamics code, GAMESS, a computational chemistry code, HYCOM, a weather code, and OOCORE, an out-of-core solver. In a stern test of the methods every year we are are allowed access to DOD machines only to gather machine profiles via low-level benchmarks. We then model these large parallel applications at several CPU counts ranging on several systems. We then predict application performance on these machines; and only after the predictions are issued are the application true runtimes independently ascertained by DOD personnel.

TABLE 7.1: Results of "blind" predictions of DoD HPCMO Workload in TI-10.

Program	Average Abs. Error	Standard Deviation
AVUS large	8.9%	4.8%
CTH large	9.0%	6.4%
CTH standard	6.6%	6.6%
HYCOM large	5.5%	5.5%
HYCOM standard	6.6%	6.6%
ICEPIC large	9.6%	9.6%
ICEPIC standard	4.6%	4.6%
Cray XT3	12.8%	7.0%
IBM PWR5+	6.9%	6.6%
DELL Woodcrest	5.4%	5.0%
Cray XT4	7.6%	4.8%
Cray XT5	4.2%	3.2%
Linux Networks Woodcrest	6.2%	4.3%
DELL Nehalem	7.6%	4.6%

Table 7.1 above gives the overall average absolute error and standard deviation of absolute average error as well as breakdowns by application/input and architecture. We conducted this "blind" test (without knowing the performance of the applications in advance) in order to subject our modeling methods to the sternest possible test and because we think it is important to report successes and failures in modeling in order to advance the science.

7.4　Performance Sensitivity Studies

Reporting the accuracy of performance models in terms of model-predicted time vs. observed time (as in the previous section) is mostly just a validating step for obtaining confidence in the model. A more interesting and useful exercise is to explain and quantify performance differences and to play "what if" using the model.

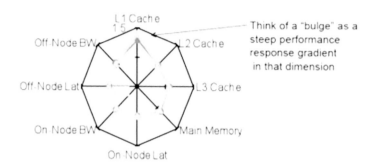

FIGURE 7.2: Performance response of WRF (Weather Research Facility) to 2x improvements in base system on several performance dimensions (relative to Power4).

A Kiviat diagram is a useful way to represent the performance response of an application to improvements in the underlying machine. In such diagrams (see Figure 7.2) a series of radial axes emanate from the center, and a series of labeled concentric polygonal grids intersect these axes. Each axis represents a performance attribute of the machine that might be individually improved, such as peak performance rate, cache bandwidth, main memory bandwidth, network latency, etc. Each polygon then represents some degree of constant performance improvement (usually interpreted here as a "speedup") of the application relative to some norm, with "1" (the degenerate polygon at the center) representing a baseline performance of the machine with no modifications. The units of the axis are normalized improvement.

In a recent exercise for the Oak Ridge National Laboratory (ORNL)'s upgrade of its Cray XT4 system we used the working example of the S3D direct numerical simulation turbulent combustion code to answer the following questions:

- How will S3D perform on upgraded Jaguar dual-core to quad-core?

- Will it take advantage of new architectural features such as SSE vector units?

- Crystal ball: how will it perform on further-out systems in 2011/2012 timeframe?

- How will its performance change as the chemistry becomes more complex, say when we move from COH2 to C2H4 to C7H16?

• Will the problem scale to hundreds of thousands of processors?

Modeling gives us a means of answering all these questions and more.

The PMaC lab at the University of California, San Diego recently completed a series of large-scale performance predictions of S3D out to full size of Jaguar using both input scaling (increasing the complexity of the chemistry) and CPU-count scaling. These predictions were 95% accurate when predicting Jaguar post-quadcore upgrade both with SSEs enable and disabled, and were independently confirmed by a team at Los Alamos National Laboratory in advance of deployment. We predicted and ORNL verified post-upgrade about a 5% "multicore tax" that is the application ran a bit slower core for core because of diminished shared bandwidth of each core accessing main memory.

Figure 7.3 shows the memory bandwidth available to a core of L1, L2, L3 and main memory relative to a Power4 processor of 12 anonymous future processors slated for availability between now and 2012. It can be seen that the main improvements will be in main memory bandwidth per core.

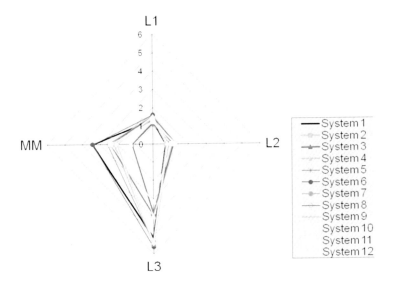

FIGURE 7.3: Memory profiles of systems out to 2012 (anonymous) (See color insert.).

Assuming the better systems are more expensive are they worth it? It depends on the application. We predict for various core counts how much execution time would shrink for S3D doing (future) C2H4 chemistry on these future systems with their different cache sizes, associativities, load store units, speeds, latencies, and main memory performance out to 2012 timeframe and sample results (on some of the systems) is presented in Table 7.2. It can be seen the best system would run S3D almost 40% faster.

TABLE 7.2: Execution time as fraction of base (Power4) for future systems on the S3D/C2H4 application.

Cores	System							
	1	2	3	4	5	6	7	8
8	1.0	0.72	0.78	0.66	0.77	0.81	0.74	0.61
64	1.0	0.73	0.78	0.67	0.77	0.81	0.73	0.62
512	1.0	0.72	0.78	0.66	0.77	0.81	0.74	0.62

For another example, it is clear from Figure 7.1 above that Lemieux, the Alpha-based system, was faster across-the-board on POP x1 than was Blue Horizon (BH), the IBM Power3 system. The question is why? Lemieux had faster processors (1GHz vs. 375 MHz), and a lower-latency network (a measured ping-pong latency of about 5 ms vs. about 19 ms), but Blue Horizons network had the higher bandwidth (a measured ping-pong bandwidth of about 350 MByte/s vs. 269 MByte/s). Without a model, one is left to conjecture "I guess POP performance is more sensitive to processor performance and network latency than network bandwidth," but without solid evidence.

With a model that can accurately predict application performance based on properties of the code and the machine, we can carry out precise modeling experiments such as that represented in Figure 7.2. Here we model perturbing the Blue Horizon system (with IBM Power3 processors and a Colony switch) into the TCS system (with Alpha ES640 processors and the Quadrics switch) by replacing components one by one. Figure 7.4 represents a series of cases modeling the perturbing from BH to TCS, going from left to right. The four bars for each case represent the performance of POP x1 on 16 processors, the processor and memory subsystem performance, the network bandwidth, and the network latency, all normalized to that of BH.

In Case 1, we model the effect of reducing the bandwidth of BH's network to that of a single rail of the Quadrics switch. There is no observable performance effect, as the POP x1 problem at this size is not sensitive to a change in peak network band-width from 350MB/s to 269MB/s. In Case 2, we model the effect of replacing the Colony switch with the Quadrics switch. Here there is a significant performance improvement, due to the five millisecond latency of the Quadrics switch versus the 20 millisecond latency of the Colony switch. This is evidence that the barotropic calculations in POP x1 at this size are latency sensitive. In Case 3, we use Quadrics latency but the Colony bandwidth just for completeness. In Case 4, we model keeping the Colony switch latency and bandwidth figures, but replacing the Power3 processors and local memory subsystem with Alpha ES640 processors and their memory subsystem. There is a substantial improvement in performance, due mainly to the faster memory subsystem of the Alpha. The Alpha can load stride-one data from its L2 cache at about twice the rate of the Power3, and

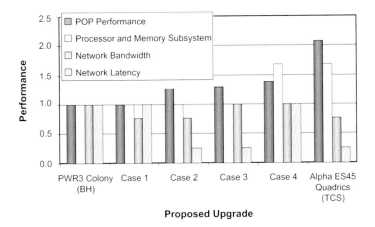

FIGURE 7.4: Performance Sensitivity study of POP applied to proposed Lemieux upgrade.

this benefits POP x1 significantly. The last set of bars show the TCS values of performance, processor and memory subsystem speed, network bandwidth and latency, as a ratio of the BH values.

The principal observation from the above exercise is that the model can quantify the performance impact of each machine hardware component.

In these studies we find that larger CPU count POP x1 problems become more network latency sensitive and remain not-very bandwidth sensitive.

7.5   Summary

We have seen that performance models enable "what-if" analyses of the implications of improving the target machine in various dimensions. Such analyses obviously are useful to system designers, helping them optimize system architectures for the highest sustained performance on a target set of applications. They are potentially quite useful in helping computing centers select the best system in an acquisition. But these methods can also be used by application scientists to improve performance in their codes, by better understanding which tuning measures yield the most improvement in sustained performance.

With further improvements in this methodology, we can envision a future wherein these techniques are embedded in application code, or even in system software, thus enabling self-tuning applications for user codes. For example, we can conceive of an application that performs the first of many iterations using numerous cache blocking parameters, a separate combination on each processor, and then uses a simple performance model to select the most favorable combination. This combination would then be used for all remaining iterations.

Our methods have reduced the time required for performance modeling, but much work needs to be done here. Also, running an application to obtain the necessary trace information multiplies the run time by a large factor (roughly 1000). The future work in this arena will need to focus on further reducing the both the human and computer costs.

7.6 Acknowledgments

This chapter is adapted from an earlier article, written by the present authors [42] This work was sponsored in part by the Performance Engineering Research Institute, which is supported by the SciDAC Program of the U.S. Department of Energy, and by the Director, Office of Science, Basic Energy Sciences, Division of Material Science and Engineering, and the Advanced Scientific Computing Research Office, of the U.S. Dept. of Energy under Contract No. DE-AC02-05CH11231.

Chapter 8

Analytic Modeling for Memory Access Patterns Based on Apex-MAP

Erich Strohmaier

Lawrence Berkeley National Laboratory

Hongzhang Shan

Lawrence Berkeley National Laboratory

Khaled Ibrahim

Lawrence Berkeley National Laboratory

8.1 Introduction

The memory behavior of a program greatly influences the overall performance observed on a specific architecture. Memory references, if not cached, involve large latencies, because of the large (and increasing) disparities between processor speeds and memory speeds, a phenomenon often termed the "memory wall." Indeed, memory wall problems have significantly influenced processor architecture for years, so that, for instance, most transistors on present-day microprocessors are dedicated to the cache subsystem.

To improve the performance of applications, computer architects and compiler designers try to exploit locality in memory reference streams. The two main form of locality that usually exist in an application are: *temporal locality*, which refers to the tendency to reuse data that were referenced a short time ago; and *spatial locality*, which refers to the tendency to reference memory locations that are either consecutive or differ by a fixed stride.

Microprocessors employ multiple hardware structures to exploit locality and to hide the memory latency, including:

- Caches that exploit temporal locality by keeping recently used memory lines. Having the cache line to hold multiple contiguous memory words implicitly exploits spatial locality as well.

- Prefetchers that exploit spatial locality by predicting the future memory references and bringing them to the cache before referencing them. Prefetching can be done either by software, hardware, or both.

- Out-of-order execution facilities that exploit, among other things, memory level parallelism by issuing multiple memory requests to reduce the impact of wait time. This technique works well when bandwidth is less critical than latency and can improve the performance significantly on a multi-level cache system.

The performance of an application relies heavily on the interaction of the application memory access behavior with the underlying architecture. Understanding memory access patterns can serve multiple purposes: first, the performance of an application can be quantified or at least the bounds of performance can be identified; second, application writers can design algorithms that exhibit better locality attributes.

This chapter discusses the characterization of memory access patterns. It also introduces the Apex-MAP model [318,319] for quantifying locality in an architecture-independent manner. This model can be used for studying the processor interaction under different memory access patterns. The model can also serve as proxy for different applications, allowing an easy and a productive characterization of machine interaction over a wide class of applications.

8.2   Memory Access Characterization

Memory access characterization is a valuable technique to help understand achieved performance on various systems. Indeed, due to the growing challenge of the "memory wall," it is increasingly for both system researchers and application programmers to understand the behavior of various types of computations in a given memory subsystem. Such understanding can help application writers design algorithms that better exploit locality of references, thereby reducing the impact of the memory wall.

In this section we discuss typical memory access patterns that most applications exhibit during execution.

8.2.1   Patterns of Memory Access

Memory access patterns can be classified based on the spatial relation between the memory references and on the reuse distance of each referenced data element:

- Regular contiguous memory access: This pattern is associated with access streams that have unit stride access to a dataset, such as while accessing a large section of a array or matrix by the most rapidly changing index.

- Regular non-contiguous (structured) memory access: This pattern is characterized by non-unit stride access, so that references can be easily be calculated by simple linear relations. This pattern often occurs most commonly in scientific applications when accessing multi-dimensional arrays by other than the most rapidly changing index.

- Random access: Random accesses are typical of applications dominated by pointer chasing algorithms, for instance in graph algorithms. Random accesses also occur in scientific code where multiple levels of indirection are needed to retrieve the data from the memory. A notable class of this type of algorithm is sparse matrix schemes, since compact memory representations for sparse matrix requires indirect addressing. This pattern is usually the most challenging pattern for microprocessor systems to handle efficiently.

The memory access pattern of an application is seldom fixed for the entire run. It is quite possible for applications to involve multiple simultaneous streams of execution, each with different memory access patterns. It is even more common for an application to have multiple phases of execution (even within a single "iteration"), each with a different pattern.

One other dimension characterizing memory access is based on the reuse of the data and how large the reuse distance is, in terms of CPU cycles or distinct memory accesses between accesses. Short reuse distance results typically results in a cache being used very effectively, whereas with a high reuse distance it is quite likely that data once in cache has been displaced before the next access.

8.2.2   Performance Dependency on Memory Access Pattern

The performance of an application can be related to the interaction between the access pattern and the architecture. It is not uncommon for an algorithm to perform very well on one architecture but rather poorly on another architecture.

Vector machines, for instance the Cray X1, typically perform quite well on applications with fixed stride access, particularly for contiguous streamed data, but typically perform rather poorly on applications exhibiting random access patterns. In other words, vector systems require high spatial locality to perform well. Caching is either not present or is a minor feature on vector systems, since they do not rely on temporal reuse.

It should be pointed out that vector processing is popular not only in scientific computing, but also in graphic processing. That is why graphic processing units often use the same hardware tradeoffs—dedicating more transistors to processing, leveraging memory level parallelism, and very few transistors to caching—as scientific vector systems.

For applications with relatively short reuse distance, in terms of distinct memory references accessed between access, cache designs are critical to performance. In general, the larger the cache, the less dependent the application becomes on referencing the slow memory system, although other features of the cache system, such as structure of the cache (cache line length, associativity, replacement policy, etc.) and the design of the translation look-aside buffer, also affect performance.

Cache-based systems, such as conventional microprocessor-based personal computer modules (and parallel systems constructed from such units) are usually preferred for applications with irregular memory accesses and moderate levels of temporal reuse.

8.2.3   Locality Definitions

Earlier in this chapter, we discussed spatial and temporal locality to characterize application memory access patterns. Spatial locality describes the relation between addresses in stream of references, while temporal locality relates to the tendency of an application to reference the same address multiple times within a certain reuse distance, measured either in cycles or in distinct in-between memory references. In most cases, locality can be quantified. For instance, for spatial locality, stride one access is typically associated with the highest locality score, while random access is typically associated with the lowest score.

In general quantifying locality is not a trivial problem. Multiple approaches exist in literature. We present here two well-known approaches.

The first formulation is based on stack reuse distance, where accesses to the same memory location are classified based on the number of distinct memory addresses referenced in between. Plotting the relation between the stack reuse distance and the cumulative percentage of memory references can provide insightful information about hit or miss rates for a certain cache size, assuming that it is fully associative with the least recently used (LRU) replacement policy. The performance for set associative caches can also be estimated as well based on a model developed by Hill and Smith [165, 309].

Figure 8.1 shows a typical stack reuse plot. For given effective cache size, one can look up the memory references that will hit in the cache as percentage of the memory references that have reuse distance of less than the effective cache size. Even though the stack reuse distance can be used to infer the cache behavior, it does not by itself provide a meaningful metric that can be used for comparing algorithms.

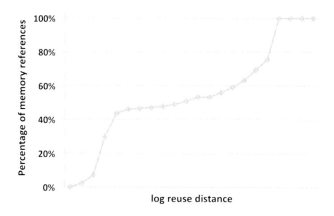

FIGURE 8.1: Typical curve for cumulative memory stack reuse distance.

The second presented formulation allows describing the locality behavior of an application by a single number. To characterize the spatial locality Weinberg et al. proposed the following metric [357]

$$\text{spatial} \;=\; \sum_{i=1}^{\infty} \text{stride}_i/i, \tag{8.1}$$

where *stride$_i$* represents the percentage of memory references of stride size i divided by the stride size. This metric for spatial locality targets characterizing an existing application in an architecture independent manner.

For the sake of evaluating particular machine, one needs to experiment with different access patterns. For the above-mentioned metrics, it is not possible to have applications that cover the whole spectrum of access patterns. Additionally, it is not easy to develop synthetic applications based on the above-presented metrics.

Apex-MAP, introduced in the next section, defines spatial and temporal locality in a way that is easier to realize in a synthetic benchmark. The patterns of memory access are parameterized to allow better coverage of the locality space while studying the performance of a specific architecture.

8.3   Apex-MAP Model to Characterize Memory Access Patterns

In this section we describe the Apex characterization framework, together with the associated performance probe. Apex-MAP was designed as a hardware- and architecture-independent application performance characterization framework. It starts with the assumption that the performance behavior of codes and/or algorithms can be characterized in an architecture-independent way based on several simple factors, whose values can be specifically selected for each of the codes. The performance probe Apex-MAP is a generic kernel that implements these factors in a parameterized fashion. With the right selection and definition of factors, Apex-MAP can thus be used as an application performance proxy for different codes on various platforms.

8.3.1   Characterizing Memory Access

Apex-MAP assumes that the data access of any code can be described as several concurrent streams of addresses which in turn can be characterized by a set of performance related factors. Currently, three major characterization parameters have been used to describe a data stream. They are the memory size of the data set accessed M, the spatial locality L and the temporal locality α:

Data Set Size (M): The data set size M is the memory size accessed by the target stream. It is evidently an important factor influencing the performance of an address stream. Its impact on performance is becoming more important with the increasing complexity of memory hierarchies in modern system architectures.

Spatial Locality (L): The spatial locality L is characterized by the number of contiguous memory locations accessed in succession (e.g., vector-length). Many modern systems benefit greatly from contiguous memory access patterns. The valid values range from 1 to M.

Temporal Locality (α): The characterization of the temporal locality in Apex-MAP is closely related with the cumulative temporal distribution function of the target stream. The cumulative temporal distribution function is defined as the following. Suppose N is the number of total memory accesses. Let X be the memory location of n-th memory access ($n <= N$). If X has been accessed earlier and the last access is the m-th memory access, then we say the temporal distance for n-th memory access is $n - m$. If X has not been accessed earlier, then we assume an infinite repetition of the whole address stream. The temporal distance is then given by $n + N - k$ and k is the location of last appearance of memory location X in the address stream. If X is only accessed one time, the temporal distance is N. The cumulative temporal distribution function is defined using this temporal distance and show for each temporal distance t the probability p, of accessing a memory location again within the next t memory accesses.

After obtaining the cumulative temporal distribution function, the next step is to find a single number to describe the cumulative temporal distribution function and use it as the temporal locality of the target stream. We find that the shape of the cumulative temporal distribution function can be approximated by using a power function with a single shape parameter α in a scale-invariant way. The address stream is generated with a power-law-distributed pseudorandom number generator (i.e., $X = r^{1/\alpha}$, where r is a uniform $(0, 1)$ pseudorandom number). The single shape parameter α of the power function is then used as a measure of the temporal locality of the target stream. Valid values of α can range between 0 and 1. A value of $\alpha = 0$ means the program will access a single address repeatedly, with the highest temporal locality, while a value of $\alpha = 1$ indicates a uniform random memory access with the lowest possible temporal locality.

8.3.2 Generating Different Memory Patterns Using Apex-MAP

Apex-MAP uses non-uniform random block access to simulate data access of a target code. The address stream is generated mainly based on Memory Size (M), Temporal Locality (α), and Spatial Locality (L) as illustrated in Figure 8.2.

The random starting addresses (X) are aligned by length L and generated

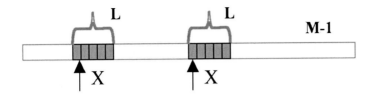

FIGURE 8.2: The random block data access in Apex-MAP.

based on uniform pseudorandom numbers r with the generating function $X = r^{1/\alpha}$. The α can take any value between 0 and 1. Once a starting address is accessed, the following $L - 1$ continuous addresses will also be accessed. The block length L is used as a measure for spatial locality; L can take any value between 1 (single-word access) and M. A non-uniform random selection of starting addresses for these blocks is used to simulate the effects of temporal locality [318]. The starting addresses cannot be dynamically generated, since the time needed to generate an address is too long compared with the memory access time. Therefore they have to be computed in advance and stored in an index buffer. The number of addresses pre-computed is controlled by a characterization parameter I. Here is the core part of the Apex-MAP code used to model the data access:

```
Repeat N Times
  GenerateIndexArray (ind)
  CLOCK(start)
  For (j=0; j < I; j++)
    For (k=0;  k < L; k++)
      sum += data[ind[j]+k]
  CLOCK(end)
  RunningTime += end - start
End Repeat
```

The *GenerateIndexArray* function will compute the random starting addresses and fill them in an index buffer *ind* whose length is set by parameter I. The parameter I is used to specify the maximum number of blocks for which no data dependencies exist and data access can occur in parallel. This parameter, together with parameter L, sets the upper limit for effective data prefetching and vectorization.

The data stream generated by Apex-MAP can be used to simulate the memory access patterns of many scientific kernels and applications. For example, a sequential access to a large array of size S can be simulated by setting $M = S$, $L = S$, $I = 1$, and α set to any value. The value of α is not important here, since the entire array is treated as one random block and thus the starting address is the first element of the array regardless of its value. For random access, such as the memory access pattern in RandomAccess of HPC

Challenge Benchmark Suite [171], the stream can be generated by setting M to be the table size, $L = 1$, $\alpha = 1$, and setting I to the number of access times. Each element is regarded as a data block in Apex-MAP.

8.4   Using Apex-MAP to Assess Processor Performance

One feature that distinguishes Apex-MAP from other performance benchmarks is that its input parameters can be varied independent of each other between extreme values. This allows one to generate continuous performance surfaces, which can be used to explore the performance effects of all potential values of the characterizing parameters. By examining these surfaces, we can understand how changes in spatial or temporal locality affect the performance and which factors are more important on a specific system. Moreover, we can compare these performance surfaces across different platforms and explore the advantages and disadvantages of each platform. Most current benchmark suites consist of a fixed set of application codes, which strongly limits the scope of performance behavior patterns they can explore. The results of such application benchmarks provide very good indications how similar applications will perform, but they are not very helpful for other applications.

As an example, Figure 8.3 and Figure 8.4 display the performance surfaces for a superscalar processor, the Itanium II, and a vector processor, the Cray X1. The Z-axis shows average CPU cycles per data access in logarithmic scale. The performance surfaces of these two kinds of processors differ significantly from each other. For the Itanium, both the temporal locality and the spatial locality affect the performance substantially. The worst performance occurs when both temporal locality and spatial locality reach the lowest point we have tested ($\alpha = 1$, $L = 1$), for which roughly 100 cycles are needed per data access. Increasing either the temporal locality or spatial locality reduces the average number of cycles per data access needed. The Itanium II needs only 0.77 cycles for $\alpha = 0.001$ and $L = 2048$. However, further increasing the spatial locality does not improve performance much. Moreover, we can find that temporal locality and spatial locality can be substituted for each other to some degree. The line of equal access speed of $Z = 10$ illustrates this as it is almost diagonal. A reduction of temporal locality could be compensated by increase in spatial locality and vice versa.

For the Cray X1, the spatial locality affects the performance much more prominently. The X1 can easily tolerate a decrease in temporal locality but is very sensitive to the loss of spatial locality (and the corresponding reduction in vector-length). In the α direction, increasing the temporal locality does not significantly change the Cycles/Access ratio, so that the performance line is fairly flat. However, increasing the spatial locality sharply reduces the number of cycles needed per data access. The differences between these two

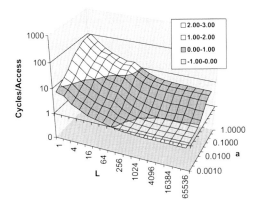

FIGURE 8.3: The performance surface map of Itanium II.

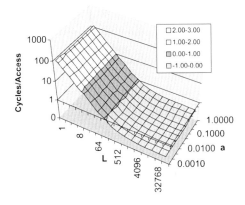

FIGURE 8.4: The performance surface map of Cray X1.

kinds of performance surfaces clearly reflect their different design concepts. Cache-based superscalar processors depend on the elaborate cache and memory hierarchies to take advantage of both temporal and spatial locality while vector processors typically have sophisticated memory access sub-systems, which makes them insensitive to the lack of temporal locality. Programmers therefore only have to focus on expressing and exploiting spatial locality.

8.5 Apex-MAP Extension for Parallel Architectures

8.5.1 Modeling Communication for Remote Memory Access

Apex-MAP is designed to use the same three main parameters for both the sequential version and parallel version. These parameters are the global memory size M, the measure of temporal locality α, and the measure of spatial locality L. For the parallel version, an important question is whether the effects of temporal locality and process locality should be treated independently of each other, by implementing different parameters and execution models for them, or if global usage of the temporal locality model can provide a sufficient first approximation of the data access behavior of scientific application kernels. For the parallel execution of a single scientific kernel, any method to divide the problem to increase process locality should also be usable to improve temporal locality in sequential execution. At the same time, any algorithm with good temporal locality should, in turn, exhibit good process locality in a parallel implementation. Thus the only cases where process and temporal localities can differ substantially would be algorithms for which the problem to be solved in each process is different from the problem between processes. One example would be the embarrassingly parallel execution of a kernel with low temporal locality by running multiple copies of the individual problem with different parameters at the same time and thus generating high process locality. For simplicity reasons we therefore decided to treat temporal and process locality in a unified way by extending the sequential temporal locality concept to global memory access in the parallel case. One important implementation detail here is that we access the global array only with load operations, which avoids any possibility of race conditions for memory update operations.

Temporal locality in actual programs arises from several sources. In some cases (e.g., a blocked matrix-matrix multiply operation), variables are not used for long periods of time but once they are used they are reused in close time proximity in numerous repetitions. Overall, however, all variables are accessed with equal frequency. In other codes (e.g., matrix-vector multiplication), some variables are simply accessed more often than others (in our example the elements of the original vector). Exploiting these different flavors of temporal locality in the sequential case requires different caching strategies such as dynamic caching in our first example and static caching in our second example. In practice, dynamic caching is used almost exclusively as it also tends to work reasonably well in many (but not all!) situations that one might think would require static caching. Apex-MAP clearly uses more frequent accesses to certain addresses to simulate the effects of temporal locality. In the parallel case, the difference between these flavors becomes more important as placement (and possibly sharing) of data and their affinity to processes becomes a performance issue. In Apex-MAP we assume that each process accesses certain variables more often and that these variables can be placed in memory closer

to this process. We do not address the different question of how to address
and exploit temporal locality in cases where overall all variables are accessed
with equal frequency and thus data placement is an ineffective strategy for exploiting temporal locality. Apex-MAP also assumes that sharing of variables
is not a performance constraint, as we are only reading global data but do not
modify them.

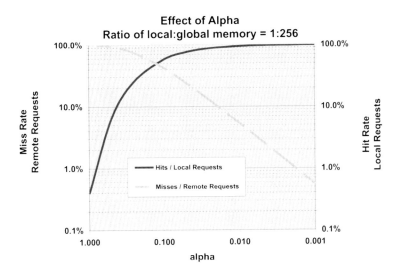

FIGURE 8.5: Effect of α on Hit/Miss rates and ratio of local to remote data
requests.

In the parallel version of Apex-MAP, the global data array of size M is
now evenly distributed across all processes. Data will again be accessed in
block mode, i.e., L continuous memory addresses will be accessed in succession and the block length L is used to characterize spatial locality. The starting
addresses X of these data blocks are computed in advance by using a non-
uniform random address generator driven by a power function with the shape
parameter α. Each process computes its own global address-stream using the
same power distribution function. However, these global addresses will be adjusted based on the rank of the process so that each process accesses its own
memory with the highest probability. Then, for each index tested, if the addressed data resides in local memory, the computation proceeds immediately,
whereas if it resides in remote memory, it is fetched into local memory first.
The following is the flowchart of a parallel implementation:

```
Repeat N Times
  GenerateIndexArray (ind)
  CLOCK(start)
  For (j=0; j < I; j++)
     If (data not in local memory)
        GetRemoteData()
     Endif
     Compute()
  CLOCK(end)
  RunningTime += end - start
End Repeat
```

Apex-MAP is designed to measure the rate at which global data can be fed into the CPU itself and not only into the memory or into cache. Therefore, it is essential that an actual computation be performed in the compute module. This computation is currently a global sum of all accessed array elements. The frequency with which remote data access occurs is mainly determined by the temporal locality parameter α. For example, for 256 processes and $\alpha = 1$ the data accesses follow a uniform random distribution and the percentage of remote accesses is $255/256 = 99.6\%$ (Figure 8.5). With the increase of temporal locality, the percentage reduces to 0.55% when $\alpha = 0.001$.

One major nontrivial issue is how the remote access is carried out. The actual implementation will be highly affected by the available parallel programming paradigm and different programming styles [319]. However, we assume that the operation for different indices is independent and multiple remote accesses can be executed independently of each other at the same time.

8.5.2   Assessing Machine Scaling Behavior Based on Apex-MAP

In this section, we first study the effectiveness of the parallel Apex-MAP. Then we show how to use Apex-MAP results to study the performance scalability of different systems.

The parallel version of Apex-MAP was developed using two-sided Message Passing Interface (MPI) calls, since that is the most popular and portable parallel programming model available today. Since in Apex-MAP the process numbers for message exchanges are generated based on a non-uniform random access, we decided to use non-blocking, asynchronous MPI functions in order to avoid blocking and deadlock. Given our non-deterministic random message pattern, it was not clear if a scalable implementation of Apex-MAP in MPI was possible. To that end, we studied the effectiveness of the parallel Apex-MAP methodology on three platforms: Seaborg, Cheetah, and Phoenix:

- *Seaborg* is an IBM Power3 based distributed memory machine at Lawrence Berkeley National Laboratory. Each node has 16 IBM Power3 processors running at the speed of 375 MHz. The peak performance of each processor is 1.5 Gflop/s. Its network switch is the IBM Colony II, which is connected to two "GX Bus Colony" network adapters per node.

- *Cheetah* is a 27-node IBM p690 system at Oak Ridge National Laboratory. It employs the IBM Federated switch, where each node has 32 Power4 processors at 1.3 GHz. The peak performance of each processor is 5.2 Gflop/s.

- *Phoenix* is a Cray X1 platform at Oak Ridge National Laboratory consisting of 512 multi-streaming vector processors (MSPs). Each MSP has four single-stream vector processors and a 2 Mbyte cache. Four MSPs form a node with 16 Gbyte of shared memory. The inter-connect functions as an extension of the memory system, offering each node direct access to memories on other nodes.

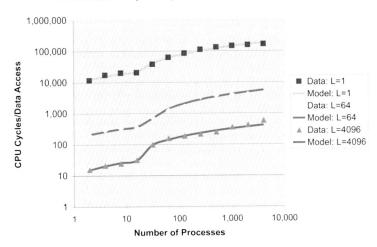

FIGURE 8.6: Scalability of Apex-MAP and its performance model for three different block sizes (*L*=1, 64, 4096) on Seaborg.

Figure 8.6 shows the increasing performance up to several processors on these three platforms, for the uniform pseudorandom distribution. We find that the Apex-MAP results on many systems follow a very simple performance model (also shown in Figure 8.6) of $A + B \log_2 P$, where A and B are parameters characterizing levels of system hierarchy and P is the number of processors. To achieve such high scalability for this type of code requires deciding carefully on numerous implementations related details [319].

We can then use Apex-MAP to analyze the scalability of different systems. We focus on low temporal locality ($\alpha = 1$) since codes with high temporal locality usually scale better. As an example, we analyze the cross-section bandwidth of the $L = 4096$ on these three systems in Figure 8.7.

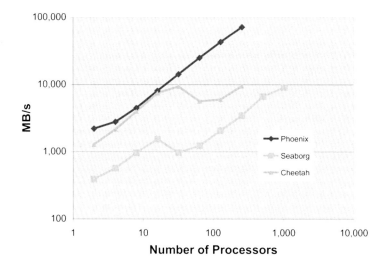

FIGURE 8.7: The performance scalability of Apex-MAP on different systems with L=4096 words and α=1 (random access).

The two SMP based IBM systems scale well within their symmetric multiprocessor (SMP) units (i.e., up to 16 processors on, and up to 32 on Cheetah), but show a performance drop if we start using more than a single SMP. For larger numbers of SMPs, cross-section bandwidth starts scaling again. The architecture of the Phoenix system is not hierarchical and shows no such effect of architectural hierarchies. Its performance scales up very well across the whole range of processors tested. The total aggregate bandwidth of 16 processors of the Phoenix system is almost equal to the total aggregate bandwidth of 256 processors on Cheetah and 1024 processors on Seaborg.

8.5.3 Using Apex-MAP to Analyze Architectural Signatures

The above IBM SP systems have a very hierarchical architecture, with several levels of cache, large SMP nodes, and distributed global memory. The Cray vector system is designed without traditional caches, no obvious SMP structure, and a comparable flat global interconnect structure. To compare the Apex-MAP performance surfaces for these different classes of architectures, we show side-by-side contour-plots of Seaborg and Phoenix in Figure 8.8 and 8.9. For the IBM systems, the area of highest performance is of rectangular shape and clearly elongated parallel to the spatial locality axis, while for the Cray system it is elongated parallel to the temporal locality axis. The IBM system can tolerate a decrease in spatial locality more easily but is much more sensitive to a loss of temporal locality. This reflects the elaborate cache and memory hierarchy on the individual nodes as well as the global system hierarchy, which also heavily relies on reuse of data as the interconnect bandwidth is substantially lower than the local memory bandwidth.

The Cray system can much more easily tolerate a decrease in temporal locality, but it is sensitive to a loss in spatial locality. This reflects an architecture that depends very little on local caching of data and an interconnect bandwidth equal to local memory bandwidth. To see such a clear signature of the Cray architecture is even more surprising considering that we us an MPI based benchmark, which does not fully exploit the capability of this system. The lines of equal performance on the Cray system are in general more vertical than diagonal as with the IBM system, which further confirms our interpretation. These differences in our performance surfaces overall clearly reflect the different design philosophies of these two different systems and demonstrate the utility of our approach.

8.6 Apex-MAP as an Application Proxy

The performance rates of most present-day scientific applications today are dominated by the cost of memory operations. However, to provide a complete understanding of the platform performance, it is not enough to characterize the data access only. Other performance factors, such as the numerical computation rate, must also be considered. In this section, we first discuss how Apex-MAP's capability can be enhanced by adding more characterization parameters. Then we show how to develop Apex-MAP performance proxies and how to validate the resulting models.

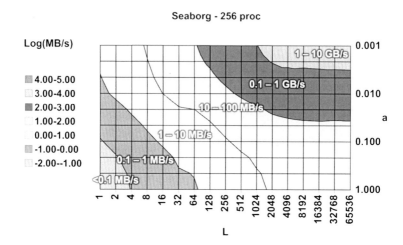

FIGURE 8.8: Contour plots of the performance surfaces for Seaborg.

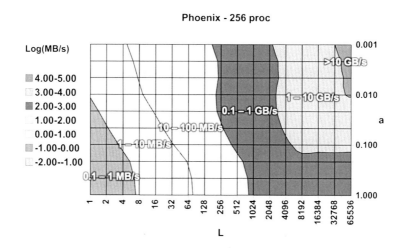

FIGURE 8.9: Contour plots of the performance surfaces for Phoenix.

8.6.1 More Characterization Parameters

In the above sections, Apex-MAP focused on data movement and used only three parameters, namely memory size, temporal locality and spatial locality, to characterize data access streams. In this section, we enhance Apex-MAP's capability to measure the performance effects of functional units by adding new concepts for characterizing computations and refining the characterization of data access to consider the effect of write operations and data prefetching. The new characterization parameters are computational intensity (C), register pressure (R), data store methods (S), and data independence (I).

- *Data Independence* (I) specifies the maximum number of blocks (the L contiguous memory locations accessed in succession is considered as one block) for which no data dependencies exist and data access can occur in parallel. This parameter sets the upper limit for effective data prefetching and vectorization.

- *Data Store Mode* (S) is used to define whether the final results will be stored in a vector or a scalar variable so that Apex-MAP can characterize both read and write operations.

- *Computational Intensity* (C) defines the ratio of floating-point operations to data access. Higher computational intensity could help to make full use of the computing units and alleviate the requirements to the memory system. Its performance effects may vary across different architectures, since it is related to the number of available computing units, the effectiveness of compilers and schedulers, and other factors.

- *Register Pressure* (R) defines the required number of registers to execute the code as a measure of computational complexity. The register pressure is increased when more variables are used, or more different sub-expressions appear.

8.6.2 The Kernel Changes for Apex-MAP

In order to incorporate these new parameters into Apex-MAP, the memory access kernel needs to be extended. The data dependence parameter is actually the parameter I we have used earlier, which controls the number of starting addresses generated in advance. Therefore, nothing needs be changed for it. To store the computational results, now we have two options. One is to use the scalar variable as in the original kernel. Another option is to use a vector in the left side of the loop statement instead of a scalar.

The computational kernel is designed to satisfy the following criteria:

- Allows Apex-MAP to exhibit close to peak performance.

- Uses only fused multiply-add operations.

- Has no common sub-expressions or the like to avoid compiler optimizations.

- Is not easily "defeated" by the compiler via loop-splitting, etc.

- Has no indirect array writes.

• Has no true dependencies in the loop.

The effect of the register pressure R is reflected by dividing the whole memory size (M) into R variables and use R statements in the loop as illustrated below, which is an example of the computational kernel for $C = 2$ and $R = 2$:

```
Repeat N times
Generate Index Array(ind);
CLOCK(start);
  for (j=0; j < I; j++) {
    for (k = 0; k < L; k++) {
      W0+=c0*(data0[ind[j]+k]  +c0*(data1[ind[j]+k]));
      W1+=c1*(data1[ind[j]+k]  +c1*(data0[ind[j]+k]));
    }
  }
CLOCK(end);
RunningTime += end - start
End Repeat
```

For different values of parameters C, R and S, the Apex-MAP kernels will be different. Therefore, we have used a code generator to produce different parameterized code segments to generate varying levels of register utilization and computational intensities. The implementation details could be find at http://ftg.lbl.gov/Apex/Apex.shtml.

Note that the original Apex-MAP model is just a special case of the current version with $C = 1$ and $R = 1$.

8.6.3 Determining Characteristic Parameters for Kernel Approximations

Ideally, a tool should be developed to automatically extract the parameters needed by Apex-MAP by analyzing the source code or based on performance profiling, such as the tools used in [227, 335]. However, such a tool is still under development and is not ready yet. In the paper [318], a statistical backfitting method had been used to find the characterization parameters L and α. With more than two parameters, this approach becomes prohibitively expensive. Therefore, for the time being, the parameters are characterized based on problem-size parameters and code inspection. For each kernel, we select five data sets which cover a wide range of the potential problem sizes so that our parameters will not be limited to one specific input data set. Parameters selected in this fashion are, in the spirit of the concepts of Apex-MAP, independent of the hardware used for our experiments. In some cases, parameters could not be easily determined this way because optimized library code was not easily analyzable or temporal locality parameters could not be deduced from the input data sets. In these cases, we determined appropriate values of some parameters by partial fitting to data, and we took extra care to check if fitted values matched our understanding of the code in question.

Typically we started by using a single data stream to approximate the performance behavior of the code. However, for some more complex applications, approximations with one stream are not sufficient. Two or more streams may be needed. If more than one stream are needed, they are either merged into one loop body if possible or they are measured by Apex-MAP separately and their performances are combined together algebraically. How to combine the measured performances for the separate streams depends on their inherit relations or profiling information.

8.6.4   Measuring the Quality of Approximation

After determining the parameters, we execute Apex-MAP with these parameter settings on several computational platforms and compare the performance results of Apex-MAP with the kernel performances and analyze their differences. We use two measures to examine the accuracy of using Apex-MAP to predict performance for the kernels. One commonly used statistical measure, the coefficient of determination R^2, quantifies the predictive capability of Apex-MAP. Our goal is for Apex-MAP to predict the performance with 90% confidence, i.e., the values of R^2 between series of application results and Apex-MAP results should not be less than 0.9.

Another simpler criterion is to examine the percentage of the performance difference between an application and Apex-MAP with corresponding parameters. Considering the generality of Apex-MAP, we consider it very satisfactory if Apex-MAP can predict the performance within 10% difference.

8.6.5   Case Studies of Applications Modeled by Apex-MAP

In this section, we will show some examples of how to use Apex-MAP to model the application performance. In particular, we picked four scientific kernels, Random Access, STREAM, DGEMM, and FFT from HPC Challenge benchmark suite. These four kernels represent the four corner cases in a two-dimensional temporal-spatial space as described in [171]: low temporal locality and low spatial locality (Random Access), low temporal locality high spatial locality (STREAM), high temporal locality high spatial locality (DGEMM), and high temporal locality low spatial locality (FFT). If the values of the characterization parameters are not discussed specifically, their default values are used. The default values are set to one, except for I, whose default value is set to 1024.

8.6.5.1   Test Environment

We tested Apex-MAP on three very different platforms: Jacquard, Bassi, and PSI:

- *Jacquard* is an Infiniband cluster built with 320 dual-socket Opteron nodes. Each node contains two single-core 2.2 GHz Opteron processors. The theoretical peak performance of a processors is 4.4 GFlop/s. The L1 data cache size is 64 Kbyte and the L2 cache size is 1 Mbyte. Processors on each node share 6 Gbyte of memory. The OS running on the nodes is Linux version 2.6.5-7.283-smp-perf. The compiler on this platform is PathScale v3.2.

- *Bassi* is a 122-node Power5-based IBM system. Each node contains eight 1.9 GHz Power5 processors with a theoretical peak performance of 7.6 GFlop/s each. Processors on each node have a shared memory pool of 32 Gbyte. The L1 data cache size of a processors is 32 Kbyte. The L2 and L3 cache sizes are 1.92 Mbyte and 36 Mbyte, respectively. The compiler is IBM XL C/C++ Enterprise Edition V8.0 for AIX.

- *PSI* is a cluster with different kinds of nodes. The nodes we used are Dell PowerEdge 1950. Each node has two quad-core Intel Xeon CPU E5345 processors running at 2.33 GHz. The primary cache (L1) size for a core is 32 Kbyte. The L2 cache size for a processor is 4 Mbyte. The shared memory on a node is 16 Gbyte. The OS system is Linux 2.6.18 (64-bit em64t). The peak performance for a core is 9.332 GFlop/s. The compiler is Intel compiler *icc/ifort* version 10.1.

8.6.5.2 Random Access

Algorithm Description: Random Access has been used to measure the peak capability of the memory system in terms of giga updates per second (GUPS). GUPS is calculated by identifying the number of memory locations that can be randomly updated in one second. Each update is a read-modify-write operation. The core part of this code is :

```
for (i=0; i<NUPDATE/128; i++) {
  for (j=0; j<128; j++) {
     ran[j] = (ran[j] << 1) ^ ((s64Int) ran[j] < 0 ? POLY : 0);
     Table[ran[j] & (TableSize-1)] ^= ran[j];
  }
}
```

There are total *TableSize* data elements and all of them are held in the *Table* array. There are total *NUPDATE* update operations. 128 target addresses are pre-computed and serve as the starting base to compute the new updated addresses. According to the description of this HPC Challenge benchmark, each thread would be permitted to look ahead up to 1024 random addresses, but the downloaded sequential code uses 128 only.

Parameter Derivation and Matching Results: The Memory size (M) for the stream is equal to TableSize. Each time only one address is updated and there is almost no relationship between the address to be updated and the next one, therefore, in Apex-MAP, the spatial locality L is defined to be one (low spatial locality), and the temporal locality α is also one (low temporal locality). The result store mode S is set to vector since after the data has been modified, all the results will be written back to the *Table* array. The computational intensity C is set as four operations are needed to compute the new address and new contents. The independent number of data blocks (I) is set as 128 since there are total 128 pre-computed addresses.

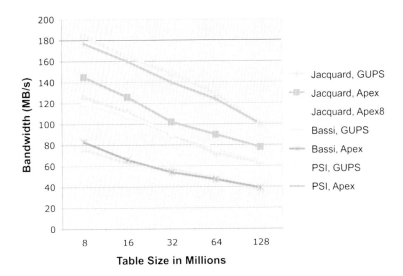

FIGURE 8.10: The performances of RandomAccess and Apex-MAP with corresponding parameters.

RandomAccess reports the final results in terms of GUPS. In order to compare with the Apex-MAP results, which reports in terms of Mbyte/s, the GUPS values are converted into Mbyte/s by timing the results with `sizeof(unsigned long long, i.e., u64Int)` × 1000. The performance results of RandomAccess and the Apex-MAP with corresponding parameters on these three platforms are shown in Figure 8.10. Bassi shows the best match and both programs deliver very similar performance. On the Intel PSI platform, the results also match quite well, especially for the largest data set, they deliver almost identical results. On Jacquard, the RandomAccess results falls below the Apex-MAP results. Further experimental results showed that if the computational intensity C was increased from four to eight, Apex-MAP could approximate RandomAccess performance much better (labeled as *Jacquard, Apex8*). A major difference between Apex-MAP and RandomAccess is the basic data type used. Apex-MAP uses eight-byte floating-point computations to orchestrate the computational intensity while in RandomAccess the computations used are logical and integer operations. These different data types and operations appear to lead to different execution conditions on different platforms and force us to adjust the value of the parameter C accordingly on Jacquard.

8.6.5.3   DGEMM

Algorithm Description: DGEMM is a subroutine in the Basic Linear Algebra Subprograms that performs the following matrix multiplication: $C = \gamma AB + \delta C$. On almost all high-performance platforms, vendors have provided optimized implementations in their libraries. One critical optimization technique for DGEMM is to decompose the input matrices into smaller blocks, so that the final result could be accumulated from the results of these smaller blocks. This decomposition provides better spatial and temporal locality for the matrix data. Therefore, a high percentage of peak performance is expected, especially after vendor's specific optimization on their platforms. In our test γ is set to 1.0, and we do not transpose any matrices.

Parameter Derivation and Matching Results: Since we do not have access to the vendor's source code, the Apex-MAP parameters are largely determined by our knowledge of the used optimization techniques and by experiments. Due to the matrix decomposition approach and vendor implementation, very high temporal locality is assumed and its value α is set to be 0.001. The spatial locality L is set to 4096, which represents the same amount of data as the commonly used 64×64 block size. Assuming fully optimized library DGEMM routines, we set the computational intensity C to the value that obtains optimal performance on a platform. Its value is set to 12 on Jacquard, and Bassi. However, on the Intel PSI platform increasing the computational intensity to such high values does not necessarily improve performance. By experiments, we find that the best performance on this platform is delivered with $C = 2$.

Figure 8.11 shows the performance of DGEMM and Apex-MAP on three different platforms across different matrix sizes. First, we notice that the DGEMM performance is consistent across different matrices. This is because the DGEMM function is provided by the vendors and has been highly optimized. Usually very high performance efficiency can be expected. On Jacquard, the Apex-MAP performance is very close to DGEMM. Similarly on Bassi, the results of Apex-MAP and DGEMM are also very close, except for the smallest matrix, where the Apex-MAP result is about 7% lower than that of DGEMM. This is probably due to the fact that for smaller data sets, other kind of optimizations in addition to the block decomposition have also been applied, making it more efficient. On PSI, for the smallest data set, Apex-MAP and DGEMM deliver almost identical results. For larger matrices, DGEMM performs better. However, the performance difference is below 9%.

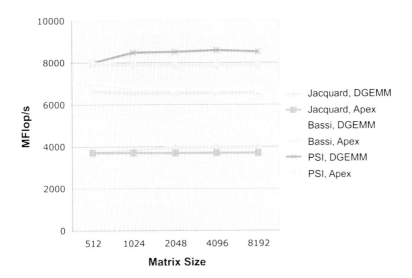

FIGURE 8.11: The performances of DGEMM and Apex-MAP with corresponding parameters.

The DGEMM subroutine is the dominant time-consuming part for HPL program when matrix size is not too small. Therefore, the parameters derived for DGEMM should also work for HPL.

8.6.5.4   STREAM

Algorithm Description: Stream [317] is a well recognized and widely used kernel to measure the memory bandwidth. It measures four operations for arrays: Copy, Scale, Add, and Triad. In our study we focus on Triad since it is more often used in the real applications. The code for Triad looks like the following:

```
for (j=0; j<N; j++)
  A[j] = B[j]+scalar*C[j];
```

Parameter Derivation and Matching Results: There are three vectors here, A, B and C. Since both B and C will be read concurrently, the number of registers R is set to be two. The memory size M is set to $2N$, the sum of the array sizes of B and C. Since the computation results will be stored in array A, the data store mode S is set as VECTOR. The spatial locality (L) is N, the length of the vector array of B and C. There is no data reuse. Therefore the temporal locality (α) is set as 1. The maximum number of independent block is also set as 1 due to the high spatial locality equals N.

The results of Stream and Apex-MAP with corresponding parameters are

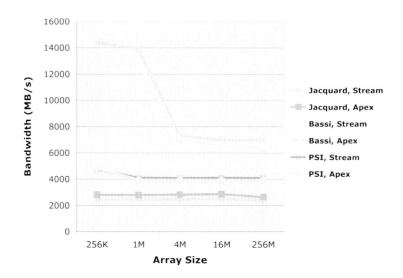

FIGURE 8.12: The performances of Stream and Apex-MAP with corresponding parameters.

shown in Figure 8.12. On Bassi, the two sets of results match well for larger data sets. For smaller data sets, the Apex-MAP performs better, but the difference is less than 10%. By analyzing the code, we find that Stream uses static memory allocation, while Apex-MAP uses dynamic memory allocation. By intuition we would expect that such a difference should not cause a large performance difference for such simple codes. However, by changing the memory allocation method in Stream from static to dynamic, we find that Stream could perform as well as Apex-MAP.

On Jacquard, Apex-MAP performs a little better than Stream. Similar to Bassi, the difference could be reduced by changing the static memory allocation to dynamic memory allocation in Stream. The best results are obtained on PSI, where they deliver almost same performance.

8.6.5.5 FFT

Algorithm Description: The code is obtained from the HPC Challenge benchmark. Our experiment is focus on the one-dimensional double complex fast Fourier transform (FFT), which named zfft1d. Although it is intended to solve one-dimensional FFT, in the implementation the one-dimensional data is actually stored into a two-dimensional matrix, so that the FFT can be done in parallel. Therefore, like many other two-dimensional FFTs, the entire solving process is done in five steps: (i) transpose the matrix, (ii) perform a one-dimensional FFTs on local rows in block mode, (iii) multiply the elements of the resulting matrix by the corresponding roots of unity and transpose the resulting matrix, (iv) perform one dimensional FFTs on rows, and finally (v) transpose the matrix again. So altogether there are three matrix transpositions and two local-node FFTs. In the original implementation, the three transposes are implemented differently, perhaps for cache reuse purpose. However, these subtle differences may cause significantly performance variations on different platforms. In such cases, we replace the slower implementation by the faster one. Here an efficient block transpose is used for performance improvement.

Parameter Derivation and Matching Results: Here we use a two-stream approach, one for the transpose (FFT_T) and another for the local FFT computation (FFT_C). For the transpose stream, the memory size M is the matrix size. The spatial locality L is the block size. and the results are stored in VECTORs S. For FFT_C, a double memory size M is selected due to additional memory used in performing the local FFTs. The spatial locality L is also selected as the block size. However, the temporal locality α is difficult to determine from the code. After some experimentation, we set to be 0.12. We set the computational intensity C to two.

After we obtained the results of these two streams separately, we combined them based on the following formula: $1/(1/P_{FFT\_C}+1/(P_{FFT\_T}*Ratio))$. Here P_{FFT_C} is the performance of the local FFT_C stream in terms of Mflop/s. P_{FFT_T} is the performance of the transpose stream in terms of Mbyte/s. The Ratio is computed by dividing the number of floating point operations needed by two local FFTs by the number of data movement in the three transposes. This is an intrinsic attribute of FFT computations and is only related to the problem size.

The well matching results between the FFT results and Apex-MAP simulated results are displayed in Figure 8.13.

8.6.6 Overall Results and Precision of Approximation

In this section, we discuss the overall ability to use Apex-MAP to predict performance for scientific kernels. First, we use a statistical measure, the coefficient of determination R^2, to see how well Apex-MAP predicts the performance for the kernels. Then, we compute the percentage performance difference between the Apex-MAP and the kernels. Considering the wide variety of HPC applications and platforms, our goal is to expect Apex-MAP to predict the performance across a range of data sets with 90% accuracy. Next we summarize the Apex-MAP parameters for these kernels and position each kernel in a temporal-spatial locality map. Finally, we discuss some lessons we learned for developing a portable performance proxy across HPC platforms.

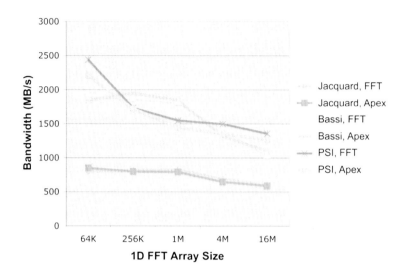

FIGURE 8.13: The performances of 1D FFT and Apex-MAP with corresponding parameters.

8.6.6.1   Coefficient of Determination R^2

The coefficient of determination is a measure used widely in statistical and modeling community to determine how certain one can be in making predictions from a model, graph, proxy, or simulation. It is the square of the linear correlation coefficient r. The mathematical formula for computing the r is:

$$r \; = \; \frac{\Sigma(x - \bar{x})(y - \bar{y})}{\sqrt{\Sigma(x - \bar{x})^2 \Sigma(y - \bar{y})^2}}, \tag{8.2}$$

where x and y are the series of Apex-MAP results and kernel results respectively, and $\bar{x}$ and $\bar{y}$ are the average of series x and y. The computed R^2 values are shown in Table 8.1. Usually, a value of over 0.8 is considered "strong" correlation [282]. A value of 1.00 indicates a perfect correlation between Apex-MAP and a kernel. For most cases, the values are at least above 0.98. In the worst case, the value is 0.92.

8.6.6.2   Percentage of Performance Difference

Another way to measure the predictive ability of Apex-MAP is to look at the relative difference between Apex-MAP results and the actual measured performance for kernels for a range of data sets.

Table 8.2 displays the absolute performance numbers and the percentage of performance difference between Apex-MAP and kernels on the three test

TABLE 8.1: The coefficients of determination between Apex-MAP results and kernel results

	Jacquard	Bassi	PSI
GUPS	0.998	0.967	0.993
DGEMM	0.966	0.965	0.959
STREAM	0.922	0.995	1.000
FFT	0.925	0.940	0.969

platforms. Positive percentages indicate that Apex-MAP performs better than the kernel, while the negative ones indicate that Apex-MAP performs worse. We find that for all cases the relative performance difference is less then 10%, satisfying our goal.

8.7 Limitations of Memory Access Modeling

- Characterization tools Currently, the automatic characterization tools for Apex-MAP is still under development. The characterization process is largely depend on code inspection and experiments. This may seriously increase the characterization difficulty. Also when more than one streams are needed, Apex-MAP is lack of a consistent way to combine the results of these multiple streams.

- Large-scale full applications

  Apex-MAP has not been applied to large-scale full applications yet.

- Computational Intensity

  The computational intensity needed for a stream may be different across different platforms. The value of this parameter is affected by the balance among hardware functional units, the compiler efficiency, and other factors.

8.8 Acknowledgment

This work was sponsored by the Director, Office of Science, Basic Energy Sciences, Division of Material Science and Engineering, and the Advanced Scientific Computing Research Office, of the U.S. Dept. of Energy under Contract No. DE-AC02-05CH11231.

TABLE 8.2: The absolute performance and percentage of performance difference

Absolute Performance						Percentage of Difference %		
Jacquard		Bassi		PSI		Jacquard	Bassi	PSI
Kernel	Apex	Kernel	Apex	Kernel	Apex			
Random Access (Mbyte/s)								
126	121	76	83	184	177	-4.23	8.97	-4.00
112	109	63	66	165	160	-2.79	4.67	-3.04
90	91	57	54	149	140	0.78	-5.36	-6.26
72	77	48	47	130	124	6.46	-2.86	-4.54
63	67	37	39	100	100	5.67	4.37	0.16
DGEMM (Mflop/s)								
3708	3716	7128	6642	7995	7968	0.21	-6.81	-0.34
3855	3708	6774	6574	8488	7922	-3.82	-2.96	-6.66
3974	3698	6690	6565	8516	7914	-6.94	-1.87	-7.07
3983	3700	6705	6578	8591	7896	-7.11	-1.89	-8.09
3984	3701	6709	6574	8523	7907	-7.09	-2.02	-7.23
STREAM (Mbyte/s)								
2677	2807	13442	14431	4747	4825	4.86	7.36	1.64
2686	2795	13342	13941	4098	3943	4.06	4.49	-3.78
2685	2822	7624	7306	4082	3940	5.10	-4.17	-3.48
2771	2859	6764	7007	4094	3937	3.18	3.59	-3.83
2792	2650	6652	7000	4092	3933	-5.09	5.23	-3.89
FFT (Mflop/s)								
815	852	1905	1850	2439	2218	4.52	-2.86	-9.07
818	801	1879	1957	1733	1723	-2.02	4.11	-0.63
829	794	1829	1846	1546	1453	-4.30	0.94	-6.01
692	645	1440	1324	1496	1361	-6.81	-8.07	-8.97
605	589	1051	1105	1355	1276	-2.60	5.15	-5.85

Chapter 9

The Roofline Model

Samuel W. Williams

Lawrence Berkeley National Laboratory

The Roofline model is a visually intuitive performance model constructed using bound and bottleneck analysis [362, 367, 368]. It is designed to drive programmers towards an intuitive understanding of performance on modern computer architectures. As such, it not only provides programmers with realistic performance expectations, but also enumerates the potential impediments to performance. Knowledge of these bottlenecks drives programmers to implement particular classes of optimizations. This chapter will focus on architecture-oriented roofline models as opposed to using performance counters to generate a roofline model.

This chapter is organized as follows. Section 9.1 defines the abstract architecture model used by the roofline model. Section 9.2 introduces the basic form of the roofline model, where Sections 9.3–9.6 iteratively refine the model with tighter and tighter performance bounds.

9.1 Introduction

In this section we define the abstract architectural model used for the Roofline model. Understanding of the model is critical in one's ability to apply the Roofline model to widely varying computational kernels. We then introduce the concept of arithmetic intensity to the reader and provide several diverse examples that the reader may find useful in their attempt to estimate arithmetic intensity for their kernels of interest. Finally, we define the requisite key terms in this chapter's glossary.

9.1.1 Abstract Architecture Model

The roofline model presumes a simple architectural model consisting of black boxed computational elements (e.g., CPUs, Cores, or Functional Units) and memory elements (e.g., DRAM, caches, local stores, or register files) interconnected by a network. Whenever one or more processing elements may access a memory element, that memory element is considered shared. In general, there is no restriction on the number or balance of computational and memory elements. As such, a large number of possible topologies exist, allowing the model to be applied to a large number of current and future computer architectures. At any given level of the hierarchy, processing elements may only communicate either with memory elements at that level, or with memory elements at a coarser level. That is, processors, cores, or functional units may only communicate with each other via a shared memory.

Consider Figure 9.1. We show two different dual-processor architectures. Conceptually, any processor can reference any memory location. However, Figure 9.1(a) partitions memory and creates additional arcs. This is done to convey the fact that the bandwidth to a given processor may depend on which

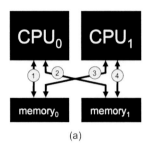 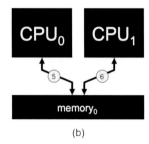

(a)   (b)

FIGURE 9.1: High-level architectural model showing two black-boxed processors either connected to separate memories (a) or to a common shared memoryi (b). The arrows denote the ISA's ability to access information and not necessarily hardware connectivity.

memory the address may lie in. As such, these figures are used to denote non-uniform memory access (NUMA) architectures.

9.1.2 Communication, Computation, and Locality

With this model, a kernel can be distilled down to the movement of data from one or more memories to a processor where it may be buffered, duplicated, and computed on. That modified data or any new data is then communicated back to those memories.

The movement of data from the memories to the processors, or communication, is bounded by the characteristics of the processor-memory interconnect. Consider Figure 9.1. There is a maximum bandwidth on any link as well as a maximum bandwidth limit on any subset of links i.e., the total bandwidth from or to $memory_0$ may be individually limited.

Computation, for purposes of this chapter, consists of floating-point operations including multiply, add, compare, etc. Each processor has an associated computation rate. Nominally, as processors are black boxed, one does not distinguish how performance is distributed among cores within a multicore chip. However, when using a hierarchical model for multicore (discussed at the end of this chapter), rather than only modeling memory-processor communication and processor computation, we will model memory-cache communication, cache-core communication, and core computation.

Although there is some initial locality of data in $memory_i$, once moved to $processor_j$ we may assume that caches seamlessly provide for locality within the processor. That is, subsequent references to that data will not generate capacity misses in 3C's (compulsory, capacity, conflict) vernacular [165]. This technique may be extended to the cache hierarchy.

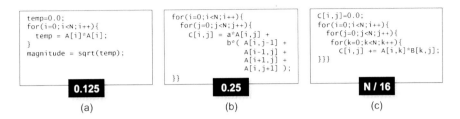

```
temp=0.0;
for(i=0;i<N;i++){
  temp = A[i]*A[i];
}
magnitude = sqrt(temp);
```

0.125

(a)

```
for(i=0;i<N;i++){
  for(j=0;j<N;j++){
    C[i,j] = a*A[i,j] +
             b*( A[i,j-1] +
                 A[i-1,j] +
                 A[i+1,j] +
                 A[i,j+1] );
}}
```

0.25

(b)

```
C[i,j]=0.0;
for(i=0;i<N;i++){
  for(j=0;j<N;j++){
    for(k=0;k<N;k++){
      C[i,j] += A[i,k]*B[k,j];
}}}
```

N / 16

(c)

FIGURE 9.2: Arithmetic Intensities for three common HPC kernels including (a) dot products, (b) stencils, and (c) matrix multiplication.

9.1.3   Arithmetic Intensity

Arithmetic intensity is a kernel's ratio of computation to traffic and is measured in flops:bytes. Remember traffic is the volume of data to a particular memory. It is not the number of loads and stores. Processors whose caches filter most memory requests will have very high arithmetic intensities. A similar concept is machine balance [77] which represents the ratio of peak floating-point performance to peak bandwidth. A simple comparison between machine balance and arithmetic intensity may provide some insight as to potential performance bottlenecks. That is, when arithmetic intensity exceeds machine balance, it is likely the kernel will spend more time in computation than communication. As such, it is likely compute bound. Unfortunately such simple approximations gloss over many of the details of computer architecture and result in performance far below performance expectations. Such situations motivated the creation of the roofline model.

9.1.4   Examples of Arithmetic Intensity

Figure 9.2 presents pseudocode for three common kernels within scientific computing: calculation of vector magnitude, a stencil sweep for a 2D PDE, and dense matrix-matrix multiplication. Assume all arrays are double precision.

Arithmetic intensity is the ratio of total floating-point operations to total DRAM bytes. Assuming N is sufficiently large that the array of Figure 9.2(a) does not fit in cache and enough to amortize the poor performance of the square root, then we observe that it performs N flops while transferring only $8 \cdot N$ bytes (N doubles). The second access to A[i] exploits the cache/register file locality within the processor. The result is an arithmetic intensity of 0.125 flops per byte.

Figure 9.2(b) presents a much more interesting example. Assuming the processor's cache is substantially larger than $8 \cdot N$, but substantially smaller than $16 \cdot N^2$, we observe that the leading point in the stencil A[i,j+1] will eventually be reused by subsequent stencils as A[i+1,j], A[i,j], A[i-1,j], and A[i,j-1]. Although references to A[i,j] only generate $8 \cdot N^2$ bytes of communication, accesses to C[i,j] generate $16 \cdot N^2$ bytes because write-allocate cache architectures will generate both a read for the initial fill on the write miss in addition to the eventual write back. As the code performs $6 \cdot N^2$ flops, the resultant arithmetic intensity is $(6 \cdot N^2)/(24 \cdot N^2) = 0.25$.

Finally, Figure 9.2(c) shows the pseudocode for a dense matrix-matrix multiplication. Assuming the cache is substantially larger than $24 \cdot N^2$, then A[i,j], B[i,j], and C[i,j] can be kept in cache and only their initial and write back references will generate DRAM memory traffic. As such, we observe the loop nest will perform $2 \cdot N^3$ flops while only transferring $32 \cdot N^2$ bytes. The result is an arithmetic intensity of $N/16$.

9.2 The Roofline

Given the aforementioned abstract architectural model and a kernel's estimated arithmetic intensity, we create a intuitive and utilitarian model that allows programmers rather than computer architects to bound attainable performance. We call this model the "Roofline Model." The roofline model is built using Bound and Bottleneck analysis [207]. As such we may consider the two principal performance bounds (computation and communication) in isolation and compare their corresponding times to determine the bottleneck and attainable performance. Consider Figure 9.1(b). Consider a simple homogeneous kernel that must transfer B bytes of data from memory_0 and perform $\frac{F}{2}$ floating-point operations on both CPU_0 and CPU_1. Moreover, assume the the memory can support *PeakBandwidth* bytes per second and combined, and the processors can perform *PeakPerformance* floating-point operations per second. Simple analysis suggests it will take $\frac{B}{\text{PeakBandwidth}}$ seconds to transfer the data and $\frac{F}{\text{PeakPerformance}}$ seconds to compute on it. Assuming one may perfectly overlap communication and computation it will take:

$$\text{Total Time} = \max \begin{cases} F \ / \ \text{PeakPerformance} \\ B \ / \ \text{PeakBandwidth} \end{cases} \tag{9.1}$$

Reciprocating and multiplying by F flops, we observe performance is bound to:

$$\text{AttainablePerformance (Gflop/s)} = \min \begin{cases} \text{PeakPerformance} \\ \text{PeakBandwidth} \times \text{ArithmeticIntensity} \end{cases} \tag{9.2}$$

Where Arithmetic Intensity is F/B.

Although a given architecture has a fixed peak bandwidth and peak performance, arithmetic intensity will vary dramatically from one kernel to the next and substantially as one optimizes a given kernel. As such, we may plot attainable performance as a function of arithmetic intensity. Given the tremendous range in performance and arithmetic intensities, we will plot these figures on a log-log scale.

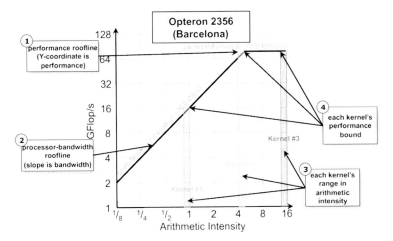

FIGURE 9.3: Roofline Model for an Opteron SMP. Also, performance bounds are calculated for three non-descript kernels.

Using the Stream benchmark [317], one may determine that the maximum bandwidth one can attain using a 2.3 GHz dual-socket × quad-core Opteron 2356 Sun 2200 M2 is 16.6 GB/s. Similarly, using a processor optimization manual it is clear that the maximum performance one can attain is 73.6 Gflop/s. Of course, as shown in Equation 9.2, it is not possible to always attain both, and in practice may not be possible to achieve either.

Figure 9.3 visualizes Equation 9.2 for this SMP via the black line. Observe that as arithmetic intensity increases, so to does the performance bound. However, at the machine's flop:byte ratio, the performance bound saturates at the machine's peak performance. Beyond this point, although performance is at its maximum, used bandwidth decreases. Note, the slope of the roofline in the bandwidth-limited regions is actually the machine's Stream bandwidth. However, on a log-log scale the line always appears at a 45-degree angle. On this scale, doubling the bandwidth will shift the line up instead of changing its perceived slope.

This Roofline model may be used to bound the Opteron's attainable performance for a variety of computational kernels. Consider three generic kernels, labeled 1, 2, and 3 in Figure 9.3, with flop:DRAM byte arithmetic intensities of about 1, 4, and 16 respectively. When mapped onto Figure 9.3, we observe that the Roofline at Kernel #1's arithmetic intensity is in the bandwidth-limited region (i.e., performance is still increasing with arithmetic intensity). Scanning upward from its X-coordinate along the Y-axis, we may derive a performance bound based on the Roofline at said X-coordinate. Thus, it would be unreasonable to expect Kernel #1 to ever attain better than 16 Gflop/s. With an arithmetic intensity of 16, Kernel #3 is clearly ultimately compute-bound. Kernel #2 is a more interesting case as its performance is heavily dependent on exactly calculating arithmetic intensity as well as both the kernel's and machine's ability to perfectly overlap communication (loads and stores from DRAM) and computation. Failure on any of these three fronts will diminish performance.

In terms of the Roofline model, performance is no longer a scalar, but a coordinate in arithmetic intensity–Gflop/s space. As the roofline itself is only a performance bound, it is common the actual performance will be below the roofline (it can never be above). As programmers interested in architectural analysis and program optimization, we are motivated to understand why performance is below the roofline (instead of on it) and how we may optimize a program to rectify this. The following sections refine the roofline model to enhance its utility in this field.

9.3   Bandwidth Ceilings

Eliciting good performance from modern SMP memory subsystems can be elusive. Architectures exploit a number of techniques to hide memory latency (HW, SW prefetching, TLB misses) and increase memory bandwidth (multiple controllers, burst accesses, NUMA). For each of these architectural paradigms, there is a commensurate set of optimizations that must be implemented to extract peak memory subsystem performance. This section enumerates these potential performance impediments and visualizes them using the concept of *bandwidth performance ceilings*. Essentially a ceiling is structure internal to the roofline denoting a complete failure to exploit an architectural paradigm. In essence, just as the roofline acted to constrain performance to be beneath it, so too do ceilings constrain performance to be beneath them. Software optimization removes these ceilings as impediments to performance.

9.3.1   NUMA

We begin by considering the NUMA issues in the Stream benchmark as it will likely be illustrative of the solution to many common optimization mistakes made when programming multisocket SMPs. As written, there is a loop designed to initialize the values of the arrays to be streamed. Subtly, this loop is also used to distribute data among the processor sockets through the combination of a OpenMP pragma (`#pragma omp parallel for`) and the use of the first touch policy [134]. Although the virtual addresses of the elements appear contiguous, their physical addresses are mapped to the memory controllers on different sockets. This optimized case is well visualized in Figure 9.4(a). We observe the array (grid) has been partitioned with half placed in each of the two memories. When the processors compute on this data they find that the pieces of the array they're tasked with using are in the memory to which

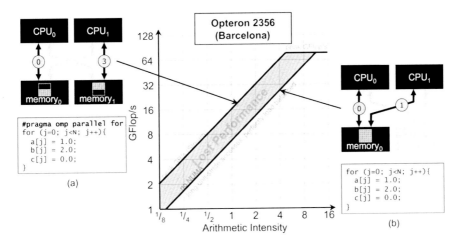

FIGURE 9.4: NUMA ceiling resulting from improper data layout. The codes shown are initialization-only (a) with and (b) without proper exploitation of a first-touch policy. Initialization is completely orthogonal to the possible computational kernels.

they have the highest bandwidth. If, on the other hand, the pragma were omitted, then the array would likely be placed in its entirety within memory$_0$. As such, not only does one forgo half the system's peak bandwidth by not using the other memory, but he also looses additional performance as link 1 likely has substantially lower bandwidth than 0 or 3, but must transfer just as much data. We may plot the resultant bandwidth on the Roofline figure to the right. We observe a 2.5× degradation in performance. Not only will this depress the performance of any memory-bound kernels, but it expands the range of memory-bound arithmetic intensities to about 10 flops per DRAM byte.

Such performance bugs can be extremely difficult to find regardless of whether one uses OpenMP, POSIX threads, or some other threading library. Under very common conditions, it can also occur even when binding threads to cores under pthreads because data is bound to a controller by the OS, not by a `malloc()` call. For example, an initial `malloc()` call followed by an initialization routine may peg certain virtual addresses to one controller or the other. However, if that data is freed, it is returned to the heap, not the OS. As such, a subsequent call to `malloc()` will use data already on the heap, and already pinned to a controller other than the one that might be desired. Unless cognizant of these pitfalls, one should strongly consider only threading applications within a socket instead of across an entire SMP node.

9.3.2   Prefetching, DMA, and Little's Law

Little's Law [41] (see also Chapter 1) states that the concurrency (independent memory operations) that must be injected into the memory subsystem to attain peak performance is the product of memory latency and peak memory bandwidth. For processors like Opterons, this translates into more than 800 bytes of data (perhaps 13 cache lines). Hardware vendors have created a number of techniques to generate this concurrency. Unfortunately, methods like out-of-order execution operate on doubles, not on cache lines. As such, it is difficult to get 100 loads in flight. The more modern methods include software prefetching, hardware prefetching, and DMA. Software prefetching and DMA are similar in that they are both asynchronous software methods of expressing more memory-level parallelism than one could normally achieve via a scalar ISA. The principal difference between the two is granularity. Software prefetching only operates on cache lines, whereas DMA operates on arbitrary numbers of cache lines. Hardware prefetchers attempt to infer a streaming access pattern given a series of cache misses. As such they don't require software modifications, and express substantial memory-level parallelism, but are restricted to particular memory access patterns.

It is conceivable that one could create a version of Stream that mimics the memory access pattern observed in certain applications. For example, a few pseudorandom access pattern streams may individually trip up any hardware or software prefetcher, but collectively allow expression of memory-level parallelism through DMA or software prefetch. As such, one could draw a series of ceilings below the roofline that denote an every decreasing degree of memory-level parallelism.

9.3.3   TLB Issues

Modern microprocessors use virtual memory and accelerate the translation to physical addresses via small highly-associative translation lookaside buffers (TLBs). Unfortunately, these act like caches on the page table (caching page table entries). If a kernel's working set, as measured in page table entries, exceeds the TLB capacity (or associativity), then one generates TLB capacity (or conflict) misses. Such situations arise more often than one might think. Simple cache blocking for matrix multiplication can result in enough disjoint address streams, which although they may fit in cache, do not fit in the TLB.

One could implement a version of Stream that scales the number of streams for operations like TRIAD. Doing so would often result in a about the same bandwidth for low numbers of streams, but would suddenly dip for an additional array. This dip could be plotted on the roofline model as a bandwidth ceiling, and labeled with the number of arrays required to trigger it.

9.3.4　Strided Access Patterns

A common solution to the above problems is to lay out the data as one multicomponent array (i.e., an array of cartesian vectors instead of three arrays one for each component). However, the computational kernels may not use all of these components at a time. Nevertheless, the data must still be transfered. Generally, small strides (less than the cache line size) should be interpreted as a decrease in arithmetic intensity, where large strides can represent a lack of spatial locality and memory-level parallelism. One may plot a ceiling for each stride with the roofline being stride-1 (unit-stride).

9.4　In-Core Ceilings

Given the complexity of modern core architectures, floating-point performance is not simply a function of arithmetic intensity alone. Rather architectures exploit a number of paradigms to improve peak performance including pipelining, superscalar out-of-order execution, SIMD, hardware multithreading, multicore, heterogeneity, etc. Unfortunately, a commensurate set of optimizations (either generated by the compiler or explicitly expressed by the user) are required to fully exploit these paradigms. This section enumerates these potential performance impediments and visualizes them using the concept of *in-core performance ceilings*. Like bandwidth ceilings, these ceilings act to constrain performance coordinates to lie beneath them. The following section enumerates some of the common ceilings. Each is examined in isolation. All code examples assume an x86 architecture.

9.4.1　Instruction-Level Parallelism

Every instruction on every architecture has an associated latency representing the time from when the instruction's operands are available to the time where the results are made available to other instructions (assuming no other resource stalls). For floating-point computational instructions like multiply or add, these latencies are small (typically less than 8 cycles). Moreover, every microprocessor has an associated dispatch rate (a bandwidth) that represents how many independent instructions can be executed per cycle. Just as Little's Law can be used to derive the concurrency demanded by the memory subsystem using the bandwidth–latency product, so too can it be used to estimate the concurrency each core demands to keep its functional units busy. We define this to be the *instruction-level parallelism*. When a thread of execution falls short of expressing this degree of parallelism, functional units will go idle, and performance will suffer [73].

As an example, consider Figure 9.5. In this case we plot scalar floating-

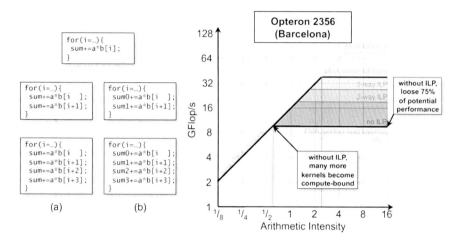

FIGURE 9.5: Performance ceilings as a result of insufficient instruction-level parallelism. (a) Code snippet in which the loop is unrolled. Note (FP add) instruction-level parallelism (ILP) remains constant. (b) Code snippet in which partial sums are computed. ILP increases with unrolling.

point performance as a function of DRAM arithmetic intensity. However, on the roofline figure we note the performance impact from a lack of instruction-level parallelism through instruction-level parallelism (ILP) ceilings. Figure 9.5(a) presents a code snippet in which the loop is naïvely unrolled either by the user or the compiler. Although this has the desired benefit of amortizing an loop overhead, it does not increase the floating-point add instruction-level parallelism — the adds to `sum` will be serialized. Even a superscalar processor must serialize these operations. Conversely, Figure 9.5(b) shows an alternate unrolling method in which partial sums are maintained within the loop and reduced (not shown) upon loop completion. If one achieves sufficient cache locality for `b[i]` then arithmetic intensity will be sufficiently great that Figure 9.5(b) should substantially outperform (a).

Subtly, without instruction-level parallelism, the arithmetic intensity at which a processor becomes compute-bound is much lower. In a seemingly paradoxical result, it is possible that many kernels may show the signs of being compute-bound (parallel efficiency), yet deliver substantially suboptimal performance.

9.4.2 Functional Unit Heterogeneity

Processors like AMD's Opteron's and Intel's Nehalem have floating-point execution units optimized for certain instructions. Specifically, although they both have two pipelines capable of simultaneously executing two floating-point instructions, one pipeline may only perform floating-point additions, while the other may only perform floating-point multiplies. This creates a potential performance impediment. For codes that are dominated by one or the other, attainable performance will be half that of a code that has a perfect balance between multiplies and adds. For example, codes that solve PDEs on structured grids often perform stencil operations which are dominated by adds with very few multiplies, where codes that perform dense linear algebra often see a near perfect balance between multiplies and adds. As such, we may create a series of ceilings based on the ratio of adds to multiplies. As the ratio gets further and further from 1, the resultant ceiling will approach one half of peak.

Processors like Cell, GPUs, POWER, and Itanium exploit what is known as fused-multiply add (FMA). These instructions are implemented on execution units where instead of performing multiplies and adds in parallel, they are performed in sequence (multiply two numbers and add a third to the result). Obviously the primary advantage of such an implementation is to execute the same number of floating-point operations as a machine of twice the instruction issue width. Nevertheless, such an architecture creates a similar performance issue to the case of separate multipliers and adders in that unless the code is entirely dominated by FMA's, performance may drop by a factor of two.

9.4.3   Data-Level Parallelism

Modern microprocessor vendors have attempted to boost their peak performance through the addition of Single Instruction Multiple Data (SIMD) operations. In effect, with a single instruction, a program may express two or four-way data-level parallelism. For example, the x86 instruction addps performs four single-precision floating-point add operations in parallel. Ideally the compiler should recognize this form of parallelism and generate these instructions. However, due to the implementation's rigid nature, compilers often fail to generate these instructions. Moreover, even programmers may not be able to exploit them due to rigid program and data structure specifications. Failure to exploit these instructions can substantially depress kernel performance.

Consider Figure 9.6. The code is a simplified version of that in Figure 9.5(b). We observe there is substantial ILP, but only floating-point adds are performed. As such, there is no data-level parallelism, and performance is bounded to less than 18.4 Gflop/s. Most x86 compilers allow the user to SIMDize their code via *intrinsics* — small functions mapped directly to one or two instructions. We observe that the first step in this process is to replace the conventional C assignments with the scalar form of these intrinsics. Of course doing so will not improve our performance bound because it has not increased the degree of data-level parallelism. However, when using the _pd form of the intrinsics we should unroll the loop 8 times so that we may simultaneously express both 2-way data level parallelism and 4-way instruction-level parallelism. Doing so improves our performance bound to 36.8 Gflop/s. As

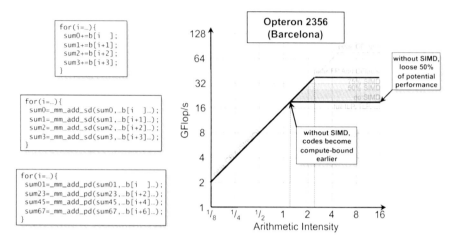

FIGURE 9.6: Example of Ceilings associated with data-level parallelism.

discussed in the previous subsection, we cannot achieve 73.6 due to the fact that this code does not perform any floating-point multiplies.

9.4.4 Hardware Multithreading

Hardware multithreading [164] has emerged as an effective solution to the memory- and instruction-level parallelism problems with a single architectural paradigm. Threads whose current instruction's operands are ready are execute while the others wait in a queue. As all of this is performed in hardware, there is no apparent context switching. There are no ILP ceilings, as typically there is enough thread-level parallelism (TLP) to cover the demanded instruction-level parallelism. Moreover, the exemplar of this architecture, Sun's Niagara [273], doesn't implement SIMD or heterogeneous functional units. However, a different set of ceilings normally not seen on superscalar processors appear: the floating-point fraction of the dynamics instruction mix. All processors have a finite instruction fetch and decode bandwidth (the number of instructions that can be fetched per cycle). On superscalar processors, this bandwidth is far greater than the instruction bandwidth required under ideal conditions to saturate the floating-point units. However, on processors like Niagara, as the floating-point fraction dips below 50%, the non-floating-point instructions begin to sap instruction bandwidth away from the floating-point pipeline. The result: performance drops. The only effective solution here is improving the quality of code generation.

More recently, superscalar manufactures have begun to introduce hardware multithreading into their processor lines including Nehalem, Larrabee, and POWER7. In such situations, SPMD programs may not suffer from ILP

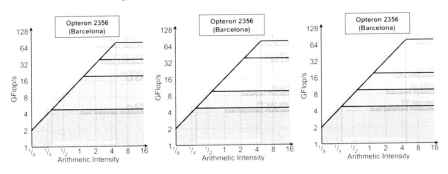

FIGURE 9.7: Equivalence of Roofline models.

ceilings but may invariably see substantial performance degradation due to data-level parallelism (DLP) and heterogenous functional unit ceilings.

9.4.5 Multicore Parallelism

Multicore has introduced yet another form of parallelism within a socket. When programs regiment cores (and threads) into bulk synchronous computations (compute/barrier), load imbalance can severely impair performance. Such an imbalance can be plotted using the roofline model. To do this, one may count the total number of floating-point operations performed across all threads and the time between the start of the computation and when the last thread enters the barrier. The ratio of these two numbers is the (load imbalanced) attained performance. Similarly, one can sum the times each thread spends in computation and divide by the total number of threads. The ratio of total flops to this number is the performance that could be attained if properly load balanced. As such, one can visualize the resultant performance loss as a load balance ceiling.

9.4.6 Combining Ceilings

All these ceilings are independent and thus may be combined as needed. For example, a lack of instruction-level parallelism can be combined with a lack of data-level parallelism to severely depress performance. As such, one may draw multiple ceilings (representing the lack of different forms of parallelism) on a single roofline figure as visualized in Figure 9.7.

However, the question of how ceilings should be ordered arises. Often, one uses intuition to order the ceilings based on which are most likely to be implicit in the algorithm or discovered by the compiler. Ceilings placed near the roofline are those that are not present in the algorithm or unlikely to be discovered by the compiler. As such, based on this intuition, one could adopt any of the three equivalent roofline models in Figure 9.7.

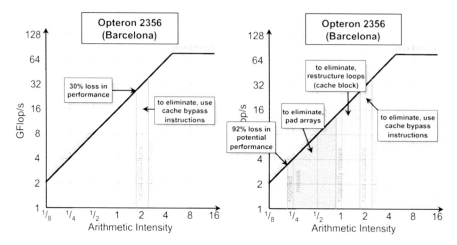

FIGURE 9.8: Performance interplay between Arithmetic Intensity and the Roofline for two different problem sizes for the same nondescript kernel.

9.5 Arithmetic Intensity Walls

Thus far, we've assumed the total DRAM bytes within the arithmetic intensity ratio is dominated by "compulsory" memory traffic in 3C's parlance [165]. Unfortunately, on real codes there are a number of other significant terms in the arithmetic intensity denominator.

$$\text{AI} = \frac{\text{Total FP Operations}}{\text{Compulsory} + \text{Allocation} + \text{Capacity} + \text{Conflict Memory Traffic} + ...} \quad (9.3)$$

In much the same way one denotes ceilings to express a lack of instruction, data, or memory parallelism, one can denote *arithmetic intensity walls* to denote reduced arithmetic intensity as a result of different types superfluous memory traffic above and beyond the compulsory memory traffic — essentially, additional terms in the denominator. As such, Equation 9.3 shows that write allocation traffic, capacity cache misses, and conflict cache misses, among others, contribute to reduced arithmetic intensity. These arithmetic intensity walls act to constrain arithmetic intensity and, when bandwidth-limited, constrain performance and is visualized in Figure 9.8. As capacity and conflict misses are heavily dependent on whether the specified problem size exceeds the cache's capacity and associativity, the walls are execution-dependent rather than simply architecture-dependent. That is, a small, non-power-of-two problem may not see any performance degradation due to capacity or conflict miss traffic, but for the same code, a large, near power-of-two problem size may result in substantial performance loss, as arithmetic intensity is constrained to be less than 0.2.

The following subsections discuss each term in the denominator and possible solutions to their impact on performance.

9.5.1   Compulsory Miss Traffic

It should be noted that compulsory traffic is not necessarily the minimum memory traffic for an algorithm. Rather compulsory traffic is only the minimum memory traffic required for a particular implementation. The most obvious example of elimination of compulsory traffic is changing data types. e.g., `double` to `single` or `int` to `short`. For memory-bound kernels, this transformation may improve performance by a factor of two, but should only be performed if one can guarantee correctness always or through the creation of special cases. More complex solutions involve in-place calculations or register blocking sparse matrix codes [346].

9.5.2   Capacity Miss Traffic

Both caches and local stores have a finite capacity. In the case of the former, when a kernel's working set exceeds the cache capacity, the cache hardware will detect that data must be swapped out, and capacity misses will occur. The result is an increase in DRAM memory traffic, and a reduced arithmetic intensity. When performance is limited by memory bandwidth, it will diminish by a commensurate amount. In the case of local stores, a program whose working size exceeds the local store size will not function correctly.

Interestingly, the most common solution to eliminating capacity misses on cache-based architectures is the same as to obtaining correct behavior on local store machines: cache blocking. In this case loops are restructured to reduce the working set size and maximize arithmetic intensity.

9.5.3   Write Allocation Traffic

Most caches today are *write-allocate*. That is, upon a write miss, the cache will first evict the selected line, then load the target line from main memory. The result is that writes generate twice the memory traffic as reads: cache line fill plus a write back vs. one fill. Unfortunately, this approach is often wasteful on scientific codes where large blocks of arrays are immediately written without being read. There is no benefit in having loaded the cache line when the next memory operations will obliterate the existing data. As such, the write fill was superfluous and should be denoted as a arithmetic intensity wall.

Modern architectures often provide a solution to this quandry either in the form of SSE's cache bypass instruction `movntpd` or PowerPC's block init instruction `dcbz`. The use of the `movntpd` instruction allows programs to bypass the cache in its entirety and write to the write combining buffers. The advantage: elimination of write allocation traffic and cache pressure is reduced. The dcbz instruction allocates a line in the cache and zeros its contents. The advantage is that write allocation traffic has been eliminated, but cache pressure has not been reduced.

9.5.4 Conflict Miss Traffic

Similarly, unlike local stores, caches are not fully associative. That is, depending on address, only certain locations in the cache maybe used to store the requested cache line — a *set*. When one exhausts this associativity of the set, one element from that set must be selected for eviction. The result: a conflict miss and superfluous memory traffic.

Conflict misses are particularly prevalent on power-of-two problem sizes as this is a multiple of the number of sets in a cache, but can be notoriously difficult to track down due to the complexities of certain memory access patterns. Nevertheless for many well structured codes, one may pad arrays or data structures cogniziant of the memory access pattern to ensure that different sets are accessed and conflict misses are avoided. Conceptually, 1D array padding transforms an array from Gird[Z][Y][X]) to Gird[Z][Y][X+pad]) regardless of whether the array was statically or dynamically allocated.

9.5.5 Minimum Memory Quanta

Naïvely, one could simple count the number of doubles a program references and estimate arithmetic intensity. However, one should be mindful that both cache- and local store-based architectures operate on some minimum memory quanta hereafter referred to as *cache lines*. Typically these lines are either 64 or 128 bytes. All loads and stores after being filtered by the cache are aggregated into these lines. When this data is not subsequently used in its entirety, superfluous memory traffic has been consumed without a performance benefit. As such another term is added to the denominator and arithmetic intensity is depressed.

9.5.6 Elimination of Superfluous Floating-Point Operations

Normally, when discussing arithmetic intensity walls, we think of adding terms to the denominator of arithmetic intensity. However, one should consider the possiblity that the specified number of floating-point operations may not be a minimum, but just a compulsory number set forth by a particular implementation. For instance, one might calculate the number of flops within a loop and scale by the number of loop iterations to calculate a kernel's flop count. However, the possibility of common subexpression elimination (CSE) exists when one or more loop iterations are inspected in conjunction. The result is that the flop count may be reduced. This has the seemingly paradoxical results of decreased floating-point performance, but improved application performance. The floating-point performance may decrease because arithmetic intensity was reduced while bandwidth-limited. However, because the total requisite work (as measured in floating-point operations) was reduced, the time to solution may have also been reduced.

Although this problem may seem academic, it has real world implications as a compiler may discover CSE optimizations the user didn't. When coupled with performance counter measured flop counts, the user may find himself in a predicament rectifying his performance estimations and calculations and the empirical performance observations.

9.6 Alternate Roofline Models

Thus far, we've only discussed a one-level processor-memory abstraction. However, there are certain computational kernel–architecture combinations for which increased optimization creates a new bandwidth bottleneck — cache bandwidth. One may construct separate roofline models for each level of the hierarchy and then determine the overall bottleneck. In this section we discuss this approach and analyze example codes.

9.6.1 Hierarchically Architectural Model

One may refine the original processor–memory architectural model by hierarchically refining the processors into cores and cache (which essentially look like another level of processors and memories). Thus, if the CPUs of Figure 9.1 were in fact dual-core processors, one could construct several different hierarchical models (Figure 9.9) depending on the cache/local store topology. Figure 9.9(a) shows it is possible for $core_0$ to read from $cache_3$ (simple cache coherency), but on the local store architecture, although any core can read from any DRAM location, $core_0$ can only read $LocalStore_0$.

Just as there were limits on both individual and aggregate processor-memory bandwidths, so too are there limits on both individual and aggregate core-cache bandwidths. As a result, what were NUMA ceilings (arising when data crossed low bandwidth/high load links) when transferring data from memory to processor, become NUCA (non-uniform cache access) ceilings when data resident in one or more caches must cross low bandwidth/high load links to particular cores

Ultimately, this approach may be used to refine cores down to the register file–functional unit level. However, when constructing a model to analyze a particular kernel, the user may have some intuition as to where the bottleneck lies — i.e., L2 cache–core bandwidth with good locality in the L2. In such a situations, there is no need to construct a model with coarser granularities (L3, DRAM, etc...) or finer granularities (register files).

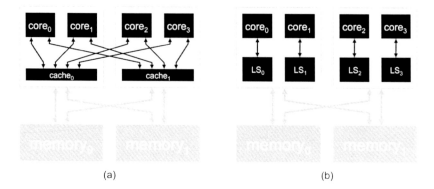

(a)   (b)

FIGURE 9.9: Refinement of the previous simple bandwidth-processor model to incorporate (a) caches or (b) local stores. Remember, arrows denote the ability to access information and not necessarily hardware connectivity.

9.6.2 Hierarchically Roofline Models

Given this memory hierarchy, we may model performance using two roofline models. First, we model the performance involved in transferring the data from DRAM to the caches or local stores. This of course means we must calculate an arithmetic intensity based on how data will be disseminated among the caches and the total number of floating-point operations. Using this arithmetic intensity and the characteristics of the processor–DRAM interconnect, we may bound attainable performance. Second, we calculate core-cache arithmetic intensity involved in transferring data to/from caches or local stores. We may also plot this using the roofline model. This bound may be a tighter or looser bound depending on architecture and kernel.

Such hierarchical models are especially useful when arithmetic intensity scales with cache capacity as it does for dense matrix-matrix multiplication. For such cases we must select a block size that is sufficiently large that the code will be limited by core performance rather than cache-core or DRAM-processor bandwidth.

9.7 Summary

The roofline model is a readily accessible performance model intended to provide performance intuition to computer scientists and computational scientists alike. Although the roofline proper is a rather loose upper bound to performance, it may be refined through the use of bandwidth ceilings, in-core ceilings, arithmetic intensity walls, and hierarchical memory architectures to provide much tighter performance bounds.

9.8 Acknowledgments

The author wishes to express his gratitude to Professor David A. Patterson and Andrew Waterman for their help in creation of this model. This work was supported by the ASCR Office in the DOE Office of Science under contract number DE-AC02-05CH11231, Microsoft (Award #024263), Intel (Award #024894), and by matching funding through U.C. Discovery (Award #DIG07-10227).

9.9 Glossary

3C's: is a methodology of categorizing cache misses into one of three types: compulsory, capacity, and conflict. Identification of miss type leads programmers to software solutions and architects to hardware solutions.

Arithmetic Intensity: is a measure of locality. It is calculated as the ratio of floating-point operations to DRAM traffic in bytes.

Bandwidth: is the average rate at which traffic may be communicated. As such it is measured as the ratio of total traffic to total time and today is measured in 10^9 bytes per second (GB/s).

Ceiling: a performance bound based on the lack of exploitation of an architectural paradigm.

Communication: is the movement of traffic from a particular memory to a particular computational element.

Computation: represents local FLOPs performed on the data transfered to the computational units.

DLP: data-level parallelism represents the number of independent data items for which the same operation can be concurrently applied. DLP can be recast as ILP.

FLOP: a floating-point operation including adds, subtracts, and multiplies, but often includes divides. It is generally not appropriate to include operations like square roots, logarithms, exponentials, or trigonometric functions as these are typically decomposed into the base floating-point operations.

ILP: instruction-level parallelism represents the number of independent instructions than can be executed concurrently.

Kernel: a deterministic computational loop nest that performs floating-point operations.

Performance: Conceptually similar to bandwidth, performance is a measure of the average rate computation is performed. As such it is calculated as the ratio of total computation to total time and today is measured in 10^9 floating-point operations per second (Gflop/s) on multicore SMPs.

Roofline: The ultimate performance bound based on peak bandwidth, peak performance, and arithmetic intensity.

TLP: instruction-level parallelism represents the number of independent threads (instruction streams) than can be executed concurrently.

Traffic: or communication is the volume of data that must be transfered to or from a computational element. It is measured in bytes. Often, we assume each computational element has some cache or internal storage capacity so that memory references are efficiently filtered to compulsory traffic.

Chapter 10

End-to-End Auto-Tuning with Active Harmony

Jeffrey K. Hollingsworth

University of Maryland, College Park

Ananta Tiwari

University of Maryland, College Park

10.1 Introduction

Today's complex and diverse architectural features require applying nontrivial optimization strategies on scientific codes to achieve high performance.

As a result, programmers usually have to spend significant time and energy in rewriting and tuning their codes. Furthermore, a code that performs well on one platform often faces bottlenecks on another; therefore, the tuning process must be largely repeated to move from one computing platform to another. Recently, there has been a growing interest in developing empirical *auto-tuning* software that help programmers manage this tedious process of tuning and porting their codes. Empirical auto-tuning software can be broadly grouped into three categories: (1) compiler-based auto-tuners that automatically generate and search a set of alternative implementations of a computation [90, 149, 388]; (2) application-level auto-tuners that automate empirical search across a set of parameter values proposed by the application programmer [93, 259]; and, (3) run-time auto-tuners that automate on-the-fly adaptation of application-level and architecture-specific parameters to react to the changing conditions of the system that executes the application [83, 345]. What is common across all these different categories of auto-tuners is the need to *search* a range of possible configurations to identify one that performs comparably to the best-performing solution. The resulting search space of alternative configurations can be very complex and prohibitively large. Therefore, a key challenge for auto-tuners, especially as we expand the scope of their capabilities, involves scalable search among alternative implementations.

While it is important to keep advancing the state-of-the-art in auto-tuning software from each of the above three categories, our experience demonstrates that full applications require a mix of these rather disjoint tuning approaches: compiler-generated code, application-level and run-time parameters exposed to auto-tuning environment. In other words, full applications demand and benefit from a cohesive environment that can seamlessly combine these different kinds of auto-tuning techniques and that employs a scalable search to manage the cost of the search process. We argue that it is important and sometimes critical to apply software tuning at all stages of application development and deployment — compile time, launch time and run-time.

To that end, in this chapter, we discuss the design and implementation of a unified end-to-end approach to auto-tuning scientific applications. We examine various sources of tunability in scientific applications. We review a set of strategies to address specific challenges that an empirical auto-tuning system faces. Finally, we present auto-tuning experience with our search-based framework, which we have named Active Harmony [168]. Active Harmony is a general-purpose infrastructure, and the auto-tuning results presented in this chapter highlight its applicability in tuning at all stages of application development and deployment.

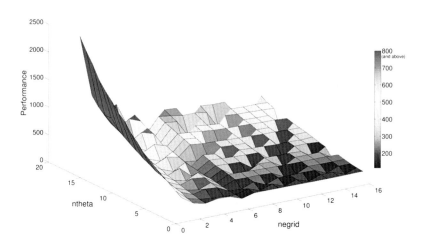

GS2 Performance as a function of two tunable parameters when the third parameter is fixed

FIGURE 10.1: GS2 performance plot with two tunable parameters. Figure from [322] [©2005 IEEE] Reprinted with Permission. (See color insert.)

10.2 Overview

To motivate the need for auto-tuning, we consider the problem of tuning a scientific simulation. GS2 [198], which is a physics application developed to study low frequency turbulence in magnetized plasma, exposes three input parameters that users can select at program launch-time. Needless to say, the task of making a good selection of the input parameters is non-trivial because this requires a concrete understanding of the interactions between the input parameters and the underlying algorithmic behaviors that they are meant to control. Figure 10.1 underscores this fact. The figure shows the performance (application run-time) of GS2, as a function of two input parameters - ntheta (number of grid points per 2π segment of field line) and negrid (energy grid). All other input parameters are fixed. Clearly, the parameter space is not smooth and contains multiple local minimums. The best and worst points are a factor of 10 different.

This example shows the need for auto-tuning systems to adapt code. The relatively large plateau of good performance is also typical of many applications we have studied. Such a topography argues for the need to have auto-tuning systems that rapidly get applications out of "bad" regions of performance and into the "good" ones. Achieving optimal performance is a secondary goal once "good" performance is reached.

A variety of tunable parameters can be exported to an auto-tuning system. These parameters can be either numeric or non-numeric (unordered lists) Numeric parameters can be defined with range and step sizes. Loop blocking and unrolling factors, input data size, buffer size used in compression algorithms, number of Openmp threads, Openmp chunk-size and MPI message size are among the most commonly seen numeric parameters in our past auto-tuning efforts. Some examples of non-numeric parameters include loop permutation and loop fusion orders for deeply nested loop-nests optimization, choice between several iterative solvers in applications such as smg2000 [63], POP [127], etc. and choices among data distribution algorithms.

10.2.1 Auto-Tuning Modes

Auto-tuning can be performed in either offline or online mode. Offline mode refers to tuning between successive full application runs. The performance data collected in this fashion can be used for training based application tuning as well. This mode is used for parameters that are read once when the program starts and must remain fixed throughout the execution of the application. For example, a parameter which governs data distribution can be tuned to minimize communication costs.

Online mode refers to run-time tuning during production runs. This mode is used when application parameters can be adapted during run-time for improved performance. For example, in a 3D Jacobi algorithm, blocking factors for triply nested stencil loops can be changed across different iterations to get better cache reuse. Online tuning has the privilege of exploiting fine-grained, up-to-date and accurate performance data that can be directly linked back to specific code sections, characteristics of input datasets, architecture specific features, changing conditions of the system, just to name a few. This information is generally not available at compile time and even when some information is available, most compilers choose not to use the information. Compilers are designed to be generic and as such, they base their optimization decisions on simple and conservative analytical models. With online tuning, more aggressive tuning strategies can be explored. Furthermore, some tuning decisions made by the compiler can be undone at run-time should they interact with more profitable optimizations negatively.

10.2.2 Need to Coordinate Auto-Tuners

Since auto-tuning can take place at many levels of a program from the user to specific libraries, an important question is how to coordinate this process. Left uncoordinated, each component of a program may try to run its own auto-tuner. Such a process would likely lead to a dissonance where multiple components change something nearly simultaneously and then try to assess if it improved the program's performance. However, without coordination it would be impossible to tell which change was actually improving the program. In fact, it is likely that one change might improve performance and the second hurt performance and the net performance benefit is little or none. Thus an important question is how to best coordinate the efforts of auto-tuners. There are a variety of approaches possible ranging from simple arbitration to ensure that only one auto-tuner is running at once to a fully unified system that allows coordinated simultaneous search of parameters originating from different auto-tuners.

In the Active Harmony project, we take the approach of a coordinated system to allow the many varied sources of auto-tuning to work together. We accomplish this by having each component of a program (library, application code, auto-tuning compiler) expose its tunable parameters via a simple, but expressive constraint language. A central search engine then uses these constraints to manage the evaluation of possible auto-tuning steps. The core search algorithm is able to decide which sources of auto-tuning information should be considered and when.

10.3 Sources of Tunable Data

Auto-tuning does not randomly create things to change in a program. Rather it provides a way optimize a program's performance based on trying out a set of possible changes. This set of possible changes comes from many sources including options that application programmers discover, library parameters and algorithmic choices that depend on library use, and from possible code transformations identified by compilers.

In addition to considering what can change, another important source of outside guidance for auto-tuning is insight in how changes might impact performance. Information from performance models that are parametrized by the tunable components can help to guide an auto-tuning system to the correct values to use. Even simple non-parametric models can help to indicate how close an auto-tuning system is to reaching the best achievable performance. Likewise, experimental data from prior runs of programs can help to guide an auto-tuning system to search-space regions that were previously successful.

Next, we briefly review these different sources of tunability and how they are similar and different.

10.3.1 Application Programmers

Application and library developers often expose a set of input parameters to allow the end-users to adjust application execution behavior to the characteristics of the execution environment. The selection of appropriate parameter values is, thus, crucial in ensuring good performance. However, the task of making a good selection of the input parameters is non-trivial because this requires a good understanding of the interactions between the input parameters, the underlying algorithmic behaviors that they are meant to control, and the specifics of target architecture. Application-level auto-tuners use offline tuning techniques to identify a set of input parameters that delivers a reasonable performance.

Nelson et al. [259] focus on empirical techniques to find the best integer value of application-level input parameters that the programmer has identified as being critical to performance. Programmers provide high-level models of the impact of parameter values, which are then used by the tuning system to guide and to prune the search for optimal input parameters. Hunold et al. [176] consider input parameter tuning for the matrix-matrix multiplication routine PDGEMM of ScaLAPACK [92]. PDGEMM is a part of PBLAS, which is the parallel implementation of BLAS(Basic Linear Algebra Subprograms) for distributed memory systems. Input parameters selected for auto-tuning include the dimension of input matrices, number of processors, logical processor grid and three blocking factors (one for each dimension).

Chung et al. [94] use short benchmarking runs to reduce the cost of full application execution incurred in application-level input parameter tuning. Short benchmarking runs exercise all aspects of the application (or at least all aspects that are influenced by the choice of input parameters). The authors use their search algorithm (modified Nelder-Mead search) to navigate input parameter space of large-scale scientific applications.

10.3.2   Compilers

Optimization decisions made by compilers tend to be general-purpose and conservative. Therefore, to give programmers and compiler-experts more freedom to make suitable optimization decisions for complex codes, a variety of compiler-based auto-tuning frameworks have been proposed (see Chapter 11). Such frameworks facilitate the exploration of large optimization space of possible compiler transformations and their parameter values.

ADAPT [345] is a compiler-supported infrastructure for high-level adaptive program optimization. It allows developers to leverage existing compilers and optimization tools by describing a run-time heuristic for applying code optimization techniques in a domain specific language, ADAPT Language. Chapter 11 describes several loop transformation tools that are based on polyhedral representation of loops. Most such tools provide a well-defined interface to describe the transformations and search space to the auto-tuner.

There are many research projects working on compiler-based empirical optimization of linear algebra kernels and domain specific libraries. Atlas [359] uses the technique to generate highly optimized BLAS routines. The Oski (Optimized Sparse Kernel Interface) [347] library provides automatically tuned computational kernels for sparse matrices. FFTW [139] and SPIRAL [386] are domain specific libraries. FFTW combines the static models with empirical search to optimize FFTs. SPIRAL generates empirically tuned Digital Signal Processing (DSP) libraries.

10.3.3 Prior Runs

Data from prior production runs also serve as another viable source for tunable items. Frameworks such as PerfExplorer [172] and Prophesy [330] can be used to mine large sets of historical performance data. The frameworks provide a variety of analysis techniques (clustering, summarization, curve fitting, etc.) that characterize application behavior and performance. This information, which helps to better understand the search space, can then be fed to auto-tuners. In addition, the historical information helps to reduce tuning time by skipping the execution of parameter configurations that have already been evaluated in the past. However, the utilization of past performance data has to carefully consider the contextual circumstances under which the data was collected. Such contextual data consist of (but not limited to) characteristics of input data and platform, version information of source code, compiler, component libraries and operating system, and finally, performance data. Ideally context is the same for both the current tuning environment and the historical data. In practice, there is almost always some difference. For mismatching contexts, interpolation techniques (described elsewhere [93, 172, 330]) can be used, where applicable, to approximate performance data.

Auto-tuners can pull historical data from performance database systems [154, 174] that support storage and retrieval of both performance data and accompanying context information. One such database system is PERI-DB [154], which thrives on accumulating performance data from a variety of performance profiling tools and on assuring interoperability between the different performance database systems.

Another source of data for tuning is to use training runs to auto-tune the software based on a well defined benchmark or typical workload. Training runs are application executions designed mainly to produce performance data that feed into the auto-tuning process. A single auto-tuning run can evaluate several configurations. More often than not, several auto-tuning runs are performed with slightly different settings (e.g., different input sizes) before the application is run under production mode. Performance data collected during such auto-tuning runs tend to have similar contextual information. Therefore, if a log of such evaluations (along with some context information) is maintained locally, tuning time can be reduced by consulting the log during consecutive auto-tuning runs.

10.3.4 Performance Models

The application performance modeling community has used various techniques [180, 187, 209, 306, 311] to study and model application behavior and predict performance. Ipek et al. [180] uses machine learning techniques to build performance models. The idea is to build the model once and query it in the future to derive performance prediction for different input configurations. The model is trained on a dataset which consists of points spread regularly across the complete input parameter space. After the training phase, they query their model at run-time for points in the full parameter space. Others [306] and [209] have also used similar machine learning techniques to build performance models. The limiting factor of the listed modeling techniques is the size of the training set. Ipek et al. report using training set of size $3K$ (for SMG2000), which amounts to $3K$ unique executions of the application just to train the model. Kerbyson et al. [187] present predictive analytical model that accurately reflects the behavior of SAGE, a multidimensional hydrodynamics code with adaptive mesh refinement. Inputs to the model include machine-specific information such as latency and bandwidth and application-specific information such as data-decomposition, problem size, etc.

Information from modeling can be fed into auto-tuning tools to determine if there are opportunities for application optimization and to gain insights as to what factors affect the application performance on a given target architecture. The Roofline model (see Chapter 9), for example, integrates architecture specific elements such as in-core performance, memory bandwidth, and locality to provide realistic expectations of performance and potential benefits and priorities for different optimizations. This information can be used by auto-tuning systems to prune the search-space, to constrain and guide the search to the interesting parts of the search space and to set stopping criteria for auto-tuning runs.

10.4   Search

A key to any auto-tuning system is how it goes about selecting the specific combinations of choices to try. We refer to this process as the search algorithm. We briefly review the unique aspects of search in an auto-tuning system. The first relates to the size of the search space. While a simple parameter space might be exhaustively searched, most systems contain too many combinations to try them all. Instead, an auto-tuning system must rely on search heuristics to evaluate only a sub-set of the possible configurations while trying to find an optimal one (or at least as nearly optimal as practical). Second, unlike traditional minimization (or maximization) algorithms, which expect well-structured and continuous search spaces, auto-tuning systems deal with constrained and often discrete parameter spaces. These spaces tend to be quite complex with multiple local minimums and maximums. As such, appropriate methods to constrain the search to allowable regions of the search space must be adopted. The third unique aspect relates to the inherent variability in performance measurements in today's architectures. In other words, the measured performance for a fixed set of parameters is not always the same. The search algorithm should, thus, be resilient and converge to good solutions even in the presence of performance variability (i.e., noise in the measured run time).

Finally, the fourth distinct factor is related to the selection of an appropriate performance metric. The final result and the convergence time (in terms of the number of evaluations of the objective function) are two common performance metrics for search algorithms. However, for auto-tuning systems, these are not the most appropriate ones. For auto-tuning systems, the overall performance of the system from the start to the end are equally important. This is particularly critical for online optimization because if the transient behavior is poor, the overall tuning time and hence the running time of the application will suffer. Therefore, the appropriate performance metric should capture the performance of the intermediate points visited in the path to the final solution.

10.4.1   Specification of Tunable Parameters

An important aspect of searching parameters is to allow the precise specification of valid parameters. Such specification could be as simple as expressing the minimum, maximum and initial values for a parameter. Sometimes not all parameters values within a range should be searched, so option to specify a step function is useful. Likewise for parameters with a large range (i.e., a buffer that could be from 1K to 100Megabytes), it is useful to specify that parameter should be searched based on the log of the value. Another critical factor is that not all parameters are independent. Frequently, there is a relationship between parameters (i.e., when considering tiling a two dimensional array, it is often useful to have the tiles be rectangles with a definite aspect ratio). To meet the needs of having a fully expressive way for developers to define parameters, we have developed a simple and extensible language (Constraint Specification Language, CSL) that standardizes parameter space representation for search-based auto-tuners. CSL allows tool developers to share information and search strategies with each other. Meanwhile, application developers can use CSL to export their tuning needs to auto-tuning tools.

CSL provides constructs to define tunable parameters and to express relationships (ordering, dependencies, constraints, and ranking) between those parameters. *Search-hints* construct can be used to provide hints to the underlying search engines. Examples of search-hints include specification for initial points to start the search, default values for parameters, soft constraints on MPI message-sizes and so on. *Optimization-plan* construct allows application developers to exert greater control over what optimizations can be applied and in what order to apply them (e.g., determine tiling parameters before unrolling). This construct can also be used to specify when to apply the optimizations; compile-time, launch-time or run-time. Finally, *parameter-group* construct can be used to assemble related parameters into groups to apply tuning at per-group granularity. This is particularly helpful when there are multiple code-sections that benefit from the same optimization.

We provide a simple example of CSL usage for matrix multiplication optimization in Table 10.1. This specification defines tiling and unrolling parameters.[1] It also defines constraint relations between the parameters to prune uninteresting points from the search space.

TABLE 10.1: Search Specification for MM Tuning (tiling + unrolling)

```
 1: search space tiling_mm {              30:    parameter unroll int {
 2:    # Now defining the search space   31:       range [1:16:1];
 3:    #  specifics for tiling and unrolling 32:    default 1;
 4:    #  define some constants          33:       region loop;
 5:    constants {                       34:    }
 6:       int l1_cache=128;              35:    # L1 cache (for array B)
 7:       int l2_cache=4096;            36:    constraint mm_l1 {
 8:       int register_file_size=16;    37:       loopK.tile*loopJ.tile
 9:    }                                 38:          <= (l1_cache*1024)/4;
10:    # code region declarations:      39:    }
11:    #    loopI,loopJ,loopK            40:    # L2 cache (for array A)
12:    code_region loopI;               41:    constraint mm_l2 {
13:    code_region loopJ;               42:       loopK.tile*loopI.tile
14:    code_region loopK;               43:          <= (l2_cache*1024)/4;
15:    # region set declaration         44:    }
16:    region_set loop [loopI, loopJ, loopK]; 45:    # unroll constraint
17:    # declare tile_size parameter and 46:    constraint mm_unroll {
18:    # associate the parameter to region 47:       (loopI.unroll*loopJ.unroll*
19:    # set loop. default value of the 48:       loopK.unroll)
20:    # parameter is set to 32.        49:          <= register_file_size;
21:    parameter tile int {             50:    }
22:       range [0:512:2];             51:    # putting everything together
23:       default 32;                   52:    specification {
24:       region loop;                  53:       mm_l1 && mm_l2
25:    }                                 54:          && mm_unroll;
26:    # declare unroll_factor parameter and 55:    }
27:    # associate the parameter to region 56: }
28:    # set loop. default unroll factor is
29:    # set to 1.
```

10.4.2   Previous Search Algorithms

In this section, we briefly review some search algorithms that have been used in various auto-tuning frameworks. While this is not a complete set of search algorithms used in the auto-tuning realm, we describe the most widely used algorithms. Atlas [359] uses orthogonal line search, which optimizes each tunable parameter independently by keeping the rest fixed to their reference values. The parameters are tuned in a pre-determined order and each successive parameter tuning uses the optimized values for parameters that precede it in the ordering. The disadvantage of using this search in a general-purpose framework is that it requires a pre-determined ordering for parameters. Atlas

---

[1]These are the same constraints that we use for MM optimization results in Section 10.5.2.

exploits years of experience in dense linear algebra tuning to determine appropriate ordering for parameters. However, such knowledge is not available for general-purpose tuning cases.

Several auto-tuners [102, 190, 202] have used genetic algorithms (GA). GA algorithm starts by randomly generating an initial population of possible configurations. Each configuration is represented as a genome and the "fitness" of the configuration is the performance metric. Based on the fitness, each successive iteration of the algorithm produces new set of configurations by using genetic operations - mutation, crossover and selection. While GA has shown its promise by converging to good configurations, the key disadvantage lies in its long convergence time. Furthermore, the transient behavior of GA is unpredictable and jittery, which makes the algorithm unsuitable for multiple auto-tuning use cases (e.g., online tuning).

Direct search methods are also popular among auto-tuners. These methods do not explicitly use function derivatives. The parameter tuning problem is a very good use-case for direct search methods, since in most cases the performance function at a given point in the search space have to be evaluated by actually running the program. The Nelder-Mead Simplex algorithm [258], is one of the most widely used direct search methods in auto-tuning systems [17, 103, 163]. For a function of N variables (tunable parameters), a set of $N + 1$ points forming the vertices of a simplex in N-dimensional space is maintained. In a two-dimensional space, the simplex is a triangle and in a three-dimensional space, the simplex is a non-degenerate tetrahedron. At each iteration, the simplex is moved towards the minimum by considering the worst point in the simplex and forming its symmetrical image through the center of the opposite (hyper) face. Thus, at each step, the worst performing point in the simplex is replaced by a better point. While the Nelder-Mead algorithm sometimes finds solutions efficiently, many studies [197, 205, 232] have described unpredictable behavior of the algorithm as the number of parameters (search space dimension) increases. Looptool [281], which is a compiler-based auto-tuning framework uses pattern-based direct search method proposed by Hookes and Jeeves [170]. The pattern-based search method have been observed to be very reliable method, however, in some cases, the convergence time of the algorithm is slow [197].

In Active Harmony, we use the parallel rank ordering (PRO) Algorithm. The algorithm belongs to a class of direct search algorithms known as the generating set search (GSS) methods. GSS methods can effectively handle high-dimensional constrained spaces and have the necessary conditions for convergence. In the next section, we provide a high-level description of the algorithm.

10.4.3   Parallel Rank Ordering

For a function of N variables, PRO maintains a set of kN points forming the vertices of a simplex in an N-dimensional space. Each simplex transformation step of the algorithm generates up to $kN - 1$ new vertices by reflecting, expanding, or shrinking the simplex around the best vertex. After each transformation step, the objective function value associated with each of the newly generated points are calculated in parallel. The reflection step is considered successful if at least one of the $kN - 1$ new points has a better f than the best point in the simplex. If the reflection step is not successful, the simplex is shrunk around the best point. A successful reflection step is followed by expansion check step. If the expansion check step is successful, the expanded simplex is accepted. Otherwise, the reflected simplex is accepted and the search moves on to the next iteration. The search stops if the simplex converges to a point in the search space (or after a pre-defined number of search steps). A graphical illustration for reflection, expansion and shrink steps are shown in Figure 10.2 for a 2-dimensional search space and a 4-point simplex.

Note that before computing all expansion points, we check the outcome of expansion for the most promising case[2] first. This seems to be counter-intuitive at first glance, since we are not taking full advantage of the parallelism. However, in our experiments, we realized there are some expansion points with very poor performance that can slow down the algorithm. Therefore, to avoid these time consuming instances and to ensure good transient behavior, we calculate the expansion point performance for the most promising case first and only if it is successful, perform the expansion for other points.

One of the unique features that distinguishes PRO from other algorithms used in auto-tuning systems is that the algorithm leverages parallel architectures to search across a set of optimization parameter values. Multiple, sometimes unrelated, points in the search space are evaluated at each timestep. For online tuning, this parallel search translates to tuning multiple sections of the code at the same time, thereby allowing the auto-tuner to explore multiple parameter interactions at each iteration. For offline tuning, it translates to tuning a parallel application in parallel by having different nodes of a cluster use different parameters and making tuning modifications between runs.

10.4.4 Parameter Space Reduction

In this section, we discuss strategies to reduce the optimization space and focus the search on only the interesting regions of the tuning space. Such strategies play a vital role in making empirical optimization a reasonable approach to auto-tuning. In our earlier work [337], we observed that parameter interaction models can be used to eliminate a large number of mediocre configurations from the search space. Furthermore, the use of programmer identified constraints on input parameters also helped reduce the search space. Other auto-tuners [87, 90, 281, 359] working on model-driven empirical optimization have also looked into an array of interesting avenues to rule out inferior code-

---

[2]Most promising point is the point in the original simplex whose reflection around the best point returns a better function value.

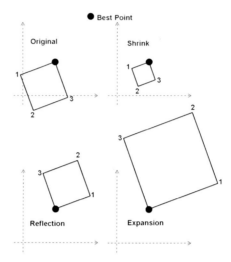

FIGURE 10.2: Original 4 point simplex in a 2-dimensional space. Figure from [337] [©2009 IEEE] Reprinted with permission.

variants. In general, information from a variety of complementary sources can be used to reduce the optimization space.

Target architecture parameters such as cache sizes, TLB size, register file size, etc. can be used intelligently to reduce the number of points in the search space. For example, the tile size can be constrained by half of the available cache. In addition, run-time tuning can take advantage of the input dataset knowledge to further reduce the search space. An example of this would be to opt out from data-copy optimization for matrix multiplication if the input matrices fit in the largest available cache.

Parameter sensitivity studies [93] for numeric parameters (defined by minimum, maximum and step size) can help understand the relative impact of parameters on performance. For each parameter, the sensitivity test involves testing all possible values the parameter can take while keeping the rest of the parameters fixed at their default values. The sensitivity is defined as the ratio of the difference between the worse and the best performance values observed during the study and the normalized distance between the values the parameter can take. Parameters with low sensitivity values can either be discarded or used later in the tuning process.

10.4.5 Constraining to Allowable Region

Parameter tuning is a constrained optimization problem. Therefore after each search iteration, we have to make sure that the points returned by the search algorithm for evaluation are admissible, i.e., they satisfy the constraints. We refer to the process of mapping points that are not admissible to nearby admissible points as "projection". We consider two types of parameter constraints: internal discontinuity constraints and boundary constraints. Internal discontinuity constraints arise because some tuning parameters can only have discrete admissible values (e.g., integer parameters). The projection operation makes sure that the computed parameters are rounded to an admissible discrete value.

We describe two methods to handle boundary constraints. The first is a distance-based method used in the Active Harmony system. We define distance between two points using the L_1 distance metric, which is the sum of the absolute differences of their coordinates. The nearest neighbor of an inadmissible point (calculated in terms of L_1) will thus be a legal point with the least amount of *change* (in terms of parameter values) summed over all dimensions.

The second method consists of adding a penalty factor to the performance metric associated with the points that violate the constraints. This approach has been used previously in the context of constrained optimization using genetic algorithms [184]. The success of this strategy will depend on how well the penalty factor is chosen. Among other strategies, the factor can be chosen based on the number of constraints a given illegal point violates. Similarly, the distance between the illegal point and the nearest legal point can also be used to determine the penalty factor.

10.4.6 Performance Variability

Besides the tunable parameters, there are many other factors affecting a program's performance. Therefore, even for a fixed set of tunable parameters, the application performance varies in time. Other applications running on the same processor, network contention, operating system jitter, and memory architecture complexity are common sources of performance variability. Attempting to minimize or control the variability is not the topic of discussion here. What auto-tuning systems need are strategies to make search algorithms as resilient as possible to such performance variabilities. The conventional method for dealing with performance variability is to use multiple samples to estimate the function value at each desired point. The widely used operator for aggregating performance values from multiple samples is an average operator.

In our earlier work [322], we studied the nature of performance variability in real systems. One of the most important observations we made was that the performance variability is heavy-tailed. Heavy-tailed performance (execution time) distributions consist of many small spikes with random occurrences of few large spikes, most likely because of external factors. What this means from the perspective of auto-tuning is that there is a non-negligible probability of observing large variations in the measured performance. The sampled performance measurements could, therefore, have infinite variance. This observation essentially renders average operator ineffective. As an alternative, we showed taking a minimum of multiple performance measurements is an effective way for performance estimation even in the presence of heavy-tail component in the performance distribution. The minimum has finite mean and variance and is not heavy-tailed.

10.5 Auto-Tuning Experience with Active Harmony

We now describe how we have used our auto-tuner, Active Harmony, to perform end-to-end tuning for scientific applications. Active Harmony takes a search-based collaborative approach to auto-tuning. Application programmers and end-users collaborate to describe and export a set of performance related tunable parameters to the auto-tuning system. These parameters define a tuning search-space. The auto-tuner monitors the program performance and suggests application adaptation decisions. The decisions are made by a central controller using the parallel rank ordering algorithm. For adaptation decisions that require new code (for example, different unroll factors), Active Harmony generates code on-the-fly. By merging the traditional feedback-directed optimization and just-in-time compilation in this fashion, the auto-tuning system can explore domain- and architecture-specific optimizations.

10.5.1 Application Parameters — GS2

Past Active Harmony tuning efforts[3] with GS2 highlight the importance of engaging directly with the application developer. GS2 [198] is a physics application developed to study low frequency turbulence in magnetized plasma. The tuning was done in two stages. The first stage looked at "data layout" input parameter. The data layout in GS2 is specified with five variables: x,y,l,e,s (where x and y are spatial coordinates; l and e are velocity coordinates and s is the particle species). The default data layout used by the application was lxyes, which indicates the order of the dimensions of the primary 5-dimensional array in the simulation. Active Harmony tuning suggested unique data layouts for two different architectures and improved the application performance by up to 3.4x. In subsequent releases of GS2, the developer changed the default layouts for the two architectures to the ones suggested by Active Harmony.

The second stage looked at three tunable parameters: ntheta (number of grid points per 2π segment of field line), negrid (energy grid), and nodes (number of nodes). The selection of these tunable parameters were made by the application developer. It should be noted that changing ntheta and negrid may affect simulation resolution. However, the search was conducted using

---

[3]We only summarize the results here. For full results, please consult an earlier paper [94] on this topic.

developer suggested parameter value ranges that generate acceptable resolutions. After tuning, execution time for a production run of GS2 improved by 37%.

The observation that there exists a trade-off between the accuracy and performance improvement is important. Scientific applications have some configurations that help improve the performance dramatically with some compromise on the accuracy of the output. The decision on how to find a balance between these two crucial factors rests with the experts. However, auto-tuning can help them quantify and make such choices.

10.5.2 Compiler Transformations — Computational Kernels

In Section 10.3.2, we noted that compiler based auto-tuning frameworks have encouraged expert participation in deciding optimization strategies for complex codes. In this section, we show how Active Harmony can use the knowledge provided by experts to successfully generate highly optimized matrix multiplication kernels.

Complex architecture features and deep memory hierarchies that characterize the state-of-art HPC platforms require applying nontrivial optimization strategies on loop nests to achieve high performance. Compounding these challenges is the fact that different loop optimizations usually have different goals, and when combined they might have unexpected (and sometimes undesirable) effects on each other. Even optimizations with similar goals but targeting different resources, such as unroll-and-jam plus scalar replacement (targeting data reuse in registers), and loop tiling plus data copy (for reuse in caches), must be carefully combined [337]. Figure 10.3 illustrates these complex interactions by showing the performance of square matrix multiplication as a function of tiling and unrolling factors. We see a corridor of best performing combinations along the $x - y$ diagonal where tiling and unrolling factors are equal.

We use Active Harmony's offline tuning mechanism to compensate for the lack of precise analytical models and to automatically handle the latent and complex interactions between the optimization strategies. The tuning involves performing a systematic search over a collection of automatically generated code-variants.

The use of a matrix multiplication (MM) kernel for the experiments presented in this section was motivated by two goals. First, the optimization of the MM kernel has been extensively studied in the past and as such, we can easily compare the effectiveness of our approach to that of well-tuned MM libraries (e.g., Atlas). And second, the MM kernel exhibits the complex parameter interactions that were discussed earlier. Therefore, the results obtained for MM can be extrapolated to generic loop-nests beyond the realm of linear algebra. CHiLL [87], a polyhedra-based framework, is used to generate code-variants. CHiLL provides a high-level script interface that allows compilers and application programmers to use a common interface to describe parametrized code

Parameter Interaction (Tiling and Unrolling for MM, N=800)

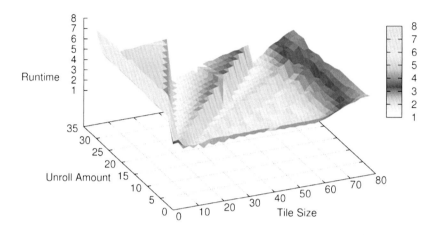

FIGURE 10.3: Parameter Search Space for Tiling and Unrolling. Figure from [337] [©2005 IEEE] Reprinted with permission.
(See color insert.)

TABLE 10.2: Matrix Multiplication Optimization. Table from [337] [©2009 IEEE] Reprinted with permission.

naive	*CHiLL − Recipe*	*Constraints*

`DO K = 1, N`
` DO J = 1, N`
`  DO I = 1, N`
`   C[I,J] = C[I,J]+A[I,K]*B[K,J]`
	`permute([3,1,2])`
	`tile(0,2,TJ)`
	`tile(0,2,TI)`
	`tile(0,5,TK)`
	`datacopy(0,3,2,1)`
	`datacopy(0,4,3)`
	`unroll(0,4,UI)`
	`unroll(0,5,UJ)`
		$TK \times TI \leq \frac{1}{2}\left(\frac{size_{L2}}{2}\right)$
		$TK \times TJ \leq \frac{1}{2}\left(\frac{size_{L1}}{2}\right)$
		$UI \times UJ \leq size_R$
		
		TI, TJ, TK
		$\in [0, 2, 4, \ldots, 512]$
		$UI, UJ \in [1, 2, \ldots, 16]$

transformations to be applied to a computation. The optimization strategy for MM reflected in the CHiLL script in Table 10.4 exploits the reuse of C(I,J) in registers, and the reuse of A(I,K) and B(K,J) in caches. Data copying is applied to avoid conflict misses. The values for the five unbound parameters TI, TJ, TK, UI, and UJ are determined by Active Harmony, which uses the PRO algorithm to navigate this five-dimensional search space.

The experiments were performed on a 64-node Linux cluster (henceforth referred to as umd-cluster). Each node is equipped with dual Intel Xeon 2.66 GHz (SSE2) processors. L1- and L2-cache sizes are 128 KB and 4096 KB re-

TABLE 10.3: MM Results - Alternate Simplex Sizes. Table from [337] [©2009 IEEE] Reprinted with permission.

	$2N$	$4N$	$8N$	$12N$
Number of Function Evals.	276	571	750	961
Number of Search Steps	49	32	22	18
Speedup over Native	2.30	2.33	2.32	2.33

spectively. To study the effect of simplex size, we considered four alternative simplex sizes - 2N, 4N, 8N, and 12N, where N(=5) is the number of unbound parameters. Figure 10.4 shows the performance of the best point in the simplex across search steps. Tuning runs conducted with 12N and 8N simplices clearly use fewer search steps than the search conducted with smaller simplices. Recall from Figure 10.3 that the loop transformation parameter space is not smooth and contains multiple local minima and maxima. The existence of long stretches of consecutive search steps with minimal or no performance improvement (marked by arrows in Figure 10.4) in the 2N and 4N cases show that more search steps are required to get out of local minimas for smaller simplices. At the same time, by effectively harnessing the underlying parallelism, 8N and 12N simplices evaluate more unique parameter configurations (see Table 10.3) and get out of local minimas at a faster rate. The results summarized in Table 10.3 also show that as the simplex size increases, the number of search steps decreases, thereby confirming the effectiveness of the increased parallelism.

The next question regarding the effectiveness of Active Harmony relates to the quality of the search result. Figure 10.5 shows the performance of the code variant produced by a 12N simplex across a range of problem sizes along with the performance of native `ifort` compiler, Atlas' search-only and full version. In addition to a near exhaustive sampling of the search space, Atlas uses carefully hand-tuned BLAS routines contributed by expert programmers. Therefore, to make a meaningful comparison, we provide the performance of the search-only version of Atlas — code generated by the Atlas Code Generator via pure empirical search. Our code version performs, on average, 2.36 times faster than the native compiler. The performance is 1.66 times faster than the search-only version of Atlas. Our code variant also performs within 20% of Atlas' full version (with processor-specific hand coded assembly).

10.5.3 Full Application — SMG2000

Full application tuning is a part of a broader effort to integrate various compilers and auto-tuning tools as part of the PERI (Performance Engineering Research Institute) project [16]. The ultimate aim of the effort is to develop a fully automated tuning tool-chain for whole applications. More details on this integration effort can also be found elsewhere [215]. The tuning process starts by first using application profiling tools such as HPCToolkit (see Chapter 4) to

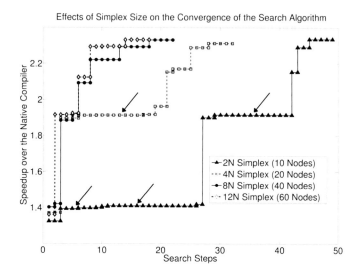

FIGURE 10.4: Effects of Different Degree of Parallelism on PRO. Figure from [337] [©2009 IEEE] Reprinted with permission.

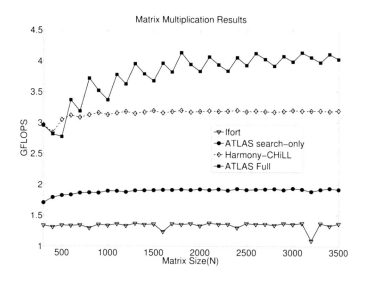

FIGURE 10.5: Results for MM Kernel. Figure from [337] [©2009 IEEE] Reprinted with permission.

identify hotspots. The hotspots are then outlined to a separate function using the ROSE compiler [15]. The actual code transformations are selected by the auto-tuning system from a recipe library. The library is created by compiler experts based on their experience working with real codes. Unlike traditional compiler optimization which must be coded into the compiler, these recipes can evolve over time.

The final stage involves using search-based auto-tuners to perform the empirical tuning. At each auto-tuning step, the auto-tuner requests code generation tools to generate code variants with a given set of code transformation parameters. The new code is compiled into a shared library. The application dynamically loads the transformed kernel by using the dlopen/dlsym mechanism. The measured performance is fed back to the auto-tuner. The use of the shared library mechanism provides advantages for both offline and online tuning modes. For offline mode, only the outlined and transformed code has to be recompiled between successive tuning runs. Furthermore, the tuning time can be kept under control by using the checkpoint restart mechanism. For online mode, the use of the shared library mechanism allows the auto-tuner to adjust application execution during a single production run. Applications can swap out poorly implemented hotspots with better performing and automatically generated code-variants.

To demonstrate the potential of this approach, we consider the SMG2000 [63] benchmark. SMG2000 is a parallel semi-coarsening multigrid solver for the linear systems arising from finite difference, finite volume, or finite element discretizations of the diffusion equation $\nabla \cdot (D \nabla u) + \sigma u = f$ on logically rectangular grids. The code solves both 2D and 3D problems with discretization stencils of up to nine-point in 2D and up to 27-point in 3D.

We use HPCToolkit to identify the most time-consuming kernel. The ROSE outliner outlines the kernel to a separate and independently compilable C source file with all dependent structures and typedef declarations preserved. Key portions of the outlined kernel is shown in Table 10.4. The kernel consists of sparse matrix vector multiply expressed in four-deep loop nest.[4] For code-generation, we used the CHiLL loop transformation tool. The optimization strategy (CHiLL-recipe) and constraints on transformation parameters are also provided in Table 10.4. The search-space is six-dimensional and includes a parameter that controls the use of gcc versus icc. For the results presented here, we use a short representative run of SMG2000. This is done to reduce the auto-tuning time. Except for the initialization/setup phase, we disable all other phases of the application and measure the time spent in the outlined kernel.

The search is conducted in an offline mode using the parallel rank ordering algorithm. New code is generated once per search-node per search-step. Transformation parameters are adjusted between successive runs of SMG2000 on umd-cluster (see Section 10.5.2). The search uses a 24-point simplex, which means up to 23 new parameter configurations are evaluated in parallel at each search-step. After the search algorithm converges, a recursive local search around the best point is performed.

---

[4]The outlined kernel shown is a simplified version. Actual code is much less clean.

TABLE 10.4: SMG2000 optimization details.

Original	CHiLL − Recipe	Constraints

`for (si = 0; si < stencil_size; si++)`
` for (kk = 0; kk < mz; kk++)`
`  for (jj = 0; jj < my; jj++)`
`   for (ii = 0; ii < mx; ii++)`
`    rp[((ri+ii)+(jj*sy3))+(kk*sz3)] -=`
`    ((Ap_0[((ii+(jj*sy1))+`
`    (kk*sz1))+(((A->indices)[i])[si])])]*`
`    (xp_0[((ii+(jj*sy2))+(kk*sz2))+`
`    (( *dxp_s)[si])])]));`
	`permute([2,3,1,4])`
	`tile(0,4,TI)`
	`tile(0,3,TJ)`
	`tile(0,3,TK)`
	`unroll(0,6,US)`
	`unroll(0,7,UI)`
		$0 \leq TI \leq 122$
		$0 \leq TJ \leq 122$
		$0 \leq TK \leq 122$
		$0 \leq UI \leq 16$
		$0 \leq US \leq 10$
		compilers $\in$
		$\{gcc, icc\}$

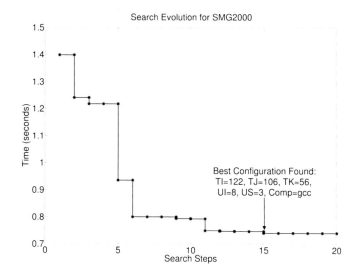

FIGURE 10.6: Search evolution.

The search converges in 20 steps. The search-evolution (performance of the best-point at each search-step) is shown in Figure 10.6. The y-axis shows the time spent in the outlined kernel in seconds. The x-axis shows the PRO steps. The configuration that PRO converges to is: $TI = 122, TJ = 106, TK = 56, UI = 8, US = 3$, Compiler=gcc. We see a performance improvement of 2.37x on the outlined kernel. We then use the code-variant associated with this parameter configuration to do a full sized run of SMG2000 (with input parameters `-n 120 120 120 -d 3`). We compile our code-variant with the following compiler optimization flags:
`-O2 -msse2 -march=pentium4`. Note the `-O2` doesn't apply loop-unrolling. This was done to prevent gcc from interfering with our unrolling. The results from production run are summarized in Table 10.5. We see that the full application execution improved by 27.2%.

TABLE 10.5: SMG2000 Production Experimental Results

Auto-tuned	Original	Improvement(%)
49.86s	68.52s	27.2

10.6 Summary

In this chapter, we motivated the need for a general-purpose end-to-end auto-tuning framework for scientific codes. To keep up with the rapidly changing HPC computing platforms, we argued that auto-tuning should be a continuous process — a process that starts at the coding stage and continues throughout the life of the program. Application developers, auto-tuners and compiler experts must work together to maximize the science done in the available computation time. Developing an effective auto-tuning framework comes with its own set of challenges, however. In this chapter, we examined the most daunting of the challenges and presented approaches to address them. We identified various sources of tunability in real codes and discussed how such sources can be exploited to steer application performance.

We presented a set of distinct characteristics and requirements that auto-tuners must consider when they design their search algorithm. We showed that parallel search algorithms can leverage available parallelism to effectively deal with constrained high-dimensional search spaces. The fact that the search algorithm converges to solutions in only a few tens of search-steps while simultaneously tuning multiple parameters demonstrates its capability of taking into account the latent interactions between tunable parameters. Finally, we presented our auto-tuning experience with Active Harmony. We showed the benefits of expert/developer participation in the overall tuning process. We presented tuning results that demonstrate how the system can provide tuning solutions at all stages of application development and deployment.

10.7 Acknowledgment

This work was sponsored in part by the Performance Engineering Research Institute, which is supported by the SciDAC Program of the U.S. Department of Energy.

Chapter 11

Languages and Compilers for Auto-Tuning

Mary Hall

University of Utah School of Computing

Jacqueline Chame

University of Southern California/Information Sciences Institute

In Chapter 10, we described a set of features of a practical auto-tuning framework. We consider in this chapter how language and compiler technology can support these features, particularly in the triage process, and in generating alternative implementations of a key computation. Application programmers are typically accustomed to interacting with compilers only through optimization flags, and then observing performance behavior of the resulting code. There is a rich history of compiler technology automatically generating high-performance code that targets specific architectural features, including vector machines, VLIW and superscalar architectures, and effective management of registers. However, the rapid changes and diversity in today's architectures, along with the complex interactions between different optimization strategies, have made it increasingly difficult for compiler technology to keep pace with application developers' needs. To enhance the role of compilers in generating high-performance code for high-end applications, we consider in this chapter new compiler technology and interfaces into the compiler that enable a more collaborative relationship with the application programmer in the autotuning process.

The specific compiler technology we describe primarily supports triage, optimization and code variant generation. In triage, compiler technology maps performance data back to code constructs and derives standalone executable programs from key computations, a process we call kernel extraction. In generating code variants, the compiler can participate in numerous ways, through program analysis, code transformation, and code generation. Program analysis can identify parallelism opportunities, understand storage requirements and reuse of data, and determine safety of code transformations. In addition, program analysis can interact with modeling to predict bottlenecks and to understand the performance impact of code transformations. Relevant code transformations rewrite a computation into an equivalent form that is expected to yield better performance, and can be used to improve utilization of memory, compute, or interconnect resources. For example, transformations can reorder the computation, reorganize data or refine communication and its relationship to computation. Finally, code generation realizes this rewriting, and can encompass specialization of code to specific input data, or other optimization parameter values. At every step of the compilation process, an appropriately structured compiler and language interface can permit the application developer to fill in gaps in the compiler's analysis results, express the set of transformations to be evaluated, describe how to search the set of possible code variants or how to dynamically select among a set of distinct implementations of a computation. This vision for current and future language and compiler technology for autotuning will be the subject of this chapter.

11.1 Language and Compiler Technology

Compilers provide a mapping from high-level programming language to executable code that targets a particular architecture. In designing new programming models, the desired goal is to abstract key features of the architecture so that programmers can express their code at a high level, possibly with the execution model of the architecture in mind. For example, MPI programmers target a distributed-memory platform where communication is considered expensive, and therefore code should have very coarse-grained parallel computations with infrequent communication. Vector architectures permit expression of array computations, and encourage repeated fine-grain operations on long vectors. Today's widely used programming models hide a lot of architectural details, sometimes making it very difficult to achieve high performance from code expressed in a way that is natural for programmers.

Today this performance gap is filled by (1) highly tuned libraries that capture common computations (*e.g.,* dense linear algebra and FFTs) and target specific architectures; (2) compiler optimization technology to automatically optimize for specific architectures; or least desirable; (3) application programmers writing low level architecture-specific code. We discuss in this section the relationship between these three approaches. First, we discuss the capabilities and limitations of current compiler technology. We then consider two cases where a combination of compiler and programming language technology can help the programmer achieve high performance in portions of their code that are not well represented by an existing library.

11.1.1 Compiler Technology: Capabilities and Limitations

Application scientists use a compiler as a black box to perform optimizations that improve the quality of the generated code beyond a straightforward translation. Today's compilers are particularly effective at diverse optimizations – for example, removing redundant computation, managing registers, and identifying instruction-level parallelism. Historically, compilers have helped support the transition to new architectures, such as the migration to vector architectures in the 1980s. Nevertheless, there are numerous challenges of targeting complex modern microprocessors, with multiple cores, deep memory hierarchies, SIMD compute engines and heterogeneity. Compilers that automatically parallelize sequential codes were the subject of intense research a decade ago [59, 161], but achieving coarse-grain parallelism across a wide range of applications is still an open problem. Codes with irregular memory access patterns and parallelism, such as applications that manipulate sparse matrices and graphs, achieve just a small fraction of peak performance due to load imbalance and poor utilization of the memory hierarchy.

Fundamentally, open issues for compilers are managing the combinatorial complexity of optimization choices and decisions that are impacted by dynamic execution behavior of applications [160]. Getting everything right in this complex optimization process is sometimes simply too hard for a compiler. Still, we would like to exploit the types of optimizations that compilers do well, and, where needed, give the programmer control they so often desire without having to write low-level code. In the next two subsections, we show how compiler technology can be applied to application and library code to achieve high performance while increasing the productivity of developers. We focus on two key ideas: (1) *auto-tuning* is used to evaluate a rich space of optimization choices, by running a series of empirical tests to measure performance and guide in optimization selection; and, (2) we engage the developer in *collaborative auto-tuning* to provide high-level guidance of the optimization process.

11.1.2 Auto-Tuning a Library in its Execution Context

Before looking at the more difficult problem of tuning general application code, let us first consider compiler optimization of libraries. Libraries are painstakingly customized by their developers to map to specific architectural features; many libraries, particularly the BLAS libraries, achieve performance that comes fairly close to architectural peak. When can compiler technology achieve even higher performance on computations that are encapsulated in a library?

Suppose that a programmer wants to use a library in an unusual way for which the library was not tuned. As an example, nek5000 is a scalable code for simulating fluid flow, heat transfer, and magnetohydrodynamics. The code is based on the spectral-element method, a hybrid of spectral and finite-element methods [135]. Nek5000 spends over 60% of its sequential execution time in what are effectively Basic Linear Algebra Subroutine (BLAS) calls, but for very small matrices (*e.g.*, $10 \times 2 \times 10$). The goal of the work we describe below was to improve the performance of nek5000 for the Cray XT5 jaguar system at Oak Ridge National Laboratory, which uses Opteron Istanbul processors. Why is it not sufficient to use standard BLAS libraries for the target architecture?

BLAS libraries are typically tuned for large square matrices. For example, for matrix-multiply on 1024×1024 square matrices, the ACML (native), ATLAS and Goto BLAS libraries all perform well, above 70% of peak performance on the target architecture. They all incorporate aggressive memory hierarchy optimizations such as *data copy*, *tiling*, and *prefetching* to reduce memory traffic and hide memory latency. Additional code transformations improve *instruction-level parallelism* (ILP). Several examples in the literature describe this general approach [55, 88, 153, 360, 388].

If we look closer at matrices of size 10 or smaller, however, those same BLAS libraries perform below 25% of peak performance. Since these matrices fit within even small L1 caches, the focus of optimization should be on managing registers, exploiting ILP in its various forms, and reducing loop overhead. For these purposes, we can use *loop permutation*, which reorders the loops in a loop nest, and aggressive *loop unrolling* to unroll the iterations for all loops in a nest. To the backend compiler, unrolling exposes opportunities for instruction scheduling, mapping array variables to registers, and eliminating redundant computations. Loop permutation may enable the backend compiler to generate more efficient *single-instruction multiple-data* (SIMD) instructions by bringing a loop with unit stride access in memory to the innermost position, as required for utilization of multimedia-extension instruction set architectures such as SSE-3. Even with this very limited set of optimizations, we must explore a search space of different loop orders and unroll factors that is still quite large. We use auto-tuning, coupled with heuristics to prune the search space, to specialize the BLAS calls to the dominant matrix sizes used by nek5000, corresponding to 99% of the matrix-multiply execution time. To demonstrate the impact of compiler optimization of library code, we show performance results in Figure 11.1. A comparison with BLAS libraries for different matrix sizes is taken from [304] and is shown in Figure 11.1(a). The use of specialization, autotuning, and optimizations focused on small matrices yields significant gains over even the best manually tuned libraries, including the native ACML BLAS (version 4.1.0), Goto BLAS (goto_barcelona-r1.26), and ATLAS (version 3.8.2 with the following architectural default: AMD64K10h64SSE3). Each performs at less than 30% of peak for the smallest matrices, which is more than 3X faster than using the native compiler on a naive implementation. The manually tuned baseline, part of Nek5000, is a little slower than the hand-tuned libraries, performing at roughly 23% of peak across all sizes. As shown in Figure 11.1(a), our automatically generated library code yields performance up to 77% of peak, more than a 2.2X improvement over the manually tuned code and the hand-tuned BLAS libraries.

Results from running the entire nek5000 on up to 256 processors of jaguar is taken from [303] and is shown in Figure 11.1(b). We see that the optimized BLAS libraries lead to a 26% performance gain over the manually-tuned versions for the same number of processors.

11.1.3   Tuning Other Application Code

We have shown an advantage to using compiler technology in tuning library code in its execution context. However, even more importantly, a compiler can apply its optimizations to application code for which there does not exist a library. We will show the SMG2000 example in Chapter 10, where compiler technology contributed to an integrated suite of tools, resulting in a 2.37X speedup on the residual computation and 27% performance improvement on the full application.

As another example, S3D is a turbulent combustion code under development at Sandia National Laboratory. Using a compiler optimization framework called LoopTool [281], the most costly loop nest in S3D was annotated with a set of transformations to be applied. In this way, the code could remain at a high level, with the annotations describing how to optimize the code. The results of this optimization were a 2.94X reduction in execution time [234].

11.2   Interaction between Programmers and Compiler

Application programmers typically interact with compilers through optimization flags. Most commercial compilers support an extensive number of optimization flags where each flag directs the compiler to enable a specific optimization or hardware mechanism (such as hardware prefetching). To achieve high performance it is often necessary to employ a number of carefully selected optimizations and mechanisms. Finding a combination of compiler flags that results in good performance often requires experimentation, due to complex interactions between code transformations, architecture features, or both. While some compilers provide feedback on the success of specific optimizations, most

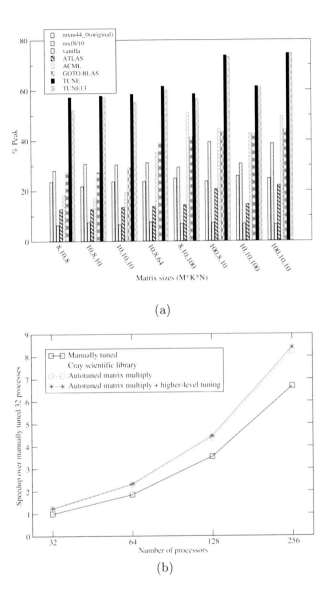

(a)

(b)

FIGURE 11.1: Performance Results. (a) Comparison of BLAS performance for small matrices from `nek5000`. (b) Nek5000 on `Jaguar` using 1 core per node.

do not provide information about negative interactions between optimizations or among optimizations and hardware mechanisms. The lack of control over the optimization strategy (code transformations, order of transformations, optimization parameters) and interactions with hardware mechanisms (such as hardware prefetching) drives application developers to manual performance tuning. Manual tuning is a time-consuming and error-prone process, and may prevent further compiler optimization as some manually optimized codes are not ammenable to compiler analysis.

We envision a compiler-based performance tuning model in which performance tuning options are exposed to developers and tools by well-defined interfaces. Under this model each performance tuning component provides an interface to its performance tuning knobs to support external control by developers or other components and tools. The key requirement of each interface is to expose the *space* of performance tuning options offered by its component. Through these interfaces a developer or tool can derive and navigate custom spaces of optimization strategies according to specific requirements and constraints. For example, a triage interface would allow customization of performance criteria for identifying performance bottlenecks; a transformation interface would allow the specification of compositions of loop transformations, and allow transformation parameters to have a range of possible values; a search interface would allow the selection of the search algorithm, evaluation criteria, stop criteria and any other search properties offered by the search component.

11.3 Triage

Code triage identifies sources of performance bottlenecks in complex applications with the goal of selecting computations, or code sections, for performance tuning. This identification is made by collecting performance data at execution time and correlating the data to code sections.

Methodologies for collecting performance data include code instrumentation and statistical sampling of hardware performance counters.

Code instrumentation directly inserts measurement code in the source program, allowing the programmer to pinpoint code sections to be profiled. Manual code instrumentation is not practical for large applications, but automatic instrumentation tools typically allow instrumentation of functions, methods, basic blocks, and statements. One of the drawbacks of code instrumentation is that the extra measurement code can substantially increase execution time. Another is that programmers typically instrument code sections which they expect to be performance bottlenecks, and may miss other opportunities for performance tuning.

Collecting performance data with statistical sampling works by sampling an execution of a program using hardware performance counters. The advantages of statistical sampling over instrumentation are that sampling does not require modification of source code and introduces much lower overheads. Sampling using performance counters captures detailed metrics such as number of operations, cache misses and pipeline stalls, which are essential for understanding performance of applications on complex architectures. Statistical sampling is used in HPCToolkit, a suite of tools described in Chapter 4 that supports measurement, analysis, attribution, and visualization of application performance for both sequential and parallel programs.

11.3.1   Code Isolation

It is often desirable to isolate each code segment selected for optimization into a standalone kernel, because tuning a standalone kernel takes much less time than tuning the same computation in the full application. The process of isolating, or outlining, a code segment can be automated. Tools that automate code outlining include Lee and Hall's *Code Isolator* [211] and the *ROSE Outliner* developed by Liao et al. [215]. The main goal of a code isolation tool is to automatically produce an isolated version of a code segment that can be compiled and executed with inputs representative of the original program. If performance tuning is the goal, the tool must also initialize the machine state so that execution of the isolated code segment mimics the performance behavior of the computation when executing in the full program.

In general, a code outlining tool must perform the following steps, (more details can be found in [211, 215]):

- Generate representative input data sets. This is achieved by instrumenting the application to capture representative input data values, just prior to executing the code to be isolated.

- Identify input parameters. This step performs an analysis of upwards exposed uses to identify which variables referenced in the code segment must be passed as input parameters. Upwards exposed uses determine what uses of variables in the outlined code may be reached by definitions outside the code.

- Initialize input parameters. The set of input parameters must be initialized prior to invoking the outline procedure in the isolated program. If the isolated code includes undefined sizes of arrays, the lower and upper bounds of array sizes are determined through instrumenting the original program. Data values are extracted from the original program, and are initialized in the isolated program. An alternative strategy to the analysis and initialization of input state, used by ROSE, is to simply checkpoint the application state on entry to the function, and use the memory state to initialize the input to outlined code.

- Capture machine state. The state of the machine from the original program is captured through instrumentation, and is set during initialization in the isolated code. The machine state describes all relevant state of the target architecture, including register, cache memory, and TLB.

- Generate a version of the isolated code with the data set and machine state initialized. The isolated code is encapsulated in a function, which can be invoked by a main program.

11.4 Code Transformation

There has been some recent research on new models where the compiler is not a monolitic block but a set of components that operate on distinct compilation phases and communicate through well-defined interfaces. These interfaces can be exposed to developers to allow external control of the tasks performed by the compiler. A major component of such a compiler framework is code transformation. Several code transformation tools that operate independently or as components of a compiler framework have been developed in the past few years. Among these tools are CHiLL [89], Graphite [279], Loop-Tool [280], Orio [162], PLuTo [60], POET [387], and URUK [149]. The ROSE compiler [15] also exposes an interface to its code transformation module.

CHiLL is a code transformation and code generation tool designed to support compiler-based autotuning and facilitate empirical search. CHiLL is based on the polyhedral model and supports a rich set of transformations that can be composed to generate custom optimization strategies. Developers or compilers interact with CHiLL using a high-level interface where an optimization strategy is specified as a *transformation recipe* [159]. A transformation recipe is a sequence of composable transformations to be applied to a code section.

Graphite is a loop transformation framework that uses the polyhedral model as its internal representation. Graphite was introduced in GCC 4.4 and is part of a polyhedral compilation package (PCP) designed to support loop optimization and analysis for GCC. PCP exposes an annotation-based transformation interface that can be used to specify transformations on a per statement basis. A single transformation can be applied per statement on each code generation cycle, and a composition of transformations to a statement must be specified over consecutive code generation cycles (this is a limitation of the transformation interface, and not of PCP.)

LoopTool is a source-to-source loop transformation tool that employs both data and computation restructuring to improve data reuse on several levels of the memory hierarchy. It uses multi-level fusion of loop nests to improve data reuse and expose opportunities for other optimizations, such as array contraction, unroll-and-jam, multi-level tiling, and iteration space splitting. Application developers can tune performance by augmenting their code with directives that control how transformations are applied. LoopTool checks the legality of the optimizations specified by the directives, performs the transformations requested and generates the transformed source code.

Orio is an annotation-based system that enables developers to augment source code with optimization and performance information. Orio is designed for portability, extensibility and automation, and is based on an extensible annotation parsing architecture. As an empirical performance-tuning system, Orio takes annotated C code as input, generates many code variants of the annotated code, and evaluates their performance empirically. In addition, Orio supports automated validation of the transformed code versions.

Pluto is an automatic parallelization tool that generates OpenMP parallel code from sequential C programs. Pluto is based on the polyhedral model and transforms code targeting both coarse-grained parallelism and locality simultaneously. Its core transformation framework supports outer, inner, and pipelined parallelization along with efficient tiling, fusion, and unrolling. Explicit tuning of parameters, such as tile sizes and unroll factors, and outer level loop fusion structure is supported by a number of options via command-line or optional files.

POET (Parameterized Optimizations for Empirical Tuning) is a dynamic scripting language and tool to express hundreds or thousands of complex code transformations and their configurations using a small set of parameters. POET was originally designed for parameterizing performance optimizations, and it is especially relevant to the evaluation of large search spaces as part of empirical tuning. Users can direct POET to apply a set of specified configurations to a code section using command line options and a configuration file.

URUK (*Unified Representation Universal Kernel*) is part of a compilation framework implemented within the Open64/ORC compiler. URUK performs code transformations in the WRAP-IT (*WHIRL-Represented As Polyhedra - Interface Tool*) polyhedra representation. Transformations are built from elementary actions called constructors, and expressed in a scripting language that supports composition of transformations.

The ROSE compiler exposes an interface that allows users to specify code transformations to CHiLL, POET, and its own loop translators. This interface was designed to support tool interopterability within PERI [271] (Performance Engineering Research Institute.)

Each of these tools exposes an interface to code transformations. The most common interface mechanisms are annotations in the code, command-line options or additional files with a description of transformations, and transformation recipes written using a transformation interface such as CHiLL's. The next section discusses transformation interfaces in more detail.

11.4.1 Transformation Interfaces

As more compiler-based transformation tools expose an interface to code developers, a few languages and high-level interfaces that allow developers to specify parameterized transformations have been proposed. The X language [117] was designed to represent parameterized programs in a compact way and to serve as an intermediate representation for library generators. X makes it easy for programmers to specify transformations and parameters, to change the order of transformations and parameter values, and to generate multiversion programs. Similarly, the interface to CHiLL permits description of high-level code transformations, parameterized by optimization variables such as unroll factors and tile sizes [159]. A recent extension to CHiLL's transformation interface, called CUDA-CHiLL, provides an even higher level of abstraction for describing transformations, with additional constructs for generating CUDA for Nvidia GPUs [287].

The transformation interfaces described in this section share some common goals: to support automatic performance tuning and user-assisted tuning with automated steps such as code generation and empirical search; and to facilitate interoperability between compilers and tools.

The key requirements of a high-level interface for transformation recipes in an auto-tuning environment are:

- a level of abstraction suitable to both compilers and application developers;

- a mechanism for naming language constructs;

- a mechanism to define transformations, such as *transformation rules* from the X language [117];

- support for empirical search with tunable optimization parameters.

High-level interface. A transformation recipe expresses an optimization strategy as a sequence of composable transformations. Each transformation has a set of input and output parameters. Input parameters are: (1) language constructs such as statements, loops, and array references that are present in the code before the transformation is applied; (2) optimization parameters such as tile size and unroll factors, which can be integer values or unbound parameters to be instantiated during empirical search. Output parameters are specifiers to new language constructs resulting from the code transformation. Output parameters allow subsequent transformations in a transformation recipe to refer to these new constructs.

Representation of language constructs. A common interface to transformation tools or compilers requires a mechanism for naming language constructs such as loops, statements, and array references. Under the DOE SciDAC PERI project, Quinlan and Liao [15] have designed a representation based on abstract handles to language constructs, to support interoperability between compiler and tools. Abstract handles represent references to language constructs in source code and optimization phases, and allow language constructs to be specified in multiple formats, including:

- Source file position information including path, filename, line and column number as in GNU standard source position [5].

- Global or local numbering of specified language construct in source file.

- Global or local names of constructs.

- Language-specific naming mechanisms.

Interoperability is achieved by writing out abstract handles as strings and reading them with other tools to build the equivalent abstract handle.

To make the discussion of transformation frameworks and transformation recipes concrete, we illustrate in Figure 11.2 the process of generating CUDA code from a sequential application and a CUDA-CHiLL recipe, taken from [287]. This transformation interface goes beyond prior systems, with the following support:

- *Parameters as variables.* Mutable variables allow for parameters to be set by auto-tuning framework or by other methods.

- *Queryable program state.* Captures the side-effects and results of commands as well as other code state as variables or through function calls.

- *Logical branching.* Reacts to captured or parameterized information with logical branching and iteration constructs.

- *Encapsulation.* Groups of commands are encapsulated and context insensitive to previous commands.

- *Readability.* References semantic constructs that carry over to the final generated program.

With these goals, it becomes immediately apparent that we are thinking of transformation recipes as programs written in a very limited programming language. For this reason, CUDA-CHiLL extends CHiLL's transformation recipes by adding a programming language frontend implemented with Lua, a lightweight, embeddable scripting language with extensible semantics and easy integration with a host program [178]. Lua has been described as "multi-paradigm" with features that allow for functional, imperative and object-oriented programming styles.

Figure 11.2 illustrates some of these features for a simple matrix-vector multiply code generation. The CUDA-CHiLL transformation recipe in Figure 11.2(b) is applied to the sequential matrix-vector multiply code in Figure 11.2(a). We use a tiling transformation (tile_by_index) to partition the sequential computation into tiles for hierarchical parallel execution of CUDA threads and blocks. The recipe introduces two new control loops to be used as CUDA block and thread indices, *ii* and *jj*. These new loops are used as handles for the CUDA mapping operation (cudaize) at line 7 and they can be seen as the loop indices of the first two loop levels of the intermediate result shown in Figure 11.2(c). The automatically-generated code for kernel invocation scaffolding and the CUDA kernel is shown in Figures 11.2(d) and (e), respectively.

11.5   Higher-Level Capabilities

The focus of this chapter, and indeed much of the auto-tuning literature, has been on tuning performance for a single core or single socket of a supercomputer. While single-socket performance is a critical factor in transitioning applications to perform efficiently on multi-core architectures, many applications fail to scale due to high communication overhead (including high latency), load imbalance, and related issues that require optimization of application behavior across sockets. Work on auto-tuning that crosses sockets is limited, but we cite a few examples here that help to frame the research challenges.

A key issue in scalable performance is how data is decomposed across the distributed memory space. In PGAS languages, this data decomposition is typically expressed by the application programmer. Zima et al. describe the design of a programming system that focuses on *productivity* of application programmers in expressing locality-aware algorithms for high-end architectures, which are then *automatically* tuned for performance. The approach combines the successes of two novel concepts for managing locality: high-level specification of user-defined data distributions and model-guided autotuning for data locality [391].

Another aspect of scalability is the overlap of communication and computation, which can be adjusted by changing the communication granularity. Danalis et al. describe how to transform code to adjust a variety of communication parameters to improve performance. They present results from a detailed study of the eect of the problem and message size, level of communication-computation overlap, and amount of communication aggregation on runtime performance in a cluster environment based on an RDMA-enabled network [107].

In general, the notion of higher-level optimization will focus on a combination of auto-tuning the partitioning of data across nodes and communication parameters, in conjunction with the auto-tuning of the single-node or single-socket computation in the context of these data partitioning and communication granularity decisions.

```
int i, j;
for (i = 0; i ¡ N; i++)
    for (j = 0; j ¡ N; j++)
        a[i] = a[i] + c[j][i] * b[j];
```

(a) Matrix-vector multiply sequential source code.

```
1 init("seqMV.suif", 0, 0)
2 dofile(hlcuda.lua)
3 N=1024
4 TI=16
5 tile_by_index("i","j", TI,TI, l1 control="ii", l2 control="jj", "ii", "jj", "i", "j")
6 print code()
7 cudaize("gpuMV", a=N, b=N, c=N*N, block="ii", thread="jj")
```

(b) A very simple transformation recipe.

```
for (ii = 0; ii < 64; ii++)
    for (jj = 0; jj < 64; jj++)
        for (i = 16 * ii; i < 16 * ii + 16; i++)
            for (j = 16 * jj; j < 16 * jj + 16; j++)
                a[i] = a[i] + c[j][i] * b[j];
```

(c) The results of output at line 6 from (b).

```
void seqMV(float (*c)[1024], float *a, float *b)
float *devO1Ptr, *devI1Ptr, *devI2Ptr;
cudaMalloc(&devO1Ptr, 1024 * 4);
cudaMemcpy(devO1Ptr, a, 1024 * 4, cudaMemcpyHostToDevice);
cudaMalloc(&devI1Ptr, 1048576 * 4);
cudaMemcpy(devI1Ptr, c, 1048576 * 4, cudaMemcpyHostToDevice);
cudaMalloc(&devI2Ptr, 1024 * 4);
cudaMemcpy(devI2Ptr, b, 1024 * 4, cudaMemcpyHostToDevice);
dim3 dimGrid(64, 1);
dim3 dimBlock(64, 1);
gpuMV<<<dimGrid,dimBlock>>>(devO1Ptr, (float(*) [1024])devI1Ptr, devI2Ptr);
cudaMemcpy(a, devO1Ptr, 1024 * 4, cudaMemcpyDeviceToHost);
cudaFree(devO1Ptr);
cudaFree(devI1Ptr);
cudaFree(devI2Ptr);
```

(d) The resulting generated CUDA scaffolding in the original host-executed function.

```
global void gpuMV(float *a, float (*c)[1024], float *b)
int bx = blockIdx.x; int tx = threadIdx.x;
for (i = 16 * bx; i < 16 * bx + 16; i++)
    for (j = 16 * tx; j < 16 * tx + 16; j++)
        a[i] = a[i] + c[j][i] * b[j];
```

(e) The resulting generated CUDA kernel.

FIGURE 11.2: A simplified example of tiled and CUDAized matrix-vector multiply. (a) Matrix-vector multiplication code, (b) Transformation recipe, (c) intermediate result, (d) CUDA scaffolding, (e) CUDA kernel.

11.6 Summary

This chapter has described how compiler and programming language technology can support the performance tuning process through automatic techniques and collaboraton with an application developer. The chapter began with examples where this technology was used to accelerate the performance of high-performance libraries and scalable production scientific applications. We described the use of this technology in the triage phase, to identify key computations using traces or sampling, and to isolate these computations into standalone executable code. We discussed the use of transformation frameworks and their associated interfaces to provide programmers with a means to automatically generate code from a high-level description. Finally, we described higher-level optimizations across sockets to optimize data layout and communication/computation overlap.

Future architectures will increase the need for language and compiler support for auto-tuning. Demands for energy efficiency and affordability are driving future exascale architectures toward heterogeneous computation and memory resources, with orders of magnitude higher flops/byte ratios. The growing trend of using GPUs as accelerators to a conventional multi-core microprocessor will continue its migration into the supercomputer platforms. Therefore, future tuning efforts will focus on computation partitioning between different heterogeneous processing units and even more attention to management of memory hierarchies of growing complexity. These trends demand a more collaborative relationship between tools and application developers.

11.7 Acknowledgments

This work was sponsored in part by the Performance Engineering Research Institute, which is supported by the SciDAC Program of the U.S. Department of Energy.

Chapter 12

Empirical Performance Tuning of Dense Linear Algebra Software

Jack Dongarra

University of Tennessee/Oak Ridge National Laboratory

Shirley Moore

University of Tennessee

Dense linear algebra (DLA) forms the core of many scientific computing applications. Consequently, there is continuous interest and demand for the development of efficient algorithms and implementations on new architectures. One response to this demand has been the development of the ATLAS (Automatic Tuning of Linear Algebra Software) system to automatically produce implementations of the BLAS (Basic Linear Algebra Subroutines) routines that underlie all of dense linear algebra. ATLAS generates efficient code by

255

running a series of timing experiments using standard techniques for improving performance (loop unrolling, blocking, etc.) to determine optimal parameters and code structures. While ATLAS has been highly successful in tuning DLA for cache-based architectures, we are developing new auto-tuning techniques for multicore and heterogeneous architectures that exploit higher levels of parallelism and asynchronous scheduling. This chapter describes the AT-LAS techniques as well as recent research on empirical tuning of dense linear algebra routines for multicore and GPU architectures.

12.1   Background and Motivation

This section begins with a a discussion of how scientific computing relies on the efficient solution of numerical linear algebra problems. It discusses the critical performance issues involved, emphasizing performance on multicore and heterogeneous architectures. It then motivates the remainder of the chapter by introducing the empirical approach to DLA performance tuning.

12.1.1   Importance of Dense Linear Algebra Software

The standard problems of numerical linear algebra include linear systems of equations, least squares problems, eigenvalue problems, and singular value problems [112]. Linear algebra software routines for solving these problems are widely used in the computational sciences in general, and in scientific modeling in particular. In many of these applications, the performance of the linear algebra operations are the main constraint preventing the scientist from modeling more complex problems, which would then more closely match reality. This then dictates an ongoing need for highly efficient routines; as more compute power becomes available the scientist typically increases the complexity/accuracy of the model until the limits of the computational power are reached. Therefore, since many applications have no practical limit of "enough" accuracy, it is important that each generation of increasingly powerful computers have optimized linear algebra routines available.

12.1.2   Dense Linear Algebra Performance Issues

Linear algebra is rich in operations that are highly optimizable, in the sense that a highly tuned code may run multiple orders of magnitude faster than a naively coded routine. However, these optimizations are platform specific, such that an optimization for a given computer architecture will actually cause a slow-down on another architecture. The traditional method of handling this problem has been to produce hand-optimized routines for a given machine. This is a painstaking process, typically requiring many man-months from personnel who are highly trained in both linear algebra and computational optimization. The incredible pace of hardware evolution makes this approach untenable in the long run, particularly so when considering that there are many software layers (eg., operating systems, compilers, etc.) that also effect these kinds of architectures.

12.1.3   Idea of Empirical Tuning

Automatic performance tuning, or auto-tuning, has been used extensively to automatically generate near-optimal numerical libraries for modern CPUs. For example, ATLAS [111, 358] and PHiPAC [55] are used to generate optimized libraries for FFT, which is one of the most important algorithms for digital signal processing. There are two general approaches to auto-tuning, namely model-driven optimization and empirical optimization. The idea of model-driven optimization comes from the compiler community. The compiler research community has developed various optimization techniques that can effectively tranform code written in high-level languages such as C and Fortran to run efficiently on modern architectures. These optimization techniques include loop blocking, loop unrolling, loop permutation, fusion, and distribution, prefetching, and software pipelining. The parameters for these transformations, such as the block size and the amount of unrolling, are determined at compile time by analytical methods. While model-driven optimization is generally effective in making programs runs faster, it may not give optimal performance for linear algebra and signal processing libraries. The reason is that analytical models used by ocmpilers are simplified abstractions of the underlying processor architecture, and they must be general enough to be applicable to all kinds of programs. Thus, the limited accuracy of analytical models makes the model-driven approach less effective for the optimization of linear algebra and signal processing kernels if the approach is solely used. In contrast to model-driven optimization, empirical optimization techniques generate a large number of parameterized code variants for a given algorithm and run these variants on a given platform to find the one that gives the best performance. The effectiveness of empirical optimization depends on the parameters chosen to optimize and on the search heuristic used. A disadvantage of empirical optimization is the time cost of searching for the best code variant, which is usually proportional to the number of variants generated and evaluated. However, this cost may be justified for frequently used code, where the cost is amortized over the total useful lifetime of the generated code.

12.2   ATLAS

This section describes the goals and approach of ATLAS. ATLAS provides highly optimized linear algebra kernels for arbitrary cache-based architectures. The initial goal of ATLAS was to provide a portably efficient implemenatation of the BLAS (Basic Linear Algebra Subroutines). ATLAS was originally released in 1997. The most recent stable release is version 3.8.3, released in February 2009. ATLAS now provides at least some level of support for all of the BLAS and has been extended to some higher level routines from LAPACK.

As explained in [358], ATLAS uses three different methods of software adaptation, described as follows:

- *parameterized adaptation:* This method involves parameterizing charactertistics that vary from machine to machine. The most important such parameter in linear algebra is probably the blocking factor used in blocked algorithms which affects data cache utilization. Not all important architectural variables can be handled by this method, however, since some of them, such as choice of combined or separate multipley and add instructions, length of floating point and fetch pipelines, etc.) can be varied only by changing the underlying source code. For these variables, the two source code adaptations described below may be used.

- *multiple implememementation:* This source code adaptation method involves searching a collection of various hand-tuned implementations until the best is found. ATLAS adds a search and timing layer to accomplish what would otherwise be done by hand. An advantage of thrs method is that multiple authors can contribute implementations without having to understand the entire package.

- *source generation:* This source code adaptation method uses a program called a source generator which takes the various source code adapatations to be made as input and produces a source code routine with the specified characteristics. This method is flexible but complicated.

The BLAS are building block routines for performing basic vector and matrix operations. The BLAS are divided into three levels: Level 1 BLAS do vector-vector operations, Level 2 BLAS do matrix-vector operations, and Level 3 BLAS do matrix-matrix operations. The performance gains from optimized implementations depends on the level of the BLAS commonly occurring in problems in numerical linear algebra. ATLAS natively provides only a handful of LAPACK routines, but the ATLAS-provided routines can be automatically added to the standard LAPACK library from netlib [10] to produce a complete LAPACK library. In the following subsections, we describe ATLAS's BLAS and LAPACK support.

12.2.1   Level 3 BLAS Support

The Level 3 BLAS perform matrix matrix operations. They have $O(N^3)$ operations but need only $O(N^2)$ data. These routines can be effectively re-ordered and blocked for cache reuse and thus made to run fairly close to theoretical peak on most architectures. All the Level 3 BLAS routines can be efficiently implemented given an efficient matrix-matrix multiply (hereafter shortened to the BLAS matrix multiplication routine name, GEMM). Hence, the main performance kernel is GEMM. GEMM itself is further reduced to an even smaller kernel, called *gemmK*, before code generation takes place. *gemmK* is blocked to constant dimensions, usually for Level 1 Cache, and then heavily optimized for both the floating point unit and memory hierarchy using parameterization combined with multiple implementation and code generation. ATLAS first empirically searches the optimization space of the *gemmK* code and then optimizes the same code using multiple implementation. The *gemmK* that is finally used is the best performing kernel from these two searches.

12.2.2   Level 2 BLAS Support

The Level 2 BLAS perform matrix-vector operations such as matrix-vector multiply, rank 1 update and triangular forward/backward solve. ATLAS requires only one kernel to support all Level 3 BLAS, but this is not true of the Level 2 BLAS. The Level 2 BLAS have $O(N^2)$ operations and $O(N^2)$ data. Two classes of kernels are needed, a tuned general matrix-vector multiply (GEMV) and a tuned rank-1 update (GER) to support a GEMV- and GER-based Level 2 BLAS. However, since the matrix cannot be copied without incurring as much cost as an operation, a different GEMV kernel is required for each transpose setting. ATLAS tunes these kernels using only parameterization for cache blocking and multiple implementation. The Level 3 BLAS performance is determined by the peak of the machine. The Level 2 and Level 1 BLAS performance, however, is usually determined by the speed of the data bus, and thus the amount gained by optimization is less.

12.2.3   Level 1 BLAS Support

The Level 1 BLAS do vector-vector operations such as dot product. These routines have $O(N)$ operations on $O(N)$ data. ATLAS tunes the Level 1 BLAS using only multiple implementation along with some simple parameterization. Essentially, the only optimizations to be done at this level involve floating point unit usage and some loop optimizations. Since these routines are very simple, a compiler can usually do an excellent job with these optimizations; hence, performance gains from auto-tuning are typically found only when a compiler is poorly adapted to a given platform.

12.2.4   LAPACK Support

ATLAS currently provides ten basic routines from LAPACK, each of which is available in all four data types. These routines all use or provide for the LU and Cholesky factorizations and are implemented using recursion rather than static blocking.

12.2.5   Blocking for Higher Levels of Cache

Note that this chapter defines the Level 1 (L1) cache as the "lowest" level of cache: the one closest to the processor. Subsequent levels are "higher": further from the processor and thus usually larger and slower. Typically, L1 caches are relatively small, employ least recently used replacement policies, have separate data and instruction caches, and are often non-associative and write-through. Higher levels of cache are more often non-write-through, with varying degrees of associativity, differing replacement polices, and combined instruction and data cache.

ATLAS detects the actual size of the L1 data cache. However, due to the wide variance in high level cache behaviors, in particular the difficulty of determining how much of such caches are usable after line conflicts and data/instruction partitioning is done, ATLAS does not detect and use an explicit Level 2 cache size as such. Rather, ATLAS employs a empirically determined value called `CacheEdge`, which represents the amount of the cache that is usable by ATLAS for its particular kind of blocking.

12.2.6   Use of Assembly Code

Hand-tuned implementations used by ATLAS that allow for extreme architectural specialization are sometimes written in assembly code. Sometimes the compiler is not able to generate efficient backend code. ATLAS also uses assembly code to achieve persistent performance in spite of compiler changes.

12.2.7   Use of Architectural Defaults

The architectural defaults provided with ATLAS are the result of several guided installations – that is, the search has been run multiple times with intervention by hand if necessary. ATLAS's empirical search is meant to be used only when architectural defaults are unavailable or have become non-optimal due to compiler changes. As pointed out in the UMD Autotuning chapter, empirical search run on real machines with unrelated load can result in high variance in the timing results. Thus, use of architectural defaults, which serve as a type of performance database of past results, can yield the best results.

12.2.8 Search Algorithm

ATLAS uses a "relaxed 1-D line search," where the "relaxed" means that interacting transforms are usually handled by 2- or 3-D searches [1]. This basic search technique is adequate, given that the ATLAS developers understand good start values and the interactions between optimizations.

12.3 Auto-Tuning for Multicore

To deliver on the promise of multicore petascale systems, library designers must find methods and algorithms that can effectively exploit levels of parallelism that are orders of magnitude greater than most of today's systems offer. To meet this challenge, the Parallel Linear Algebra Software for Multicore Architectures (PLASMA) project is developing dense linear algebra routines for multicore architectures [278]. In PLASMA, parallelism is no longer hidden inside Basic Linear Algebra Subprograms (BLAS) [2] but is brought to the fore to yield much better performance. PLASMA relies on tile algorithms, which provide fine granularity parallelism. The standard linear algebra algorithms can be represented as a Directed Acyclic Graph (DAG) where nodes represent tasks and edges represent dependencies among them. Asynchronous, out of order scheduling of operations is used as the basis for a scalable and highly efficient software framework for computational linear algebra applications. PLASMA is currently statically scheduled with a tradeoff between load balancing and data reuse. PLASMA performance depends strongly on tunable execution parameters, the outer and inner blocking sizes, that trade off utilization of different system resources, as illustrated in Figures 12.1–12.3.

12.3.1 Tuning Outer and Inner Block Sizes

The outer block size (NB) trades off parallelization granularity and scheduling flexibility with single core utilization, while the inner block size (IB) trades off memory load with extra-flops due to redundant calculations. Only the QR and LU factorizations use inner blocking. If no inner blocking occurs, the resulting extra-flops overhead may represent 25% and 50% for the QR and LU factorization, respectively [71]. Tuning PLASMA consists of finding the (NB,IB) pairs that maximize the performance depending on the matrix size and on the number of cores. An *exhaustive search* is cumbersome since the search space is huge. For instance, in the QR and LU cases, there are 1352 possible combinations for (NB,IB) even if we constrain NB to be an even integer between 40 and 500 and if we constrain IB to divide NB. All these combinations cannot be explored for large matrices ($N >> 1000$) on effective factorizations in a reasonable time. Knowing that this process should be re-

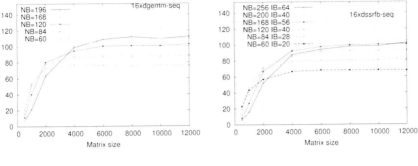

(a) DPOTRF - Intel64 - 16 cores   (b) DGEQRF - Intel64 - 16 cores

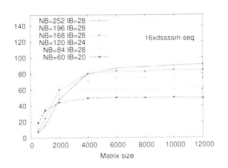

(c) DGETRF - Intel64 - 16 cores

FIGURE 12.1: Effect of (NB,IB) on PLASMA performance (Gflop/s) on (a) DPOTRF, (b) DGEQRF, and (c) DGETRF.

peated for each number of cores and each matrix size motivates us to consider a *pruned search*. The idea is that tuning the serial level-3 kernel (dgemm-seq, dssrfb-seq, and dssssm-seq) is not time-consuming since peak performance is reached on relatively small input matrices ($NB < 500$) that can be processed fast. Therefore, we first tune those serial kernels. As illustrated in Figure 12.4 and Figure 12.5, not all the (NB,IB) pairs result in a high performance.

For instance, the (480,6) pair leads to a performance of 6.0 Gflop/s whereas the (480,96) pair achieves 12.2 Gflop/s, for the dssrfb-seq kernel on Power6 (Figure 12.5(b)). We select a limited number of (NB,IB) pairs (*pruning* step) that achieve a local maximum performance on the range of NB. We have selected five or six pairs on the Intel64 machine (Figure 12.4) for each factorization and eight on the Power6 machine (Figure 12.5). We then benchmark the performance of PLASMA factorizations only with this limited number

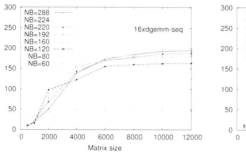

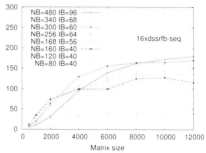

(a) DPOTRF - Power6 - 16 cores   (b) DGEQRF - Power6 - 16 cores

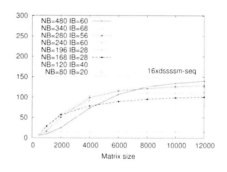

(c) DGETRF - Power6 - 16 cores

FIGURE 12.2: Effect of (NB,IB) on PLASMA performance (Gflop/s) on (a) DPOTRF, (b) DGEQRF, and (c) DGETRF.

of combinations (as seen in Figures 12.1–12.3). Finally, the best performance obtained is selected.

The dssssm-seq efficiency depends on the amount of pivoting performed. The average amount of pivoting effectively performed during a factorization is matrix-dependent. Because the test matrices used for our LU benchmark are randomly generated with a uniform distribution, the amount of pivoting is likely to be important. Therefore, we have selected the (NB,IB) pairs from dssssm-seq executions with full pivoting (Figures 12.4(c) and 12.5(c)). The dssssm-seq performance drop due to pivoting can reach more than 2 Gflop/s on Power6 (Figure 12.5(c)).

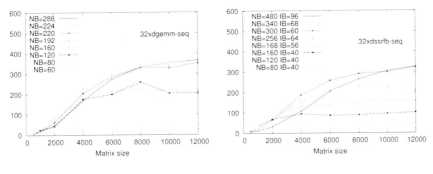

(a) DPOTRF - Power6 - 32 cores (b) DGEQRF - Power6 - 32 cores

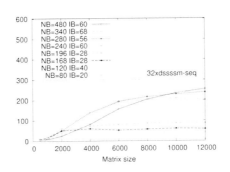

(c) DGETRF - Power6 - 32 cores

FIGURE 12.3: Effect of (NB,IB) on PLASMA performance (Gflop/s) on (a) DPOTRF, (b) DGEQRF, and (c) DGETRF.

12.3.2 Validation of Pruned Search

We have validated our *pruned search* methodology for the three one-sided factorizations on Intel64 16 cores. To do so, we have measured the relative performance overhead (percentage) of the pruned search (PS) over the exhaustive search (ES), that is: $100 \times \left(\frac{ES}{PS} - 1\right)$. Table 12.1 shows that the pruned search performance overhead is bounded by 2%. Because the performance may slightly vary from one run to another on cache-based architectures [308], we could furthermore observe in some cases higher performance (up to 0.95%) with pruned search (negative overheads in Table 12.1). However, the (NB,IB) pair that leads to the highest performance obtained with one method consistently matches the pair leading to the highest performance obtained with the other method.

We expect that the results will generalize to other linear algebra prob-

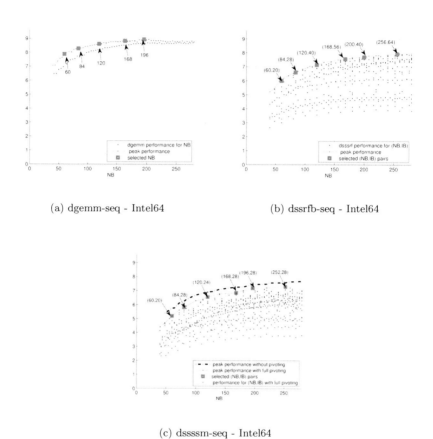

(a) dgemm-seq - Intel64

(b) dssrfb-seq - Intel64

(c) dssssm-seq - Intel64

FIGURE 12.4: Effect of (NB,IB) on the performance (in GFlop/s) of the serial PLASMA computational intensive level-3 BLAS kernels: (a) DGEMM, (b) DSSRFB, (c) DSSSSM.

TABLE 12.1: Overhead (in %) of Pruned search (Gflop/s) over Exhaustive search (Gflop/s) on Intel64 16 cores

Matrix Size	DPOTRF			DGEQRF			DGETRF		
	Pruned Search	Exhaustive Search	Over-head	Pruned Search	Exhaustive Search	Over-head	Pruned Search	Exhaustive Search	Over-head
1000	53.44	52.93	-0.95	46.35	46.91	1.20	36.85	36.54	-0.84
2000	79.71	81.08	1.72	74.45	74.95	0.67	61.57	62.17	0.97
4000	101.34	101.09	-0.25	93.72	93.82	0.11	81.17	80.91	-0.32
6000	108.78	109.21	0.39	100.42	100.79	0.37	86.95	88.23	1.47
8000	112.62	112.58	-0.03	102.81	102.95	0.14	89.43	89.47	0.04

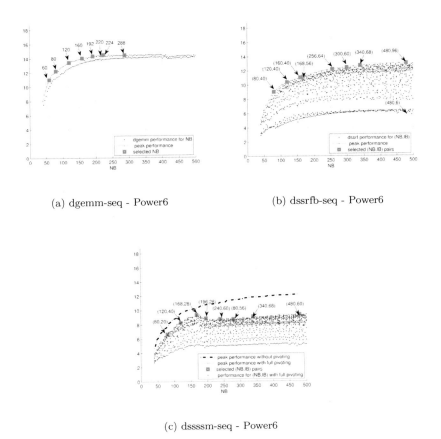

(a) dgemm-seq - Power6   (b) dssrfb-seq - Power6

(c) dssssm-seq - Power6

FIGURE 12.5: Effect of (NB,IB) on the performance (in GFlop/s) of the serial PLASMA computational intensive level-3 BLAS kernels: (a) DGEMM, (b) DSSRFB, (c) DSSSSM.

lems and even to any algorithm that can be expressed by a DAG of fine-grain tasks. Compiler techniques allow for the DAG of tasks to be generated from the polyhedral model applied to code that is free of runtime dependeces such as Cholesky or LU factorization without pivoting [44]. In comparison, by working at a much higher abstraction layer (a whole matrix tile as opposed to individual matrix elements) and with semantic knowledge of functions called from within the loop nests, PLASMA is able to produce highly-tuned kernels.

12.4 Auto-Tuning for GPUs

As mentioned above, the development of high performance dense linear algebra (DLA) depends critically on highly optimized BLAS, and especially on the matrix multiplication routine (GEMM). This statement is especially true for Graphics Processing Units (GPUs), as evidenced by recently published results on DLA for GPUs that rely on highly optimized GEMM [338, 344]. However, the current best GEMM performance on GPUs, e.g., up to 375 Gflop/s in single precision and up to 75 Gflops/s in double precision arithmetic on NVIDIA's GTX 280, is difficult to achieve. The software development requires extensive GPU knowledge and even backward engineering to understand undocumented aspects of the architecture. This section describes preliminary work on some GPU GEMM auto-tuning tecnhiques that allow keeping up with changing hardware by rapidly reusing existing ideas. Preliminary results show auto-tuning to be a practical solution that, in addition to enabling easy portability, can achieve substantial speedups on current GPUs (e.g., up to 27% in certain cases for both single and double precision GEMM on the GTX 280).

12.4.1 GEMM Auto-Tuner

This section presents preliminary work on the design of a GEMM "auto-tuner" for NVIDIA CUDA-enabled GPUs. Here auto-tuner means a system that automatically generates and searches a space of algorithms. More details may be found in [214].

In [344], Volkov and Demmel presents kernels for single-precision matrix multiplication (SGEMM) that significantly outperforms CUBLAS on CUDA-enabled GPUs, using an approach that challenges those optimization strategies and programming guidelines that are commonly accepted. In this chapter, we will focus on the GEMM kernel that computes $C = \alpha A \times B + \beta C$. Additionally, we will investigate auto-tuning on both single precision and double precision GEMM kernels (i.e., SGEMM and DGEMM). The SGEMM kernel proposed in [344] takes advantage of the vector capability of NVIDIA CUDA-enabled GPUs. The authors argue that modern GPUs should be viewed as

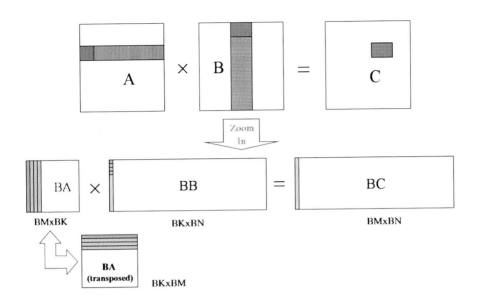

FIGURE 12.6: The algorithmic view of the code template for GEMM.

multi-threaded vector units, and their algorithms for matrix multiplication resemble those earlier ones developed for vector processors. We take their SGEMM kernel for computing $C = \alpha A \times B + \beta C$ as our code template, with modifications to make the template accept row-major input matrices, instead of column major used in their original kernel.

Figure 12.6 depicts the algorithmic view of the code templates respectively for both SGEMM and DGEMM. Suppose A, B, and C are M×K, K×N, and M×N matrices, and that M, N, and K are correspondingly divisible by BM, BN, and BK (otherwise "padding" by zero has to be applied or using the host for part of the computation). Then the matrices A, B, and C are partitioned into blocks of sizes BM×BK, BK×BN, and BM×BN, respectively (as illustrated on the figure). The elements of each BM×BN block of the matrix C (denoted by BC on the figure, standing for 'block of C') are computed by a $t_x \times t_y$ thread block. Depending on the number of threads in each thread block, each thread will compute either an entire column or part of a column of BC. For example, suppose BM = 16 and BN = 64, and the thread block has 16×4 threads, then each thread will compute exactly one column of BC. If the thread block has 16 × 8 threads, then each thread will compute half of a column of BC. After each thread finishes its assigned portion of the computation, it writes the results (i.e., an entire column or part of a column of BC back to the global memory where the matrix C resides). In each iteration, a BM×BK block BA of the matrix A is brought into the on-chip shared memory and kept there until the computation of BC is finished. Similarly to the matrix C, matrix B always resides in the global memory, and the elements of each block BB are brought from the global memory to the on-chip registers as necessary in each iteration. Because modern GPUs have a large register file within each multiprocessor, a significant amount of the computation can be done in registers. This is critical to achieving near-optimal performance. As in [344], the computation of each block BC = BC + BA×BB is fully unrolled. It is also worth pointing out that in our SGEMM, 4 `saxpy` calls and 4 memory accesses to BB are grouped together, as in [344], while in our DGEMM, each group contains 2 `saxpy` and 2 memory accesses to BB. This is critical to achieving maximum utilization of memory bandwidth in both cases, considering that the different widths between `float` and `double`.

As outlined above, 5 parameters (BM, BK, BN, t_x, and t_y) determine the actual implementation of the code template. There is one additional parameter that is of interest to the actual implementation. This additional parameter determines the layout of each block BA of the matrix A in the shared memory, i.e., whether the copy of each block BA in the shared memory is transposed or not. Since the share memory is divided into banks and two or more simultaneous accesses to the same bank cause the so-called bank conflicts, transposing the layout of each block BA in the shared memory may help reduce the possibility of bank conflicts, thus potentially improving the performance. Therefore, the actual implementation of the above code template is determined or parametrized by 6 parameters, namely BM, BK, BN, t_x, t_y, and a flag *trans* indicating whether to transpose the copy of each block BA in the shared memory.

We implemented code generators for both SGEMM and DGEMM on NVIDIA CUDA-enabled GPUs. The code generator takes the 6 parameters as inputs, and generates the kernel, the timing utilities, the header file, and the Makefile to build the kernel. The code generator first checks the validity of the input parameters before actually generating the files. By validity we mean (1) the input parameters confirm to hardware constraints, e.g., the maximum number of threads per thread block $t_x \times t_y \leq 512$, and (2) the input parameters are mutually compatible, e.g., $(t_x \times t_y)\%\mathrm{BK} = 0$, $\mathrm{BM}\%t_y = 0$, and $\mathrm{BN}\%t_x = 0$. By varying the input parameters, we can generate different variants of the kernel, and evaluate their performance, in order to identify the best variant. One way to implement auto-tuning is to generate a small number of variants for some matrices with typical sizes during installation time, and choose the best variant during run time, depending on the input matrix size.

12.4.2 Performance Results

The performance results in this section are for NVIDIA's GeForce GTX 280.

First, we evaluate the performance of the GEMM autotuner in both single and double precision. Figure 12.7 (left) compares the performance of the

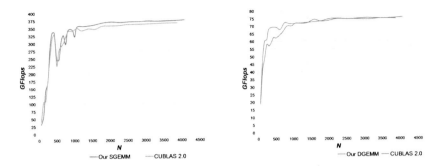

FIGURE 12.7: Performance comparison of CUBLAS 2.0 *vs* auto-tuned SGEMM (left) and DGEMM (right) on square matrices.

GEMM autotuner in single precision with the CUBLAS 2.0 SGEMM for multiplying square matrices. We note that both CUBLAS 2.0 SGEMM and our auto-tuned SGEMM are based on V.Volkov's SGEMM [344]. The GEMM autotuner selects the best performing one among several variants. It can be seen that the performance of the autotuner is apparently slightly better than the CUBLAS 2.0 SGEMM. Figure 12.7 (right) shows that the autotuner also performs better than CUBLAS in double precision. These preliminary results demonstrate that auto-tuning is promising in automatically producing near-optimal GEMM kernels on GPUs. The most attractive feature of auto-tuning is that it allows us to keep up with changing hardware by automatically and rapidly generating near-optimal BLAS kernels, given any newly developed GPUs.

The fact that the two performances are so close is not surprising because our auto-tuned code and CUBLAS 2.0's code are based on the same kernel, and this kernel was designed and tuned for current GPUs (and in particular the GTX 280), targeting high performance for large matrices. In practice though, and in particular in developing DLA algorithms, it is very important to have high performance GEMMs on rectangular matrices, where one size is large, and the other is fixed within a certain block size (BS), e.g., BS = 64, 128, up to about 256 on current architectures. For example, in an LU factorization (with look-ahead) we need two types of GEMM, namely one for multiplying matrices of size N×BS and BS×N−BS, and another for multiplying N×BS and BS×BS matrices. This situation is illustrated on Figure 12.8, where we compare the performances of the CUBLAS 2.0 *vs* auto-tuned DGEMMs occurring in the block LU factorization of a matrix of size 6144 × 6144. The graphs show that our auto-tuned code significantly outperforms (up to 27%) the DGEMM from CUBLAS 2.0.

Using the new DGEMM for example in the block LU (of block size BS

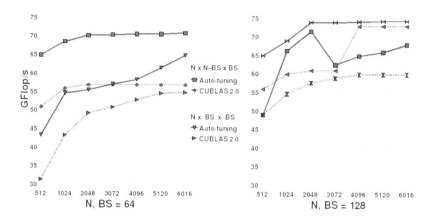

FIGURE 12.8: Performance comparison of the auto-tuned (solid line) *vs* CUBLAS (dotted line) DGEMMs occurring in the block LU factorization (for block sizes BS = 64 on the left and 128 on the right) of a matrix of size 6144×6144. The two kernels shown are for multiplying N×BS and BS×N−BS matrices (denoted by N×N−BS×BS), and N×BS and BS×BS matrices (denoted by N×BS×BS).

= 64) with partial pivoting [344] improved the performance from 53 to 65 Gflop/s on a matrix of size 6144 × 6144.

We highlighted the difficulty in developing highly optimized codes for new architectures, and in particular GEMM for GPUs. On the other side, we have shown an auto-tuning approach that is very practical and can lead to optimal performance. In particular, our auto-tuning approach allowed us

- To easily port existing ideas on quickly evolving architectures (e.g., demonstrated here by transferring single precision to double precision GEMM designs for GPUs), and

- To substantially speed up even highly tuned kernels (e.g., up to 27% in this particular study).

These results also underline the need to incorporate auto-tuning ideas in our software. This is especially needed now for the new, complex, and rapidly changing computational environment. Therefore our future directions are, as we develop new algorithms (e.g., within the MAGMA project), to systematically define their design/search space, so that we can easily automate the tuning process.

12.5 Summary

Auto-tuning is crucial for the performance and maintenance of modern numerical libraries, especially for algorithms designed for multicore and hybrid architectures. It is an elegant and very practical solution for easy maintenance and performance portability. While an empirically based exhaustive search can find the best performance kernels for a specific hardware configuration, applying performance models can often effectively prune the search space.

12.6 Acknowledgments

This work was sponsored in part by the Performance Engineering Research Institute, which is supported by the SciDAC Program of the U.S. Department of Energy, by the National Science Foundation and by Microsoft Research.

Chapter 13

Auto-Tuning Memory-Intensive Kernels for Multicore

Samuel W. Williams

Lawrence Berkeley National Laboratory

Kaushik Datta

University of California at Berkeley

Leonid Oliker

Lawrence Berkeley National Laboratory

Jonathan Carter

Lawrence Berkeley National Laboratory

John Shalf

Lawrence Berkeley National Laboratory

Katherine Yelick

Lawrence Berkeley National Laboratory

In this, chapter, we discuss the optimization of three memory-intensive computational kernels — sparse matrix-vector multiplication, the Laplacian differential operator applied to structured grids, and the `collision()` operator with the lattice Boltzmann magnetohydrodynamics (LBMHD) application. They are all implemented using a single-process, (POSIX) threaded, SPMD model. Unlike their computationally-intense dense linear algebra cousins, performance is ultimately limited by DRAM bandwidth and the volume of data that must be transfered. To provide performance portability across current and future multicore architectures, we utilize automatic performance tuning, or auto-tuning.

The chapter is organized as follows. First, we define the memory-intensive regime and detail the machines used throughout this chapter. Next, we discuss the three memory-intensive kernels that we auto-tuned. We then proceed with a discussion of performance optimization and automatic performance tuning. Finally, we show and discuss the benefits of applying the auto-tuning technique to three memory-intensive kernels.

13.1   Introduction

Arithmetic Intensity is a particularly valuable metric in predicting the performance of many single program multiple data (SPMD) kernels. It is defined as the ratio of requisite floating-point operations to total DRAM memory traffic. Often, on cache-based architectures, one simplifies total DRAM memory traffic to include only compulsory reads, write allocations, and compulsory writes.

Memory-intensive computational kernels are characterized by those kernels with arithmetic intensities that are constant or scale slowly with data size. For example, BLAS-1 operations like the dot product of two N-element vectors perform $2 \cdot N$ floating-point operations, but must transfer $2 \cdot 8N$ bytes. This results in an arithmetic intensity $(1/8)$ that does not depend on the size of the vectors. As this arithmetic intensity is substantially lower than most machines' Flop:Byte ratio, one generally expects such kernels to be memory-bound for any moderately-large vector size. Performance, measured in floating-point operations per second (GFlop/s), is bound by the product of memory bandwidth and arithmetic intensity. Even computational kernels whose arithmetic intensity scales slowly with problem size, such as out-of-place complex-complex FFT's, roughly $0.16 \log(n)$, may be memory-bound for any practical size of n for which capacity misses are not an issue.

Unfortunately, arithmetic intensity (and thus performance) can be degraded if superfluous memory traffic exists (e.g., conflict misses, capacity misses, speculative traffic, or write allocations). The foremost goal in optimizing memory-intensive kernels is to eliminate as much of this superfluous memory traffic as possible. To that end, we may exploit a number of strategies that either passively or actively elicit better memory subsystem performance. Ultimately, when performance is limited by compulsory memory traffic, reorganization of data structures or algorithms is necessary to achieve superior performance.

13.2 Experimental Setup

In this section, we discuss the three multicore SMP computers used in this chapter — AMD's Opteron 2356 (Barcelona), Intel's Xeon E5345 Clovertown, and IBM's Blue Gene/P (used exclusively in SMP mode). As of 2009, these architecture's dominate the top500 list of supercomputers [13]. The key features of these computers are shown in Table 13.1 and detailed in the following subsections.

13.2.1 AMD Opteron 2356 (Barcelona)

Although superseded by the more recent Shanghai and Istanbul incarnations, the Opteron 2356 (Barcelona) effectively represents the future x86 core and system architecture. The machine used in this work is a 2.3 GHz dual-socket × quad-core SMP. As each superscalar out-of-order core may complete both a single instruction-multiple data (SIMD) floating-point add and a SIMD floating-point multiply per cycle, the peak double-precision floating-point performance (assuming balance between adds and multiplies) is 73.6 GFlop/s. Each core has both a private 64 Kbyte L1 data cache and a 512 Kbyte L2 victim cache. The four cores on a socket share a 2 Mbyte L3 cache.

Unlike Intel's older Xeon, the Opteron integrates the memory controllers on chip and provides an inter-socket network (via HyperTransport) to provide cache coherency as well as direct access to remote memory. This machine uses DDR2-667 DIMMs providing a DRAM pin bandwidth of 10.66 Gbyte/s per socket.

13.2.2   Xeon E5345 (Clovertown)

Providing an interesting comparison to Barcelona, the Xeon E5345 (Clovertown) uses a modern superscalar out-of-order core architecture coupled with an older frontside bus (FSB) architecture in which two multichip modules (MCM) are connected with an external memory controller hub (MCH) via two frontside buses. Although two FSBs allows a higher bus frequency, as these chips are regimented into a cache coherent SMP, each memory transaction on one bus requires the MCH to produce a coherency transaction on the other. In effect this eliminates the parallelism advantage in having two FSBs. To rectify this, a snoop filter was instantiated within the MCH to safely eliminate as much coherency traffic as possible. Nevertheless, the limited FSB bandwidth (10.66 Gbyte/s) bottlenecks the substantial DRAM read bandwidth of 21.33 Gbyte/s.

Each core runs at 2.4 GHz, has a private 32 KB L1 data cache, and, like the Opteron, may complete one SIMD floating-point add and one SIMD floating-point multiply per cycle. Unlike the Opteron, the two cores on a chip share a 4 Mbyte L2 and may only communicate with the other two cores of this nominal quad-core MCM via the shared frontside bus.

13.2.3   IBM Blue Gene/P (Compute Node)

IBM's Blue Gene/P (BGP) takes a radically different approach to ultrascale system performance compared to traditional superscalar processors, as it relies more heavily on power efficiency to deliver strength in numbers instead of maximizing performance per node. To that end, the compute node instantiates four PowerPC 450 embedded cores in its one chip. These cores are dual-issue, in-order, SIMD enabled cores that run at a mere 850 MHz. As such, each node's peak performance is only 13.6 GFlop/s — a far cry from the x86 superscalar performance. However, the order of magnitude reduction in node power results in superior power efficiency.

Each of the four cores on a BGP compute chip has a highly associative 32 Kbyte L1 data cache, and they collectively share an 8 Mbyte L3. As it is a single chip solution, cache-coherency is substantially simpler, as all snoops and probes are on chip. The chip has two 128-bit DDR2-425 DRAM channels providing 13.6 Gbyte/s of bandwidth to 4 Gbyte of DRAM capacity. Like Opterons and Xeons, Blue Gene/P has hardware prefetch capabilities.

13.3   Computational Kernels

In this section, we introduce the three memory-intensive kernels used as exemplars throughout the rest of the chapter: the Laplacian stencil,

TABLE 13.1: Architectural summary of evaluated platforms.

Core Architecture	AMD Barcelona	Intel Core2	IBM PowerPC 450
Type	superscalar out-of-order	superscalar out-of-order	dual issue in-order
Clock (GHz)	2.3	2.40	0.85
DP Peak (GFlop/s)	9.2	9.60	3.4
Private L1 Data Cache	64 Kbyte	32 Kbyte	32 Kbyte
Private L2 Data Cache	512 Kbyte	—	—

Socket Architecture	Opteron 2356 Barcelona	Xeon E5345 Clovertown	Blue Gene/P Chip
Cores per Socket	4	4 (MCM)	4
Shared Cache	2 Mbyte L3	2×4 Mbyte L2	8 Mbyte L2
Memory Parallelism Paradigm	HW prefetch	HW prefetch	HW prefetch

Node Architecture	Opteron 2356 Barcelona	Xeon E5345 Clovertown	Blue Gene/P Node
Sockets per SMP	2	2	1
DP Peak (GFlop/s)	73.69	76.80	13.60
DRAM Bandwidth (GB/s)	21.33	21.33 (read) 10.66 (write)	13.60
DP Flop:Byte Ratio	3.45	2.40	1.00

the `collision()`–`stream()` operators extracted from Lattice Boltzmann Magnetohydrodynamics (LBMHD), and sparse matrix-vector multiplication (SpMV).

13.3.1 Laplacian Differential Operator (Stencil)

Partial differential equation (PDE) solvers constitute a large fraction of scientific applications in such diverse areas as heat diffusion, electromagnetics, and fluid dynamics. Solutions to these problems are often implemented using an explicit method via iterative finite-difference techniques. Computationally, these approaches sweep over a spatial grid performing *stencils* — a linear combinations of each a point's nearest neighbor. Our first kernel is the quintessential finite difference operator found in many partial difference equations — the 7-point Laplacian stencil.

This kernel is implemented as an out-of-place Laplacian stencil and is visualized in Figure 13.1. Since it uses Jacobi's method, we maintain a copy of the grid for both the current and next time steps and thereby avoid any data hazards. Conceptually, the stencil operator in Figure 13.1(b) is simultaneously

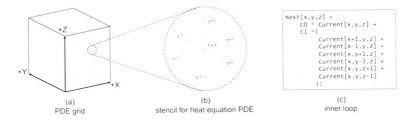

FIGURE 13.1: Visualization of the data structures associated with the heat equation stencil. (a) The 3D temperature grid. (b) The stencil operator performed at each point in the grid. (c) Pseudocode for stencil operator.

applied to every point in the 256^3 scalar grid shown in Figure 13.1(a). This allows an implementation to select any traversal of the points.

This kernel is exemplified by an interesting memory access pattern with seven reads and one write presented to the cache hierarchy. However, there is possibility of 6-fold reuse of the read data. Unfortunately this requires substantial per-thread cache capacity. Much of the auto-tuning effort for this kernel is aimed at eliciting this ideal cache utilization through the elimination of cache capacity misses. Secondary efforts are geared toward the elimination of conflict misses and write allocation traffic. Thus, with appropriate optimization, memory bandwidth and compulsory memory traffic provide the ultimate performance impediment. To that end, in-core performance must be improved trough various techniques only to the point where it is not the bottleneck. For further details on the heat equation and auto-tuning approaches, we direct the reader to [108].

13.3.2   Lattice Boltzmann Magnetohydrodynamics (LBMHD)

The second kernel examined in this chapter is the inner loop from the Lattice Boltzmann Magnetohydrodynamics (LBMHD) application [223]. LBMHD was developed to study homogeneous isotropic turbulence in dissipative magnetohydrodynamics (MHD) — the theory pertaining to the macroscopic behavior of electrically conducting fluids interacting with a magnetic field. The study of MHD turbulence is important in the physics of stellar phenomena, accretion discs, interstellar and intergalactic media, and plasma instabilities in magnetic fusion devices [56].

In Lattice methods, the macroscopic quantities (like density or momentum) at each point in space are reconstructed through operations on a momentum lattice — a discretization of momentum along 27 vectors. As LBMHD couples computational fluid dynamics with Maxwell's equations, the momentum lattice is augmented with a 15-velocity (cartesian vectors) magnetic lattice as shown in Figure 13.2. Clearly, this creates very high memory capacity requirements — over 1 Kbyte per point in space.

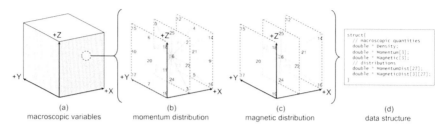

FIGURE 13.2: Visualization of the data structures associated with LBMHD. (a) The 3D macroscopic grid. (b) The D3Q27 momentum scalar velocities. (c) D3Q15 magnetic vector velocities. (d) C structure of arrays datastructure. Note: each pointer refers to a N^3 grid, and X is the unit stride dimension.

Lattice Boltzmann Methods (LBM) iterate through time calling two functions per time step: a `collision()` operator, where the grid is evolved one timestep, and a `stream()` operator that exchanges data with neighboring processors. In a shared memory, threaded implementation, `stream()` degenerates into a function designed to maintain periodic boundary conditions.

In serial implementations, `collision()` typically dominates the run time. To ensure that an auto-tuned `collision()` continues to dominate runtime in a threaded environment, we also thread-parallelize `stream()`. We restrict our exploration to a 128^3 problem on the x86 architectures, but only 64^3 on Blue Gene as it lacks sufficient main memory. For further details on LBMHD and previous auto-tuning approaches, we direct the reader to the following papers [363, 364].

The `collision()` code is far too large to duplicate here. Superficially, the `collision()` operator must read the lattice velocities from the current time step, reconstruct the macroscopic quantities of momentum, magnetic field, and density, and create the lattice velocities for the next time step. When distilled, this involves reading 73 doubles, performing 1300 floating point operations, and writing 79 doubles per lattice update. This results in a compulsory-limited arithmetic intensity of about 0.7 Flops per byte on write-allocate architectures, but may be improved to about 1.07 through the use of cache bypass instructions.

Conceptually, the `collision()` operator within LBMHD comprises both a 15 and a 27 point stencil similar to the previously discussed Laplacian Stencil. However, as lattice methods utilize an auxiliary grid that stores a distribution of velocities at each point, these stencil operators are different in that they reference a different velocity from each neighbor. As such, there is no inter-lattice update reuse. Proper selection of data layout (structure-of-arrays) transformes the principal optimization challenge from cache blocking to translation lookaside buffer (TLB) blocking. When coupled with a code transformation, one may reap the benefits of good cache and TLB locality simultaneously with effective SIMDization.

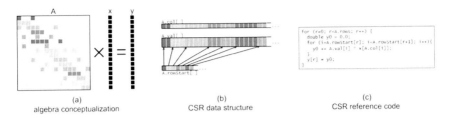

(a)   (b)   (c)
algebra conceptualization   CSR data structure   CSR reference code

FIGURE 13.3: Sparse matrix-vector multiplication (SpMV). (a) Visual-ization of the algebra: $y \leftarrow Ax$, where A is a sparse matrix. (b) Standard compressed sparse row (CSR) representation of the matrix. This structure of arrays implementation is favored on most architectures. (c) The standard im-plementation of SpMV for a matrix stored in CSR. The outer loop is trivially parallelized without any data dependencies.

As this application was designed for a weak-scaled message passing (MPI), single program, multiple data (SPMD) environment, we may simply tune to optimize single node performance, then integrate the resultant optimized im-plementation back into the MPI version.

13.3.3   Sparse Matrix-Vector Multiplication (SpMV)

Sparse matrix-vector multiplication (SpMV) dominates the performance of diverse applications in scientific and engineering computing, economic modeling and information retrieval; yet, conventional implementations have historically performed poorly on single-core cache-based microprocessor sys-tems [346]. Compared to dense linear algebra kernels, sparse kernels like SpMV suffer from high instruction and storage overhead per floating-point opera-tions, and a lack of instruction- and data-level parallelism in the reference implementations. Even worse, unlike the implicit (arithmetic) addressing pos-sible in dense linear algebra and structured grid calculations (stencils and lattice methods), indexing neighboring points in a sparse matrix requires an indirect access. This can result in potentially irregular memory access pat-terns (jumps and discontinuities). As such, achieving good performance on these kernels often requires selection of a compact data structure, reorder-ing of the computations to favor regular memory access patterns, and code transformations based on runtime knowledge of the sparse matrix. This need for run-time optimization and tuning is a major distinction from most other computational methods.

In this chapter, we consider the SpMV operation $y \leftarrow Ax$, where A is a sparse matrix, and x, y are dense vectors. A sparse matrix is a special case of the matrices found in linear algebra in which most of the matrix entries are zero. In a matrix-vector multiplication, computation on zeros does not change the result. As such, they may be elliminated from both the representation

	Dense	Protein	Spheres	Cantilever	Wind Tunnel	Harbor	QCD
Spyplot (Sparsity)		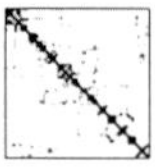		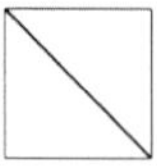	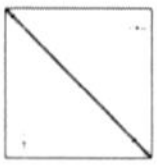	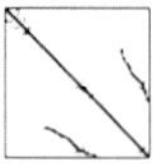	
Rows	2K	36K	83K	62K	218K	47K	49K
Columns	2K	36K	83K	62K	218K	47K	49K
Nonzeros (NNZ)	4.0M	4.3M	6.0M	4.0M	11.6M	2.4M	1.9M

	Ship	Economics	Epidem- iology	Accelerator	Circuit	Webbase	LP
Spyplot (Sparsity)		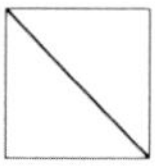	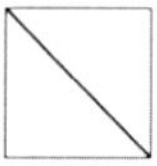		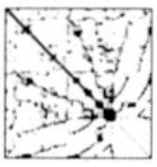	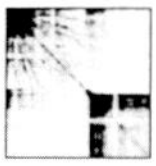	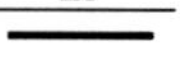
Rows	141K	207K	526K	121K	171K	1M	4K
Columns	141K	207K	526K	121K	171K	1M	1M
Nonzeros (NNZ)	4.0M	1.3M	2.1M	2.6M	0.9M	3.1M	11.3M

FIGURE 13.4: Benchmark matrices used as inputs to our auto-tuned SpMV library framework.

and the computation, leaving only the *nonzeros*. Although the most common data structure used to store a sparse matrix for SpMV-heavy computations is compressed sparse row (CSR) format, illustrated with the corresponding kernel in Figure 13.3, we will explore alternate representations of the compute kernel. CSR requires a minimum overhead of 4 bytes (column index) per 8 byte nonzero. As microprocessors only have sufficient cache capacity to cache the vectors in their entirety, we may define the compulsory memory traffic as 12 bytes per nonzero. SpMV will perform 2 Flops per nonzero. As such, the ideal CSR arithmetic is only 0.166 Flops per byte; making SpMV heavily memory-bound. Capacity misses and sub-optimal bandwidth will substantially impair performance.

Unlike most of dense linear algebra, stencils on structured grids, and Fourier transforms, matrices used in sparse linear algebra are not only characterized by their dimensions but also by their *sparsity* pattern — a scatter plot of nonzeros. Figure 13.4 presents the spyplot and the key characteristics associated with each matrix used in this chapter. Observe that for the most part, the vectors are small, but the matrices (in terms of nonzeros) are large. Remember, 12 bytes are required per nonzero. As such, a matrix with four million nonzeros requires at least 48 Mbyte of storage — far larger than most caches. We selected a set of matrices that would exhibit several classes of sparsity: dense, low bandwidth (principally finite element method), unstructured, and extreme aspect ratio. Such matrices will see differing cache capacity issues on multicore SMPs. In addition, we ensured the matrices would run the gambit of nonzeros per row — a key component in CSR performance. Finally, although some matrices are symmetric, we convert all of them to non-symmetric format and do not exploit this characteristic.

For further details on the sparse matrix-vector multiplication and previous auto-tuning efforts, we direct the reader to [362, 365, 366].

13.4   Optimizing Performance

Broadly speaking, we may either classify optimizations by their impact on implementation and usage, or by the performance bottleneck they attempt to eliminate. For example, an implementation-based categorization may delineate optimizations into four groups based on what changes are required: only code structure, data structures, the style of parallelism, or algorithms. On the other hand, if we categorize optimizations by bottleneck, we may create groups that more efficiently exploit parallelism, minimize memory traffic, maximize memory bandwidth, or maximize in-core performance. That being said, in this section, we describe the optimizations employed by our three auto-tuners grouped using the bottleneck-oriented taxonomy. Moreover, as we're focused on memory-intensive kernels, we will prioritize the optimizations accordingly.

13.4.1   Parallelism

Broadly speaking parallelism encompasses approaches to synchronization, communication, use of threads or processes, and problem decomposition.

Synchronization and Communication: Although a number of alternate strategies are possible (including DAG-based schedulers [278]), we adopted a POSIX thread-based, SPMD, bulk-synchronous approach to exploit multicore parallelism. Unlike process-based, shared memory-optimized message passing approaches, we exploit the ever-present cache coherency mechanisms for both efficient communication as well as to eliminate system calls. We enforce bulk synchronous semantics via a shared memory spin barrier.

Problem Decomposition: We utilize two different approaches to problem decomposition. First, the structured grid codes spatially decompose the stencil sweep into subdomains by partitioning the problem in two dimensions (not the unit stride). We ensure there are at least as many subdomains as there are threads. Subdomains may then be assigned to threads in chunks in a round-robin ordering. For LBMHD, the subdomains are not perfect rectahedral volumes. Rather, within each plane the boundaries are aligned to cache lines. In effect this performs only loop parallelization through blocking. No part of the data structure is changed.

Conversely, we apply a very different technique when parallelizing sparse matrix-vector multiplication. To ensure there are no data dependencies, we only parallelize by rows, creating a number of submatrices that contain roughly

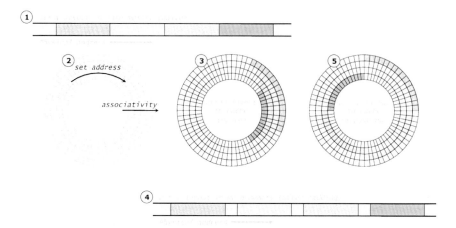

FIGURE 13.5: Conceptualization of array padding in both the physical (linear) and cache (periodic) address spaces. (See color insert.)

the same number of nonzeros, but may span wildly different numbers of rows. Each of these submatrices is as if it were its own matrix (hierarchical storage). We may now optimize each submatrix with a unique register blocking (see below). That is, some submatrices may be encoded using 1×1 COO while others are encoded with 8×4 CSR. Clearly, SpMV optimization went well beyond simple loop parallelization and subsumes data structure transformations as well.

13.4.2 Minimizing Memory Traffic

When a kernel is memory-bound, there are two principal optimizations: reduce the volume of memory traffic or increase the attained memory bandwidth. For simple memory access patterns, modern superscalar processors often achieve a high fraction of memory bandwidth. As such, our primary focus should be on techniques that minimize the volume of memory traffic. Broadly speaking, we may classify memory traffic using the Three C cache model's compulsory, conflict, and capacity misses [165] augmented with speculative (prefetch) and write-allocate traffic. We implemented a set of optimizations that attempt to minimize each class of traffic. Not all optimizations are applicable to all kernels.

Array Padding: Caches have limited associativity. When too many memory references map to the same set in the cache, a conflict miss will occur and useful data will be evicted. These conflicts may arise from intra-thread conflicts or, when shared caches are in play, from inter-thread conflicts.

An example of this complex behavior may be illustrative. In physical memory, subsets of an array assigned to different threads (red, orange, green, blue) are disjoint: Figure 13.5(1). Unfortunately, a given set in a cache may map many different addresses in physical memory (there is a 1 to many relationship). As such, in order to properly visualize caches, we must abandon the linear memory concept in favor of the periodic coordinate system shown in Figure 13.5(2). In this case, every 64th physical address maps to the same cache set. Figure 13.5(3) shows that when those disjoint physical addresses are mapped to the cache set address space, the segments overlap. This can put substantial pressure on cache associativity (number of elements mapped to the same set address). Execution on any cache less than 4-way associative will generate conflict misses and limit the working set size to 2 elements. Array padding is the simplest and most common remediation strategy. Simply put, dummy elements (placeholders) are inserted between each thread's subset of the array (4). As a result, when mapped to the cache coordinate system, Figure 13.5(5), we see there is no longer any overlap of data at a given set address, and the demands on cache associativity have been reduced from 4-way to 1-way (direct mapped).

Due to the limited number of memory streams and reuse, inter-thread conflict misses predominate on SpMV and the Laplacian stencil. On LBMHD, where the `collision()` operator attempts to keep elements from 150 different arrays in the cache, eliminating intra-thread conflict misses is key. As such, we implemented two different array padding strategies to mitigate cache conflict misses. For SpMV and stencils, we pad each array so that the address of the first element maps to a unique set in the cache. Moreover, the padding is selected to ensure that the set addresses of the threads' first elements are equally spaced (by set address) in the last level cache. The ideal padding may be either calculated arithmetically or obtained experimentally. For SpMV, we simply `malloc` each thread block independently with enough space to pad by the cache size. We then align to a 4 MB boundary and pad by the thread's fraction of the cache. Array padding for LBMHD is somewhat more complex. We pad each velocity's array so that when referenced with the corresponding stencil offset (and corresponding address offset) the resultant physical address maps to a unique, equally-spaced cache set. Although this sounds complicated in practice, its relatively easy to implement. For further details on how these kernels exploit array padding, we direct the reader to [108, 362–364].

Cache Blocking: The reference implementations of many kernels demand substantial cache working sets. In practice, as processor architects cannot implement caches that are large enough to avoid capacity misses. The volume of memory traffic is increased. LBMHD does not exhibit any inter-stencil reuse. That is, there is no two stencils reuse the same values. As such, cache capacity misses are nonexistent. However, the Laplacian stencil shows substantial reuse. Like dense matrix-vector multiplication, SpMV will also show reuse on vector accesses. In either case, we must restructure code, and possibly data, to eliminate capacity miss traffic. Like cache blocking in dense linear algebra, we may apply a simple loop blocking technique to the Laplacian stencil to ensure an implementation generates relatively few capacity misses. In practice, this is implemented the same as problem decomposition for parallelization. However, defining dense blocks (of source vectors) for SpMV often yields a dramatically suboptimal solution. As such, we employ a novel sparse blocking technique that ensures that each cache block touches the same number of cache lines regardless of how many rows or columns the block spans. In practice, for a given thread's block of the matrix, we cache block it by simply adding columns of the sparse matrix until the number of unique cache lines touched reaches a preset number. Clearly, this requires substantial data structure changes.

Cache Bypass: Based on consumer applications, most cache architectures implement a *write-allocate* protocol. That is, if a store (write) misses in the cache, an existing line will be selected for eviction, the target line will be loaded into the cache, and the target word will be written to the line in the cache. Such an approach is based on the implicit assumption that if data is written, it will be promptly read and modified many times. Note, usage of such a policy is orthogonal to the *write-back* or *write-through* choice. Unfortunately, many computational kernels found in HPC read and write to separate arrays or data structures. As such, most writes allocate a line in the cache, completely obliterate its previous contents, and eventually write it back to DRAM. This makes write-allocate not only superfluous, but expensive as it will generate twice the memory traffic as a read — an obvious target for optimization when memory-bound.

Modern write-allocate cache architectures provide a means of eliminating this superfluous memory traffic via a special store instruction that bypasses the cache hierarchy. In the x86 ISA, this is implemented with the `movntpd` instruction. Unfortunately, most compilers cannot resolve the complex decision as to when to use this instruction; improper usage can reduce performance by an order of magnitude, where correct usage can improve performance by 50%. As such, in practice, we may only exploit this functionality through the use of SIMD *intrinsics* — a language construct with the interface of a function that the compiler will map directly to one instruction.

Often, the vectors used in SpMV are small enough to fit in cache. As such, the totality of DRAM memory traffic is reads and there is no need to use cache bypass. However, Jacobi stencils and lattice methods read and write to separate arrays. For other finite-difference operators like gradient or divergence, cache bypass may only improve performance by 75% or 25% respectively. Given this variability in benefit and the human effort required to implement this optimization, one should analyze the code before proceeding with this optimization. Nevertheless, usage of cache bypass on the Laplacian stencil or LBMHD can reduce the total memory traffic by 33% and improve performance by 50%.

Register blocking for SpMV: For most matrices, SpMV is dominated by compulsory misses. As such, neither cache blocking nor cache bypass will provide substantial benefits. That is not to say nothing can be done. Rather, a radical solution has emerged that eliminates compulsory miss traffic. Broadly speaking, sparse matrices require substantial meta data per nonzero — perhaps a 50% overhead. However, we observe that many nonzeros are clustered in relatively small regions. As such, the optimization known as *register block-ing* reorganizes the sparse matrix of nonzeros into a sparse matrix of small R×C dense matrices. Meta data is now needed only for each register block rather than each nonzero. If the zero fill required to make those R×C register blocks dense is less than the reduction in meta data, then the total memory traffic has been reduced — a clear win for a memory-bound kernel. Similarly, we may note that a range of column or row indices can be represented by a 16-bit integer instead of a 32-bit integer. This can save 2 bytes (or 17%) per nonzero. In this chapter we explore the product of 16 different register blockings, two matrix formats: compressed sparse row (CSR) and coordinate (COO), and the two index sizes. Register blocked CSR and COO are noted as BCSR and BCOO, respectively.

Please note, the term register blocking, when applied to sparse linear al-gebra refers to a hierarchical restructuring of the data, but when applied to dense linear algebra refers to a unroll and jam technique. In this chapter, our SpMV code heuristically explores these matrix compression techniques to find the combination that minimizes the matrix footprint.

13.4.3   Maximizing Memory Bandwidth

Now that we've discussed optimizations designed to minimize the volume of memory traffic, we may examine optimizations that maximize the rate at which said volume of data can be streamed into the processor. Basically, these optimizations aim to either avoiding memory latency or hide memory latency.

Blocking for the Translation Lookaside Buffer: All modern micro-processors use virtual memory. To translate the virtual address produced by the program's execution into the physical address required to access the cache or DRAM, the processor must inspect the page table to determine the map-ping. As this is a slow process and page table entries rarely change, page table entries may be placed in a very fast, specialized (page) cache on chip — the *translation lookaside buffer* or TLB. Unfortunately, TLBs are small and thus may not be able to cache all the pages referenced by an application (regardless of page size). As such, it is possible to generate TLB capacity misses. These typically don't generate superfluous DRAM traffic like normal cache capacity misses because evicted page table entries may land in the L2 or L3 cache. The performance difference (resulting from an increase in average memory la-tency) between translations that hit in the TLB and those that hit in the L3 is substantial. We may recast the cache blocking technique (which eliminated cache capacity misses) to eliminate TLB capacity misses and avoid memory latency. In LBMHD, we used a loop interchange technique coupled with an auxiliary data structure. This allowed us to trade cache capacity for increased page locality (and reduced TLB capacity misses). This technique is detailed in [362–364].

Prefetching: Memory latency is high. To satisfy Little's Law [41] and maximize memory bandwidth, the processor must express substantial memory-level parallelism. Unfortunately, superscalar execution may be insufficient. As such, hardware designers have incorporated both hardware and software prefetching techniques into their processors. The goal for either is to hide memory latency.

A software prefetch is an instruction like a load without a target address. As such, the processor will not stall waiting for it to complete. The user simply prefetches one element from each cache line to initiate the entire line's load. Unfortunately, such a practice requires the programmer to tune for the optimal "prefetch distance" — how far ahead prefetch addresses should be from load addresses. If he aims too low, latency will not be completely hidden. If he aims too high, cache capacity will be exhausted. More recently, a hardware prefetchers have begun to supplant software prefetching. Typically, they detect a series of cache misses, speculate as to future addresses, and prefetch them into the cache without requiring any user interaction.

In this chapter, we structure our auto-tuned codes to synergize with hardware prefetchers (long unit-stride accesses) but supplement this with software prefetching. This general approach provides performance portability as we make no assumptions as to whether a processor implements software prefetching, hardware prefetching, or both.

13.4.4 Maximizing In-Core Performance

For memory-intensive computations our primary focus should be on minimizing memory traffic and maximizing memory bandwidth. However, it is important not to overlook in-core(cache) performance. Code written without thought to the forms of parallelism required to attain good in-core performance may actually be compute-bound rather than memory-bound. The most common techniques are unroll and jam, permuting or reordering the computation given an unrolling, and SIMDization. We explored all of these via auto-tuning on all three kernels.

13.4.5 Interplay between Benefit and Implementation

The bottleneck that an optimization attempts to alleviate is orthogonal to the scope of the software implementation effort that is required to achieve it. For example, Table 13.2 lists the optimizations used when auto-tuning our three memory-intensive kernels. Loop or code structure transformations have perennially been the only changes allowed by an auto-tuner as they preserve the input and output semantics. Nevertheless, we see many optimizations require an abrogation of this convention as changes to data structures are required for ideal performance.

TABLE 13.2: Interplay between the bottleneck each optimization addresses (parallelism, memory traffic, memory bandwidth, in-core performance) and the primary impact on implementation (code-only, data structures, styles of parallelism). Obviously, changing data or parallelism structure will mandate some code changes. †Efficient SIMD requires data structures be aligned to 128-byte boundaries.

Optimization		Loop/Code Structure	Data Structure	Style of Parallelism
BS SPMD	(pthreads)			✓
Decomposition	(loop-based)	✓		
	(hierarchical)		✓	
Array Padding			✓	
Cache Blocking	(loop-based)	✓		
	(sparse)		✓	
Cache Bypass	(movntpd)	✓		
Reg. Blocking	(sparse)		✓	
TLB Blocking	(loop-based)	✓		
	(sparse)		✓	
Prefetching	(software)	✓		
Unroll and Jam		✓		
Reordering		✓		
SIMDization		✓	†	

13.5   Automatic Performance Tuning

Given this diversity of computer architectures, performance optimization has become a challenge as optimizing an application for one microarchitecture may result in a substantial performance loss on another. When coupled with the demands to optimize performance in a shorter timeframe than architectural evolution (several new variants of the x86 processor lines appear every year), hand optimizing for each is not practical. To that end, automatic performance tuning, or *auto-tuning* has emerged as an productive approach to tune key computational kernels and even full applications in minutes instead of months [140, 314, 346, 361]. In essence, auto-tuning is built on the premise that if one can enumerate all possible implementations of a kernel, the performance of modern computers allows for the exploration of these variants in less time than a human would require to optimize for one. Moreover, once this auto-tuner has been constructed it can be reused on any evolution of these architectures. The best choice or parameterization for the optimizations in question may be either architecture-dependent, input-dependent, or both. If it is neither, simple optimization will suffice, and auto-tuning is not needed.

Typically, auto-tuning a kernel is divided into three phases: enumeration

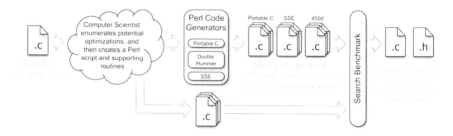

FIGURE 13.6: Generic visualization of the auto-tuning flow.

of potentially valuable optimizations, implementation of a code generator to produce functionally equivalent implementations of said kernel using different combinations of the enumerated optimization space, and implementation of a search component that will benchmark these variants (perhaps using real problem data) in an attempt to find the fastest possible implementation. We may visualize the auto-tuning flow in Figure 13.6, and will discuss the principal components in the following sections.

13.5.1 Code Generation

For purposes of this chapter, we use a simple auto-tuning methodology in which we use a Perl script to generate a few hundred potentially viable parameterizable implementations of a particular kernel. An implementation is a unique code representation that may be parameterized with a run time configuration. For example, cache blocking transforms a naive three nested loop implementation of matrix-matrix multiplication into a six nested loop implementation that is parameterized at runtime with the sizes of the cache blocks (the range of the inner loops). This is still just one variant. However, when one register blocks matrix multiplication, the inner three nested loops are so small (less than 16) it is common to simply fully unroll all loops and create perhaps a few thousand different code variants. When combined with cache blocking, we may have hundreds of individually parameterizable code variants.

Code variants are also needed when dealing with different data structures (i.e., hierarchical instead of flat), styles of parallelism (dataflow instead of bulk synchronous), or even algorithms. Each of these may in turn be parameterized.

As the differences between clusters of code variants may easily be expressed algorithmically, the Perl scripting language provides a pragmatic and productive means of tackling the intellectually-uninspiring task of producing the tens of thousands of lines of code. In essence, the simplest Perl code generation techniques are nothing more than a for loop over a series of `printf`'s. Every line in the resultant C code (function declarations, variables, statements, etc.) maps to a corresponding `printf` in the Perl script.

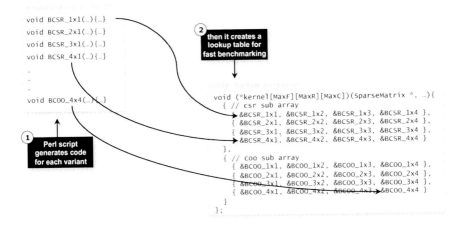

FIGURE 13.7: Using a pointer-to-function table to accelerate auto-tuning search.

13.5.2 Auto-Tuning Benchmark

A Perl script may generate thousands of code variants. Rather than trying to compile an auto-tuning benchmark for each, we integrate and compile all of them into one auto-tuning benchmark. This creates the challenge of selecting the appropriate variant without substantial overhead. To that end, we create an N-dimensional pointer-to-function table indexed by the code variants and possible parameters.

For example, in SpMV we use a 3-dimensional table indexed by kernel type (BCSR, BCOO, etc.) and the register block sizes as measured in rows and columns. As shown in Figure 13.7, the Perl script first generates the code for each kernel variant. It then creates a pointer-to-function table that provides a very fast means of executing any kernel. During execution, one simply calls `kernel[BCSR][4-1][3-1](...)` to execute the 4×3 BCSR kernel. The auto-tuning benchmark can be constructed to sweep through all possible formats and register blockings (nested for loops). For each combination, the matrix is reformatted and blocked, and the SpMV is benchmarked through a simple function call via the table lookup. This provides a substantial tuning time advantage over the naïveapproach of compiling and executing one benchmark for every possible combination. Moreover, it provides a fast runtime solution as well as easy library integration. We've demonstrated this technique when auto-tuning dense linear algebra, sparse linear algebra, stencils, and lattice methods.

13.5.3   Search Strategies

Given an auto-tuning benchmark, we must select a traversal of the optimization–parameter space that finds good performance quickly. Over the years, a number of strategies have emerged. In this chapter, we employ three different auto-tuning strategies: exhaustive, greedy, and heuristic. When the optimization–parameter space is small, an exhaustive search implemented as a series of nested loops is often acceptably fast. However, in recent years we've observed a combinatoric explosion in the size of the space. As a result, exhaustive search is no longer time- and resource-efficient. As a result, a number of new strategies have emerged designed to efficiently search the space. Greedy algorithms assume the locally optimal choice is also globally optimal. As such, it assumes the best parameterization for each optimizations may be explored independently. As such, they may transform a N^D optimization–parameter space of D optimizations each of N possible parameters into a sequential search through D optimizations each of N parameters ($N \times D$ points). Often, with substantial architectural intuition, we may express the best (or very close to best) combination in $O(1)$ time through an arithmetic approach that combines machine parameters and kernel characteristics.

Due to the size of the search space for the Laplacian stencil, we were forced to perform a greedy search algorithm after ordering the optimizations with some architectural intuition. This reduced the predicted tuning time from three months to 10 minutes. Conversely, LBMHD almost essentially uses an exhaustive strategy across seven code variants, each of which could accept over one hundred different parameter combinations. Typical tuning time was less than 30 minutes. SpMV used a combination of heuristics and exhaustive search. The none/cache/TLB blocking variant space was search exhaustively. That is, we benchmarked performance not blocking for either the cache or TLB, blocking for just the cache, and blocking for both the cache and TLB. Unlike the typical dense approach, the parameterization for cache and TLB blocking was obtained heuristically. Similarly, unlike the approach used by Berkeley's Optimized Sparse Kernel Interface (OSKI) [346], the register blocking was obtained heuristically by examining the resultant memory footprint size for each power-of-two register blocking. However, like LBMHD, the prefetch distance was obtained through an exhaustive search.

13.6   Results

In this section, we present and discuss the results from the application of three different custom auto-tuners to the three benchmarks used in this chapter. Previous papers have performed a detailed performance analysis for these three kernels [108, 363–366].

13.6.1   Laplacian Stencil

Figure 13.8 shows the benefits of auto-tuning the 7-point Laplacian stencil on a 256^3 grid on our three computers as a function of thread concurrency and increasing optimization. Threads are ordered to fully exploit all the cores within a socket before utilizing the second socket. We have condensed all optimizations into two categories: those that may be expressed in a portable C manner, and those that are ISA-specific. The former is a common code base that may be used on any cache-based architecture, not just these three. The latter includes optimizations like explicit SIMDization, and cache bypass. Thus Barcelona and Clovertown use the same x86 ISA-specific auto-tuner, and Blue Gene/P uses a different one. If an architecture with a third ISA were introduced (e.g., SPARC or Cell) it too would need its own ISA-specific auto-tuner. The auto-tuning search strategy uses a problem size-aware, greedy search algorithm in which the optimizations are searched one a time for the best parameterizations.

Clearly, the reference implementation delivers substantially suboptimal performance. As expected on the bandwidth-starved x86 processors, we see the reference implementation shows no scalability as one core may come close to fully saturating the available memory bandwidth.

When the portable C auto-tuner is applied to this kernel, we see that optimizations like cache blocking dramatically reduce superfluous memory traffic allowing substantially better performance. In general, on the x86 processors, we see a saturation of performance at around four cores (one socket), but a jump in performance at eight cores as using the second socket doubles the useable memory bandwidth. However, on NUMA architectures, like Barcelona, this boost is only possible if data is allocated in a NUMA-aware manner.

Rather than hoping the compiler, the non-portable, ISA-specific auto-tuner explicitly SIMDizes the kernel via intrinsics. Unfortunately, this is only useful on compute-bound platforms like Blue Gene. Unfortunately, despite the simplicity of this kernel, the lack of unaligned SIMD loads in the ISA results in less than perfect ($2\times$) benefit. Although explicit SIMDization was not beneficial on the x86 architectures, a different ISA-specific optimization, cache bypass, was useful as it reduces the memory-traffic on a memory-bound kernel. Doing so can substantially improve performance. Unfortunately, compilers will likely never be able to determine when this instruction is useful as it requires run-time knowledge.

In the end, auto-tuning improved the performance at full concurrency by $6.1\times$, $1.9\times$, and $4.5\times$, for Barcelona, Clovertown, and Blue Gene/P respectively. Moreover, thread-parallelizing and auto-tuning the 7-point stencil improved performance by $6.8\times$, $2.3\times$, and $17.8\times$, for Barcelona, Clovertown, and Blue Gene/P respectively. Although substantial effort was required in implementing a x86-specific auto-tuner, it may be reused on all subsequent x86 architectures like Nehalem or Magny-Cours thus amortizing the up front productivity cost.

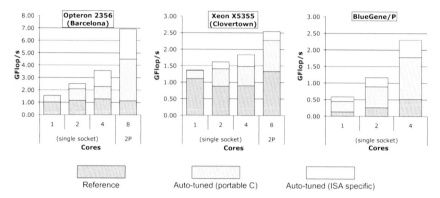

FIGURE 13.8: Benefits of auto-tuning the 7-point Laplacian stencil. Note: this Clovertown is actually the 2.66 GHz E5355 incarnation. As it has the same memory subsystem as the E5345, we expect nearly identical performance on this memory-bound kernel.

13.6.2 Lattice Boltzmann Magnetohydrodynamics (LBMHD)

Figure 13.9 shows the benefits of auto-tuning LBMHD on our three computers as a function of thread concurrency and increasing optimization. Unfortunately, Blue Gene/P does not have enough DRAM to simulate the desired 128^3 problem. As such, it only simulates a 64^3 problem. Once again, we have condensed all optimizations into two categories: those that may be expressed in a portable C manner, and those that are ISA-specific. The auto-tuning search strategy is exhaustive although vectorization is quantized into cache lines.

Although the reference implementation delivers good scalability, simple performance modeling using the Roofline Model (see Chapter 9) suggests it was delivering substantially suboptimal performance. Such a model also explains why even after auto-tuning the bandwidth-starved Clovertown shows poor scalability despite the performance boosts.

The biggest boosts derived from the portable C auto-tuner are NUMA-aware allocation, lattice-aware array padding, and vectorization (to eliminate TLB capacity misses). The Opteron and Blue Gene/P, with moderate machine balance (Flops:bandwidth), see good scaling on this kernel. Conversely, Clovertown, with a high machine balance, sees poor multicore scalability. Whether compute-bound or memory-bound, the non-portable, ISA-specific auto-tuner provided tremendous performance boosts either through explicit SIMDization or cache bypass.

In the end, auto-tuning improved the full concurrency performance by 3.9×, 1.6×, and 2.4×, for Barcelona, Clovertown, and Blue Gene/P respectively. Moreover, the coupling of thread-parallelization and auto-tuning improved LBMHD performance by an impressive 28.3×, 8.9×, and 9.2×,

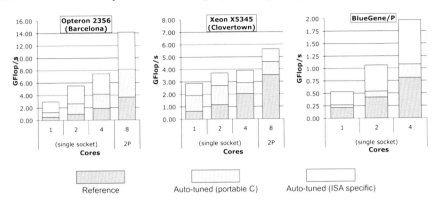

FIGURE 13.9: Benefits of auto-tuning the lattice Boltzmann Magnetohydrodynamics (LBMHD) application.

for Barcelona, Clovertown, and Blue Gene/P respectively. Clearly, the ISA-specific auto-tuners were critical in achieving these speedups.

13.6.3   Sparse Matrix-Vector Multiplication (SpMV)

Figure 13.10 shows the benefits of auto-tuning SpMV on our three computers. Unlike the previous two figures, where optimization and benefit is basically independent of problem size, the horizontal axis in Figure 13.10 represents different problems (matrices). The ordering preserves that in Figure 13.4. The lowest bar is the untuned serial performance, while the middle bar represents untuned performance using the maximum number of cores. The top bar is the tuned (portable C) performance using the maximum number of cores. Unlike the other kernels, a non-portable ISA-specific auto-tuner was not implemented for SpMV.

The auto-tuning search strategy is somewhat more complex. Register, cache and TLB blocking use a footprint minimization heuristic based on cache and TLB topology, where prefetching is based on an exhaustive search quantized into cache lines.

We observe a trimodal performance classification: problems where both the vectors and matrix fit in cache, problems where only the vectors can be kept in cache, and problems where neither the matrix nor the vectors can be kept in cache. Clearly, on Barcelona, no matrix ever fits in the relatively small cache, but the performance differences between the problems where the vectors fit can be clearly seen. On Clovertown, where the cache grows from 4 Mbyte to 16 Mbyte using all eight cores, we can see the three matrices that get a substantial performance boost through utilization of all the cache and compression of the matrices. Blue Gene/P, like Barcelona can never fit any matrix in cache, but we can see the matrices (*Economics* through *Linear Programming*) where the vectors don't fit.

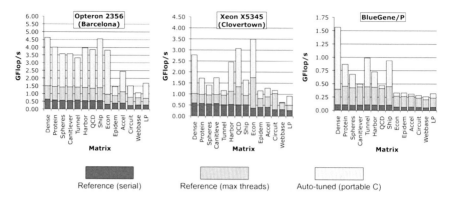

FIGURE 13.10: Auto-tuning Sparse Matrix-Vector Multiplication. Note: horizontal axis is the matrix (problem) and multicore scalability is not shown.

As it turns out, on Barcelona, NUMA-aware allocation was an essential optimization across all matrices. Across all architectures, matrix compression delivered substantial performance boosts on certain matrices. Interestingly, on Blue Gene/P, matrix compression improved performance by a degree greater than the reduction in memory traffic — an effect attributable to an initially compute-bound reference implementation. Interestingly, TLB blocking delivered substantial performance boosts on only one matrix, the extreme aspect ratio linear programming problem.

We observe that threading alone provided median speedups of 1.9×, 2.5×, and 4.2× on Barcelona, Clovertown, and Blue Gene/P. Clearly, only Blue Gene/P showed reasonable scalability. However, when coupled with auto-tuning, we observe median speedups of 3.1×, 6.3×, and 5.1× and maximum speedups of 9.5×, 11.9×, and 14.6×. Ultimately, performance is hampered by memory bandwidth on both Barcelona and Clovertown, leading to sublinear scaling. As a result, auto-tuning strategies targeted at reducing memory traffic are critical.

13.7 Summary

Naïvely, one might expect that nothing can be done to improve the performance of memory-intensive or memory-bound kernels like stencils, LBMHD, or SpMV. However, in this chapter, we discussed a breadth of useful optimizations applicable not only to our three example kernels but to many others domains. Unfortunately, no human could explore all the parameterizations of these optimizations by hand. To that end, we showed how automatic performance tuning, or *auto-tuning*, can productively tune code and thereby dramatically improve performance across the breadth of architectures that currently dominate the top500 list. Unfortunately, to achieve the best performance, non-portable ISA-specifc auto-tuners that generate explicitly SIMDized code are required.

13.8 Acknowledgments

We would like to express our gratitude to Forschungszentrum Jülich for access to their Blue Gene machine. This work was supported by the ASCR Office in the DOE Office of Science under contract number DE-AC02-05CH11231, by NSF contract CNS-0325873, and by Microsoft and Intel Funding under award #20080469.

Chapter 14

Flexible Tools Supporting a Scalable First-Principles MD Code

Bronis R. de Supinski

Lawrence Livermore National Laboratory

Martin Schulz

Lawrence Livermore National Laboratory

Erik W. Draeger

Lawrence Livermore National Laboratory

14.1 Introduction

As discussed in previous chapters, performance analysts have many tools at their disposal. The performance issues presented by any particular application determine the specific utility of each of these tools, including which we should even use. However, experience frequently shows that issues arise that require features not available in any existing tool.

The optimization process undertaken for *Qbox*, a first principle molecular dynamics (FPMD) code developed at Lawrence Livermore National Laboratory (LLNL) in collaboration with the University of California at Davis, clearly illustrates this point. *Qbox* won the ACM Gordon Bell Prize for peak performance in 2006 [157] with a sustained performance of 207 Tflop/s running on 65,536 nodes/131,072 cores of LLNL's Blue Gene/L (BG/L) system, which equals a peak performance rate of 56.4%. Even though *Qbox* was designed for scalability and had already exhibited good parallel efficiency on large runs using conventional cluster architectures, it only achieved this record breaking performance at such an unprecedented scale after a series of performance optimization steps.

The major changes during this processes included carefully optimizing base operations like fast Fourier transforms (FFTs) and dense matrix multiplications to fully exploit the underlying hardware, tuning node placements to achieve a good embedding of the application's communication into the machine's network topology, and integrating an adaptive type scheme into low-level communication to reduce overhead. Each of these steps required a set of sophisticated performance analysis tools to identify the problem and to devise the matching optimization strategy. However, existing tools did not always support the required analysis so we modified them to suit our requirements.

This custom modification of tools proved cumbersome. As our experience demonstrated a general need for a flexible tool infrastructure, we developed P^NMPI [293,294]. This tool infrastructure directly supports the analyses that we required to optimize *Qbox* by allowing selective appplication of existing tools. It employs a stack-based exension of the MPI profiling interface that supports multiple instantiations of the same tool built on that interface. It also provides modular tool components that support rapid development of tools targeting the analysis requirements that arise during the optimization of any MPI application. Current efforts are extending P^NMPI to provide a general interposition mechanism that will support an even wider range of analyses.

In this chapter, we first provide a general description of *Qbox*. We then discuss the system and experiments that led to its 2006 ACM Gordon Bell Prize. Next, we detail the optimizations that we employed to achieve its record setting performance and the tools that we required to identify them. Finally, we detail the customization of existing tools using P^NMPI in the context of the analyses that led to *Qbox*'s outstanding peak performance efficiency.

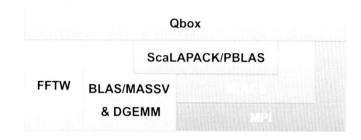

FIGURE 14.1: *Qbox* software structure.

14.2 *Qbox*: A Scalable Approach to First Principles Molecular Dynamics

Qbox is an implementation of FPMD, an accurate atomistic simulation approach used to predict the properties of materials without the need for experimental data or variational parameters. It is a widely used simulation tool in a variety of areas, including solid-state physics, chemistry, biochemistry, and nanotechnology. The FPMD approach combines a quantum mechanical description of electrons with a classical description of atomic nuclei. To reduce the exponential complexity of the many-body Schrodinger equation to a more tractable $O(N^3)$ scaling, a plane wave, pseudopotential density functional theory (DFT) formalism is used [155], where N is the number of chemically-active valence electrons. To simulate the dynamic properties of physical systems at finite temperature, the forces acting on the nuclei are computed from derivatives of the Kohn-Sham (KS) equations at each discrete time step of the trajectory. These forces are then used to move the atoms according to Newton's equations of motion, using a standard velocity Verlet molecular dynamics algorithm. Various authors have extensively discussed the solution of the KS equations within FPMD [75, 267].

14.2.1 Code Structure

Qbox, written entirely in C++, uses the highly optimized FFTW library [141] and the widely known ScaLAPACK and BLACS libaries [58] for the bulk of its computation and communication [156]. It also relies on hand optimized double precision general matrix multiplication (DGEMM) routines, which we will discuss in more detail in Section 14.4. The design of *Qbox* yields good load balance through an efficient data layout and a careful management of the data flow during time consuming operations.

Figure 14.1 gives an overview of the central software structure of *Qbox* and the dependencies between the software modules. The main application uses high-level libraries like ScaLAPACK, but also directly accesses underlying libraries such as BLACS and MPI. This structure provides additional execution efficiency and implementation flexibility. However, it complicates the software optimization process since the access patterns for the underlying libraries are interleaved from multiple sources, which reduces the ability to specialize or to fine-tune individual routines.

Qbox performs many three dimensional (3D) FFTs simultaneously (one for each electronic state). The 3D FFT is notoriously difficult to scale to large task counts, as multiple data transposes require `MPI_Alltoallv` calls. However, because *Qbox* distributes electronic states across subpartitions of the machine, only moderate scalability of the 3D FT kernel is required. *Qbox* uses a custom parallel implementation of 3D FFT that shows excellent scaling properties on up to 512 tasks. Scaling beyond this size is not necessary since a sufficient number of transforms can occur simultaneously to use the entire machine. The 3D FFT kernel requires an efficient implementation of single-node, one-dimensional complex FFTs. We used the FFTW-GEL library that was optimized for BG/L.

14.2.2   Communication Structure

Qbox distributes its central data structure, a dense matrix, across the available nodes in both dimensions. In addition to several global data exchanges, *Qbox* performs most of its communication along rows or columns of this matrix using both point-to-point and collective operations. The latter mostly consist of broadcasts, which tend to dominate the application's execution time, particularly at large scale. To enable and to simplify this communication structure, *Qbox* creates separate row and column communicators along the two dimensional (2D) matrix, such that each node shares a communicator with all nodes that share data in the same matrix column or row.

The 2D matrix in *Qbox* does not represent a traditional homogeneous 2D grid. Instead it stores the coefficients of a Fourier series expansion of the electronic wave functions for all atoms. Hence, access patterns for rows and columns are radically different, which our optimizations must reflect.

14.3   Experimental Setup and Baselines

In this section, we detail the architecture of BG/L. Next, we discuss how we used its hardware performance counters to calculate floating point operation rates. We then describe the test problems used in our experiments. Finally, we present the initial performance of *Qbox* on BG/L in order to establish a baseline for our optimization efforts.

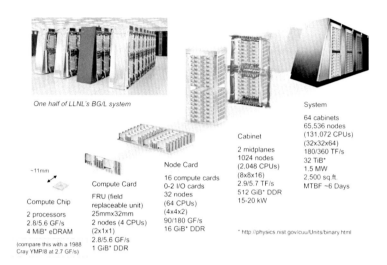

One half of LLNL's BG/L system

~11mm

Compute Card

Compute Chip

2 processors
2.8/5.6 GF/s
4 MiB* eDRAM

(compare this with a 1988
Cray YMP/8 at 2.7 GF/s)

FRU (field
replaceable unit)
25mmx32mm
2 nodes (4 CPUs)
(2x1x1)
2.8/5.6 GF/s
1 GiB* DDR

Node Card

16 compute cards
0-2 I/O cards
32 nodes
(64 CPUs)
(4x4x2)
90/180 GF/s
16 GiB* DDR

Cabinet

2 midplanes
1024 nodes
(2,048 CPUs)
(8x8x16)
2.9/5.7 TF/s
512 GiB* DDR
15-20 kW

System

64 cabinets
65,536 nodes
(131,072 CPUs)
(32x32x64)
180/360 TF/s
32 TiB*
1.5 MW
2,500 sq.ft.
MTBF ~6 Days

* http://physics.nist.gov/cuu/Units/binary.html

FIGURE 14.2: BG/L's scalable system architecture.

14.3.1 The Blue Gene/L Architecture

BG/L is a tightly-integrated large-scale computing platform jointly developed by IBM and LLNL. The original system installed at LLNL consisted of 65,536 compute nodes, each with a dual-core (PowerPC 440) application-specific integrated circuit (ASIC), resulting in an overall peak performance of 367 Tflop/s. At the time of the experiments described here, it headed the Top500 [13] list with a Linpack performance of 280.6 Tflop/s.

BG/L has a modular and hierarchical design, as Figure 14.2 illustrates and an SC2002 paper describes [24]. The base element of the hierarchy is the compute node ASIC. It includes all networking and processor functionality; in fact, a compute node uses only that ASIC and nine dynamic random access memory (DRAM) chips. This system-on-a-chip design results in extremely high power and space efficiency.

Each compute card houses two compute ASICs and their associated DRAM; 16 compute cards (32 nodes or 64 processors) form a node card. 16 node cards form a midplane and two midplanes make up one cabinet with 1024 nodes or 2048 CPUs. The BG/L installation at LLNL at the time of our experiments had 64 cabinets, totaling 65,536 (64 K) nodes or 131,072 (128 K) processors. The system has since then be expanded to 104 racks with 106,496 nodes (212,992 processors).

Each compute ASIC has two 32 bit embedded PowerPC 440 processor cores that run at only 700 MHz. Each core is enhanced with a single instruction, multiple data (SIMD)-like extension of the PowerPC floating-point unit [36], and integrated L1 data and instruction caches (32 KB each) for which no hardware coherence support is provided. The compute ASIC includes hardware prefetching based on the detection of sequential data accesses. The prefetch buffer for each core (its L2 cache) holds 64 L1 cache lines (16 128 byte L2/L3 cache lines). Each compute ASIC also has a four MB embedded DRAM L3 cache that the cores share and an integrated double data rate (DDR) memory controller. The BG/L nodes used in our experiments had 512 MB memory.

BG/L has five different networks, with support for each integrated into the compute ASIC. A 3D torus supports all point-to-point communication as well as collective operations involving subpartitions. These collective operations can exploit torus functionality that deposits packets at each node along a particular direction that it traverses to its overall destination. A tree-based collective network and a fast barrier network provide fast partition-wide MPI collectives. I/O nodes use an Ethernet connection for external communication. Finally, a JTAG[1] network is used for maintenance and bootstrapping.

BG/L supports two node-operation modes. *Co-processor mode* uses one core for computation and typically dedicates the second to communication. *Virtual node mode* uses both cores for computation and communication. The mode that an application employs depends on its computation and communication characteristics as well as memory requirements. Due to the large problem sizes per node of most of our experiments, we ran *Qbox* in co-processor mode although some optimizations (for DGEMM and ZGEMM in particular) employed a special mechanism that allows threaded computation to run on the second core despite the lack of cache coherence.

Each compute node runs a custom light weight kernel (CNK or Compute Node Kernel). This kernel does not include any support for scheduling or multi-tasking and directly implements only lightweight system calls. A larger subset of standard Unix system calls is *function shipped* to external I/O nodes, which execute them as a proxy for the compute nodes and return the results. As a consequence of this operating system (OS) design, the machine is virtually noise free [109], which leads to low overheads and high reproducibility of individual experiment run timings.

14.3.2   Floating Point Operation Counts

We counted floating point operations using the APC[2] performance counter library. This library provides access to the hardware performance counters on the compute ASIC. APC tracks several floating point events including single floating point pipe operations, some SIMD operations, and load and store operations. APC limits counting to selected code regions when calls to the `ApcStart` and `ApcStop` functions delimit them. It saves operation counts in binary format in separate files for each task at the end of the run. The post-processing program *apc_scan* then produces a cumulative report of operation

---

[1] JTAG originally stands for joint test action group and the protocol it defined, but is now more commonly used to describe debug ports for embedded devices in general.

[2] An IBM library to access hardware performance.

counts and total cycles. By default, the APC library limits the number of data files. We used the `APC_SIZE` environment variable to obtain one file per task, since the operation counts of different nodes can vary, especially during ScaLAPACK calls.

We instrumented *Qbox* with APC calls that delimited its main iteration loop. We did not include the initialization phase that involves opening and reading input files and allocating large objects. Since FPMD simulations are typically run for thousands of iterations, our performance measurements are representative of the use of *Qbox* in actual simulations. BG/L's hardware counters do not include events for SIMD addition, subtraction or multiplication, although the fused multiply-add operations can be counted. Thus, our counts can omit some SIMD operations. For this reason, we cannot measure floating point performance from a single run. Instead, we obtain the cycle and operations counts through the following procedure:

1. Compile the code without SIMD instructions (i.e., use $qarch = 440$ with the xlC compiler and non-SIMD versions of the FFTW, DGEMM and ZGEMM libraries);

2. Measure the total floating point operation count with this executable;

3. Recompile the code, enabling the SIMD instructions (i.e., using $qarch = 440d$ and the SIMD libraries);

4. Measure the total cycle count and, thus, the total time with this executable; do **not** use its floating point operation count since it likely omits SIMD operations;

5. Divide the total floating point operation count from the first run by the total time from the second.

Although this procedure requires two simulation runs to get a single measurement, it rigorously measures floating point performance through BG/L's hardware counters.

14.3.3   Test Problem

The relative cost of each term in the Kohn-Sham equation can vary significantly for different materials. Heavy transition metals are among the most challenging, as not only are more valence electrons needed per atom, but their complex electronic structure requires larger basis sets and more pseudopotential projectors to achieve numerical convergence. We chose a benchmark system of 1000 molybdenum atoms with 12 electrons/atom (12,000 electrons total) in a rectangular simulation cell with a large plane wave basis (corresponding to a 112 Ry cutoff energy) and semi-local pseudopotential integration scheme (64 projectors/atom). This system is computationally representative of a typical high accuracy heavy transition metal simulation.

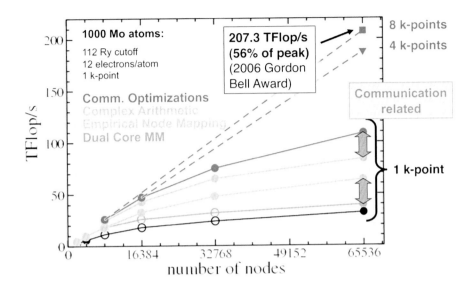

FIGURE 14.3: Strong scaling of *Qbox* molybdenum simulation.

Metallic systems add an additional level of complexity due to the need to explicitly sample different points in the Brillouin zone ("k-points") to compensate for finite-size effects. The Kohn-Sham equation must be solved for all k-points simultaneously, but for the self-consistent iterative approach used by *Qbox*, communication between k-points needs to happen only once each iteration. This allows for straightforward parallization over k-points. The need for multiple k-points also precludes the use of symmetry to cancel out all imaginary components in the Fourier expansion of the electronic wavefunction, and thus requires the use of complex linear algebra throughout. For our benchmark molybdenum calculation, two k-point grids were used: four k-points ($2 \times 2 \times 1$ grid) and eight k-points ($2 \times 2 \times 2$ grid).

14.3.4 Baseline Performance

Figure 14.3 shows the results for strong scaling runs (i.e., constant total problem size) using Co-Processor mode on our test problem. We report the aggregate floating point rate for each partition size measured by hardware performance counters as described in Section 14.3.2. The black (lowest) line captures the baseline performance of porting *Qbox* to BG/L. We observe superlinear scaling from 2048 to 4096 nodes due to communication benefits from changes in the MPI rank to node mapping and to the reduced problem size per node, which led to better cache efficiency. For the largest partition (65,536 nodes), we achieve approximately 32 Tflop/s.

14.4   Optimizing QBox: Step by Step

We describe the series of optimizations that led to *Qbox*'s record setting performance in this section. In particular, we discuss four key optimizations: using a dual core DGEMM implementation; empirical optimization of the MPI rank to node mapping; the use of complex double precision general matrix multiplication (ZGEMM) to support multiple k-points; and fully optimized communication, including further refinements of the node mapping as supported through a custom tool set. We highlight the tools that we used, as well as their deficiencies throughout. Figure 14.3 shows the progressive impact of these optimizations, which led to an overall performance of 207.3 Tflop/s with eight k-points for our test problem, an improvement of a factor of 6.46.

14.4.1   Dual Core Matrix Multiply

Qbox makes extensive use of dense matrix multiplication. Thus, a highly optimized version of this routine is essential for the code's overall performance. In particular, the implementation must use both cores on BG/L despite that we run *Qbox* in Co-processor mode. We employed a special co-routine mode that allows threaded computation to run on the second core when it is not engaged in communication.

BG/L's architecture allows this co-routine execution but provides minimal support since it does not include hardware support for coherence of the L1 caches. Thus, we had to carefully design the matrix multiplication routine to avoid stale cache contents and to ensure proper synchronization. This aspect is in addition to the usual need to limit pipeline stalls by properly blocking the routine so that the working set fits into the local cache resources.

Design and implementation of these routines was a manual process based on detailed knowledge of the underlying architecture. We worked closely with hardware and software engineers at IBM, who provided the specialized co-routine library, as well as the basic DGEMM implementation. For greatest possible efficiency, the implementation blocks and tiles the matrix at multiple levels. The smallest level is a 5x5 matrix multiplication, which can be handled entirely in registers, avoiding additional memory transactions. Overall, this specialized matrix multiplication boosted *Qbox*'s performance for the computation of a single k-point from approximately 32 Tflop/s to about 39.5 Tflops, which is roughly a 25% performance increase.

While the hand-coded matrix multiplication routine significantly improved

TABLE 14.1: Five most time consuming communication routines

	number of calls	avg. message size (bytes)	time (s)	% MPI	% total
MPI_Bcast	3250	710503.9	263.7	67.7%	14.1%
MPI_Allreduce	54168	21306.2	71.4	18.3%	3.8%
MPI_Alltoallv	3381	399.1	27.0	6.9%	1.4%
MPI_Barrier	2355	0.0	15.3	3.9%	0.8%
MPI_Reduce	564	186007.1	11.4	2.9%	0.6%

Qbox's performance, this strategy is not easily extended to large regions of loop-based application code. However, it demonstrates the need to encapsulate fundamental kernels that many applications can reuse, such as dense matrix multiplication, into standard libraries for which complex optimizations are worthwhile. Thus, the highly optimized code specific to low-level architecture details can provide wider benefits and we can reduce the need for general application programmers to consider those details.

14.4.2 Empirical Node Mapping Optimizations

Overall optimization of large scale applications must consider communication as well as computation. Most communication in *Qbox* occurs in existing, optimized libraries and, thus, was already fairly efficient. However, opportunities exist to optimize the mapping of the communication to the target system. In particular, this mapping requires special attention for BG/L, with its multiple networks and its three dimensional torus.

BG/L includes a simple mechanism to specify the node mapping at application load time, which facilitates its optimization. Users can choose from three default mappings (XYZ, YZX, or ZXY) that specify which directions to vary fastest in assigning MPI processes to torus nodes. Alternatively, they can provide a mapping file that specifies the exact torus coordinates of each process.

Unlike stencil-based applications, for which mapping their nearest neighbor communication along a 3D domain decomposition to a 3D torus is relatively straightforward, *Qbox* has a complex communication structure spread throughout several software levels. Since the interactions of the different communications required to solve the KS equations is unclear, we first explored various node mappings that intuitively should perform well. We repeated this process for each partition size, since the shape of a BG/L partition depends on its size. For example, a 4096-node partition consists of an $8 \times 16 \times 32$ block of nodes, whereas a 16,384 partition is a $16 \times 32 \times 32$ block. Thus, an efficient mapping for one partition size may perform relatively poorly at another.

We first measured the communication times in *Qbox* with mpiP, a low-

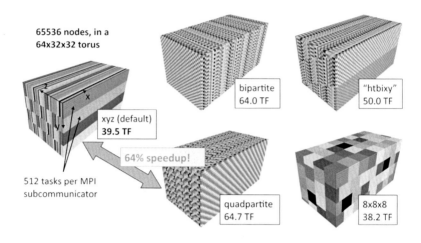

FIGURE 14.4: Performance on different node mappings. (See color insert.)

overhead profiler [343]. Table 14.1 shows the five largest contributors to *Qbox* communication on an 8192 node partition. `MPI_Bcast` and `MPI_Allreduce` dominate (by total time) the communication cost of the code. However, mpiP provides limited information in terms of which communicators are involved in these operations. Thus, we examined the source more closely in order to discern key issues for optimizing this communnication.

The *Qbox* data layout distributes the degrees of freedom that describe electronic wave functions on a two dimensional process grid similar to the BLACS process grids [58]. Most *Qbox* communication occurs as collective operations on MPI communicators that correspond to rows and columns of that process grid. While consistent with the mpiP data, this source code analysis indicates that these collectives involve only a fraction of the total nodes. Thus, these operations do not use BG/L's collective network since it can only support collectives over the entire job, i.e., the `MPI_COMM_WORLD` communicator. Therefore, these collectives use the torus network and the node mapping can substantially impact their performance.

As the ideal mapping is not obvious, we empirically tested a range of node mappings that we expected to provide a balance between the row and column communication. For a 65,536-node calculation of 1000 molybdenum atoms with a single k-point, i.e., a single set of KS equations, we found that the mapping dramatically impacted performance, as Figure 14.4 illustrates. The colors in each configuration represent the nodes belonging to one 512-node column of the process grid. Our empirical node mapping experiments found that the default XYZ, YZX, or ZXY node orderings poorly interleaved the nodes for row or column communication (or both). For example, the default XYZ mapping (left-most configuration in Figure 14.4) resulted in a sustained floating-point performance of 39.5 Tflop/s. Intuitively, compactly packing the columns (lower right configuration in Figure 14.4) should minimize intra-column communication. However, this choice actually reduced performance to 38.2 Tflop/s. Overall, we found that distributing each process column over a torus slice in a bipartite (upper middle configuration in Figure 14.4) or quadpartite (lower middle configuration in Figure 14.4) arrangement provided the highest performance for the 65,536-node partition: 64.0 Tflop/s and 64.7 Tflop/s. We found that the bipartite mapping resulted in the highest performance for partition sizes of 8192, 16,384, and 32,768 nodes. These mappings provide a balance between efficient communication on the rows and the columns of the *Qbox* process grid. Overall, we derived an improvement of 1.64X through empirical optimization of the node mapping, which illustrates the critical importance of the task layout on a machine like BG/L.

14.4.3   Use of Complex Arithmetic

We gained the next large jump in performance when we extended *Qbox* to simultaneously solve multiple k-points. This change requires the use of complex arithmetic in order to handle the phase differences between the k-points. Thus, handling multiple k-points not only increased *Qbox*'s capabilities but also increased the computational density of its core routines, thus increasing the computation to communication ratio and decreasing relative parallel overhead.

Complex arithmetic increased the workload in the matrix multiplications at the heart of *Qbox*, thus making it more profitable to execute them as a co-routine on the second core. However, this change in arithmetic changed the computation algorithm significantly and doubled the working set. Thus, we had to develop a new optimized implementation that specifically targeted complex matrix multiplication. Similarly, other libraries, like the FFT computation, also required new implementations. Overall, direct support for complex arithmetic that multiple k-points required increased performance by 31% from under 65 Tflop/s to around 85 Tflop/s for the computation of one k-point.

14.4.4   Systematic Communication Optimization

After optimizing the use of complex arithmetic, we revisited *Qbox*'s communication. We facilitated a more detailed investigation by modifying mpiP [343] to provide a detailed breakdown between row, column and global collective operations. Further, we developed simple benchmarks that we combined with BG/L hardware performance counters related to its integrated network queues to understand the link-level optimization of the torus of different mapping strategies. Finally, we designed a task mapping that reflected this detailed knowledge, in addition to implementing other communication optimizations suggested by the mpiP data.

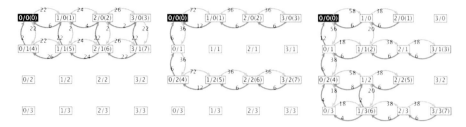

FIGURE 14.5: MPI_Bcast packet counts for different task mappings.

Our empirical task mapping optimizations significantly decreased the overall execution time. However, the best mappings were unintuitive, since they did not pack either rows or columns of the processor grid compactly. Thus, we observed hardware performance counters of network queue usage in order to understand how BG/L's MPI implementation maps collective operations to the torus network. We focus on the performance of MPI_Bcast, since our performance profile, taken with our hand-modified version of mpiP, showed that this call dominates *Qbox* communication costs under the empirically best quadpartite mapping. Figure 14.5 shows the torus packet counts of a program that issues one 4000 byte MPI_Bcast using an eight node communicator with three different task mappings. In these graphs, node labels indicate the coordinates of the node within the grid. Gray nodes are part of the communicator, with the rank indicated in parenthesis, while the black node is the root of the broadcast. White nodes are not part of the communicator on which the broadcast is performed, although they may forward packets of the operation. Edge labels indicate the number of packets transmitted on the associated torus link as measured using BG/L's hardware counters. Edge colors indicate the direction of the torus traffic.

When using a compact rectangle of nodes (Figure 14.5, left), the broadcast exploits the ability to deposit torus packets at each node traversed to the final destination. Thus, we observe similar packet counts at each node. While compact mappings provide the optimal packet count for torus broadcasts, *Qbox* communicates along both rows and columns, which implies we cannot use a compact mapping along at least one direction. Further, the resulting mapping for the other direction, when one is placed compactly, requires many more packets. The middle of Figure 14.5 shows that broadcasts for the resulting mapping for the second direction when the other uses compact rectangles not only cannot use the deposit functionality but also poorly utilize the torus links, leading to overall lower performance.

Figure 14.5 (right) shows that a checkerboard mapping highly utilizes the torus links. Thus, task mappings that use this type of layoput for both rows and columns provide high overall efficiency. In particular, the quadpartite mapping that Figure 14.4 shows, should provide good performance based on this analysis. More importantly, this analysis suggests that we can further improve its performance by careful placement of the tasks within that mapping to reflect the expected binomial tree communication pattern of the MPI_Bcast implementation. Thus, we designed a new layout that is based on a space filling curve. This new mapping significantly improved performance when combined with other optimizations that our closer examination of the mpiP data suggested.

The mpiP data revealed other optimization opportunities related to *Qbox*'s layered design. Besides directly calling MPI functions, it also directly calls BLACS routines, which are implemented by MPI functions, and ScaLAPACK routines, which use BLACS routines. We found that this layered implementation led to many datatype creation operations within the BLACS library. The generic version of BLACS creates new datatypes for each message transfer that might access strided data. However, all such calls within *Qbox* send data regions that were contiguous, which can simply use the message count, thus avoiding the type creation (and destruction) costs. We modified the BLACS library to detect possible strided accesses that result in contiguous memory regions, in which case we used the message count parameter to save the datatype costs without changing the interface between *Qbox* and ScaLAPACK or BLACS.

Our more systematic optimization of *Qbox*'s communication further improved performance for one k-point from 85 Tflop/s to 110 Tflop/s, an additional improvement of 29.4%. Our analysis indicated that the final task mapping provided most of this benefit although the BLACS optimization and other minor communication improvements accounted for a few percent of it. When we apply this version of the code to the full eight k-point configuration, we obtained the overall sustained performance of 207.3 Tflop/s.

14.5   Customizing Tool Chains with P^NMPI

Our experiences optimizing *Qbox* demonstrated the need for a flexible tool infrastructure. In particular, we created a custom version of mpiP to guide our communication optimizations since no existing tool provides the information that we needed. Tools built using the PMPI interface are invoked on every instance of any MPI call that they intercept. It is difficult to restrict them to calls with specific properties, such as the use of a particular communicator, without explicitly adding this capability.

The limitations of the PMPI interface led us to develop P^NMPI, a flexible infrastructure for MPI tools [293, 294]. P^NMPI provides a variety of capabilities based on a simple abstraction of stacking the wrapper functions that PMPI tools use. For example, it supports generic MPI tool functionality, such as request tracking. Relative to our experiences optimizing *Qbox*,

```
#define default stack
module commsize-switch
argument sizes 8 4
argument stacks row column
module mpiP

#define tool stack for rows
stack row
module mpiP1

#define tool stack for columns
stack column
module mpiP2
```

FIGURE 14.6: Example P^NMPI configuration file.

P^NMPI includes *switch modules*, which directly support selective application of PMPI-based tools. A switch module examines the relevant state, such as values of parameters to the MPI call, and then invokes a separate stack with an independent copy of the tool for calls that share that state.

14.5.1 Applying Switch Modules to *Qbox*

We use P^NMPI to target independent copies of mpiP [343] to *Qbox* communication. Specifically, we separate the use of row and column communicators and MPI_COMM_WOLRD, instead of aggregating the communication behavior across all communicators, as mpiP does by default. Thus, we can distinguish the distinct communication patterns and target optimizations to the semantic breakdown.

We use a P^NMPI switch module in order to obtain the separated measurements without having to hardcode it into a profiling tool. This module intercepts each communication call and then forwards the call into separate tool stacks depending on the size of the communicator used for that particular MPI call. We use the communicator size instead of its handle since our *Qbox* simulations had distinct sizes for row and column communicators and some of its underlying libraries, like BLACS, create the associated communicators multiple times, which results in different handles for the same row or column in a single execution while the size remains constant. Alternatively, a more complicated switch module could perform calculations when communicators are created in order to partition them appropriately. We then run an unmodified, separate copy of mpiP in each tool stack, which collects a profile only for MPI calls forwarded to that particular stack. This strategy results in separate mpiP data for the row, column and job-wide classes of communicators.

TABLE 14.2: Global and communicator-specific mpiP results

Count	Global	Sum	COMM_WOLRD	Row	Column
Send	317365	317245	31014	202972	83259
Allreduce	319028	319028	269876	49152	0
All2allv	471488	471488	471488	0	0
Recv	379355	379265	93034	202972	83259
Bcast	401312	401042	11168	331698	58176

Figure 14.6 shows the P^NMPI configuration file that targets these communicator classes. We load the switch module with a `sizes` argument that applies different stacks to communicators of sizes eight and four (assuming a 32 task job mapped to a 8x4 processor grid). The former size is associated with the `row` stack, while the latter is associated with the `column` stack.

The rest of the configuration file then loads mpiP on the default stack as well as the `row` and `column` stacks. The different names of the mpiP instantiations represent copies of the mpiP library. These copies guarantee that the dynamic linker treats them as distinct libraries and allocates a separate global state for each. Thus, we obtain a separate mpiP report for each communicator type.

14.5.2   Experiments and Results

In this section, we demonstrate that the flexible tool support provided by P^NMPI significantly simplifies gathering of the information that we needed to optimize *Qbox*'s communication. While our actual optimization efforts had to distinguish row and column communicators manually, we recreated the measurements simply using the approach described in Section 14.5.1.

Table 14.2 shows the profiling results for the top five MPI routine counts observed during an execution of *Qbox* (using a working set size of 54 molecules on 32 tasks). The second column shows the results of a global profiling run, while the last three columns represent the results from our multiplexed tool stack. The third column shows the sum of the last three, which we provide for comparison purposes and is not gathered by the tool stack. Nondeterministic differences during the computation cause the minor difference between that sum and results from the separate global profiling run.

The results clearly show the different behavior for the three types of communicators used in the code: all-reduces and all-to-all calls are mainly restricted to `MPI_COMM_WORLD`, while rows and columns dominate point-to-point communication. Further, the code uses many broadcasts along rows, which the standard profiling run does not reveal. We made significant changes to mpiP to gather this data directly during the original effort to optimize *Qbox*. With P^NMPI we were able to replace this coding effort with just a single

TABLE 14.3: Overhead of different profiling approaches

	Execution Time (seconds)	Overhead (percent)
Native	446.62	0.0%
mpiP	499.40	11.8%
P^NMPI & mpiP	499.88	11.9%
Multplexed	500.03	12.0%

switch module that only required a few lines of code. Hence, P^NMPI significantly reduces the necessary effort to distinguish the states for which we need distinct profiling information and thereby enables an easy and fast creation of application specific tools or tool adaptations.

Importantly, the reduced effort to gather the needed information does not come at a significant cost in accuracy or overhead. Our experiments above confirm that the approach is accurate since we see the same overall data while obtaining the breakdown by communicator classes. Further, Table 14.3 shows the execution times and overhead numbers for running those experiments. We compare the native execution without tool support to using mpiP as a standard PMPI tool, using mpiP as a transparent P^NMPI module, and running our complete multiplexed tool stack. The results show that mpiP adds nearly 12% overhead to the execution. We see small additional overheads when we use mpiP within the basic P^NMPI infrastructure and when we target different copies to the different communicator classes.

14.6 Summary

In this chapter, we presented the optimizations of *Qbox* that led to its ACM Gordon Bell Prize winning performance of 207 Tflop/s on BG/L. We achieved this record setting performance of 56.4% of the peak performance of the then world's largest supercomputer through optimizations of both the application's computation and communication. This effort was guided not only by detailed knowledge of the application and the architecture but also by performance analysis tools that identified specific optimization opportunities. Overall, our results show that all three of these information sources are required in order to exploit large-scale systems effectively.

We learned several lessons from optimizing *Qbox* for BG/L. Some help guide future architecture designs. For example, we had to use the second processor for computation as well as communication. However, large memory applications, such as *Qbox*, could not benefit from virtual node mode and we had to rely on low-level coding to use that processor since the system lacked L1 cache coherence. Overall, processor architectures are complex and features such as prefetching effectively limit the ability of application programmers to maintain coherence for themselves. Thus, Blue Gene/P, the next generation Blue Gene system, included L1 coherence.

Other lessons help guide future tool development. In particular, we found that no tool provided the ability to measure MPI communication based on the communicator used. We hand modified mpiP in order to gather the needed information that allowed us to derive a node mapping that reflected the actual communication pattern of *Qbox*. This process was time consuming and difficult. Further, our general lesson is that tool writers cannot anticipate every specific separation of performance information that optimization might require. Thus, we have designed and implemented P^NMPI, a flexible tool infrastructure that greatly simplifies targeting tools to the needs of a specific optimization effort. Our results demonstrate that it can substantially reduce the effort to gather the information needed to optimize *Qbox* node mappings at little additional overhead. We are currently extending P^NMPI so that it can apply to optimization scenarios beyond MPI communication.

14.7   Acknowledgment

This work was sponsored in part by the Performance Engineering Research Institute, which is supported by the SciDAC Program of the U.S. Department of Energy. This manuscript and its illustrations have been authored by contractors of the U.S. Government under the auspices of the U.S. Department of Energy by Lawrence Livermore National Laboratory under contract DE-AC52-07NA27344. Accordingly, the U.S. Government retains a nonexclusive, royalty-free license to publish or reproduce the published form of this contribution, or allow others to do so, for U.S. Government purposes.

Chapter 15

The Community Climate System Model

Patrick H. Worley

Oak Ridge National Laboratory

15.1   Introduction

The Community Climate System Model (CCSM) is a modern world-class climate code consisting of atmosphere, ocean, land, and sea-ice components coupled through exchange of mass, momentum, energy, and chemical species [98, 101]. Investigating the impact of climate change is a computationally expensive process, requiring significant computational resources [263]. Making progress on this problem also requires achieving reasonable throughput rates for individual experiments when integrating out to hundreds or thousands of simulation years. Climate models employ time-accurate numerical methods, and exploitation of significant parallelism in the time-direction has yet to be demonstrated in production climate models. For the CCSM this leaves functional parallelism between the component models, parallelizing over

315

the spatial dimensions, and loop-level parallelism exploited within a shared-memory multi-processor compute node or a single processor. Due to as yet unavoidable parallel inefficiencies, the size of the spatial computational grids that can be used and still achieve the required throughput rates for long time integrations is small compared to other peta- and exa-scale computational science. As a consequence the maximum number of processors that can be applied in a single experiment is also relatively small. Parallel algorithms need to be highly optimized for even a modest number of computational threads to make best use of the limited amount of available parallelism.

The CCSM is also a community code that is evolving continually to evaluate and include new science. Thus it has been very important that the CCSM be easy to maintain and port to new systems, and that CCSM performance be easy to optimize for new systems or for changes in problem specification or processor count [122].

This chapter describes some of the performance issues encountered, solutions developed, and methodologies utilized during the performance engineering of the CCSM and its predecessors over the past 20 years. The exposition alternates between descriptions of specific case studies and general observations drawn from a larger body of work. Topics discussed include interprocess communication, load balance, data structures, system performance characteristics and idiosyncrasies, and performance portability.[1] Of note is that many of the performance issues today are similar to those identified 20 (or more) years ago.

The emphasis here is on diagnosing and addressing performance "bugs." There is only limited discussion of the design and evaluation of parallel algorithms, which is arguably the most critical aspect of the performance engineering of the CCSM. Moreover, while the CCSM is the context in which the discussion takes place, the issues and solutions are not specific to the CCSM.

There have been, and continue to be, many contributors to the performance analysis and optimization of the CCSM. References to some of this other work is provided here, but this chapter is in no way meant to be a comprehensive discussion of performance engineering in the CCSM.

15.2   CCSM Overview

The CCSM consists of a system of four parallel geophysical components (atmosphere, land, ocean, and sea ice) that exchange boundary data (flux and state information) periodically through a parallel flux coupler [124]. The flux coupler serves to remap the boundary-exchange data in space and time. The

---

[1] Performance portability refers to the ability to optimize performance easily upon porting to a new system, changing problem specification, and/or changing process or thread counts.

concurrency of the CCSM components is not perfect; the atmosphere, land and sea-ice models are partially serialized in time, limiting the fraction of time when all five CCSM components can execute simultaneously. For this reason, it is typical for the atmosphere, land, and sea-ice model to run on a common set of processors, while the ocean model runs concurrently on a disjoint set of processors. This is not a requirement, however, and the CCSM can run all components on disjoint processor subsets, all on the same processors, one after the other, or any combination in between. (This flexibility became available with version 4 of the CCSM [101], released on April 1, 2010.)

The atmosphere model is the Community Atmosphere Model (CAM) [100]. The ocean model is the Parallel Ocean Program (POP) [310]. The land model is the Community Land Model (CLM) [114]. The sea ice model is the Community Ice Model (CICE) [57,175]. With the CCSM4 release, all components are hybrid parallel application codes, using MPI [238] to define and coordinate distributed-memory parallelism and OpenMP [106] to define and coordinate shared-memory parallelism. Each component model has its own performance characteristics, and the coupling itself adds to the complexity of the performance characterization [124].

The first step in optimizing CCSM performance is the optimization of the individual component models. Once this is complete, the number and distribution of processors for each component must be determined, based on some performance or resource requirements. This exposition discusses performance issues arising in the two three-dimensional components, the ocean and the atmosphere, which have in the past been the most expensive in CCSM simulations. For a discussion of performance engineering in the other components and of the load balancing of the full CCSM, see [122,124] and all of the articles in [123].

15.3 Parallel Computing and the CCSM

The Community Climate Model (CCM), the predecessor to CAM, was developed at the National Center for Atmospheric Research (NCAR) in 1983. CCM0 was run initially on the Cray-1 vector system [369]. CCM1, released in 1987, also supported the Cray-XMP shared memory parallel vector processing system [370]. CCM2, released in 1992, added support for the Cray-YMP [158]. The Parallel Ocean Program (POP) was developed at Los Alamos National Laboratory (LANL) in 1992, written initially in a data parallel formulation using CM Fortran and targeting the Thinking Machines CM-2 and CM-5 systems [331].

As distributed-memory parallel systems became viable platforms for production climate simulations, new parallel algorithms were developed and codes were redesigned and ported. Since the early 1990s, kernel codes extracted from the component models (for evaluation and as the first step in porting), component models, or the full CCSM have been ported and performance evaluated on a large number of systems. The author has first hand knowledge of work on systems from Convex (SPP-1200, SPP-2000), Cray (T3D, T3E, X1, X1E, XT3, XT4, XT5), HP/Compaq (AlphaServer SC), IBM (clusters of Power1 (SP1), Power2 (SP2), Power3 (SP3), Power4, Power5, and Power6 processor-based SMP compute nodes, BG/L, BG/P), Intel (iPSC/860, Delta, and Paragon), nCUBE (nCUBE 2), NEC (SX-5, SX-6), SGI (a number of different Origin and Altix systems), as well as the Earth Simulator and a variety of commodity clusters. The point of this list is to emphasize the effort that has been taken in porting to and evaluating potential target platforms and the diversity of these targets, thus motivating the need for performance portability.

15.4 Case Study: Optimizing Interprocess Communication Performance in the Spectral Transform Method

One of the early efforts in porting the atmosphere model to distributed-memory parallel computers targeted an Intel iPSC/860 hypercube with 128 processors [348,380]. The initial goal of this work was to develop and evaluate parallel algorithms for the spectral transform method that was used in the CCM2. The initial approach utilized a decomposition of both the physical and spectral domains and a ring algorithm when transforming between the two, as described below.

When transforming from spectral space to physical space, the spectral coefficients defining the spectral representation of the solution are moved between processors that are viewed as being being arranged in a ring topology. At each step a processor sends its current block of spectral coefficients to the processor on its left (in the ring topology), uses these data to calculate contributions to the physical quantities it is responsible for, then receives a new block of coefficients from the processor on the right. After P steps, where P is the length of the ring, the calculation is complete and each block of spectral coefficients has been returned to its source processor.

When transforming from physical to spectral space, the spectral coefficients being calculated are circulated around the ring. Each processor calculates its contribution to the block of coefficients it knows that it will receive next, receives that block from the *right*, adds in its contribution, sends the updated block to the *left*, and begins the calculation for the next block.

This approach has many advantages, including the ability to overlap communication with computation. The initial implementation performed poorly

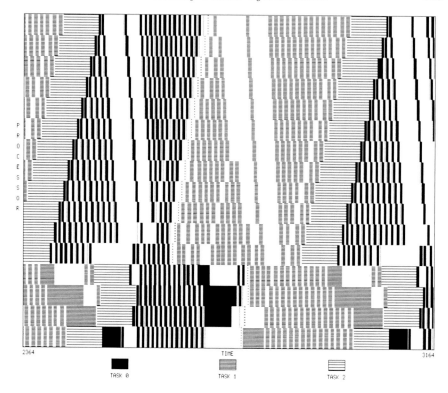

FIGURE 15.1: Task graph showing effect on ring algorithm of load imbalances. (From [380]. With permission.)

however. See [380] for the details, but the sources of the problems are still issues today or have analogues.[2]

   a) load imbalance. See Figure 15.1. The task graph shows that certain instances of *Task 0* on processors 0-3 take 20 times longer than on the other processors, introducing large idle times in the interprocess communication (unlabeled regions in the task graph). Here the X-axis is wallclock time and the Y-axis is the process ID, 0 to 15. The shading indicates what logical task a given process was engaged in at a given point in time.

   b) blocking send/receive communication protocol preventing communication/computation overlap. See Figure 15.2. The Gantt chart shows that a send on a *source* processor does not complete until the receive is posted

---

[2]The iPSC/860 system used the NX message passing library for interprocess communication [274]. The MPI standard had not yet been defined, but NX had a large number of capabilities that were included in MPI and all of the commonly used NX commands have close analogues in MPI.

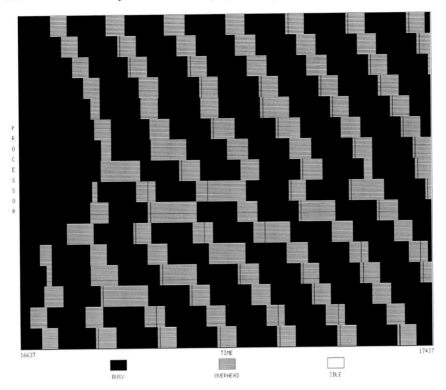

FIGURE 15.2: Gannt chart showing effect on ring algorithm of blocking communications without system buffering. (From [380]. With permission.)

on the *destination* processor.[3] The source processor is always immediately above the destination processor in this figure. Here *busy* denotes computation, *idle* denotes time blocked waiting for a message to arrive, and *overhead* denotes other time spent in the communication library routines. X-axis and Y-axis are again time and process ID, respectively.

c) handshaking message used in library implementation of nonblocking send/receive of large messages preventing communication/computation overlap. See Figure 15.3. Here the system handshaking message from the left neighbor is sometimes blocked from being received because the processor is receiving a large application message from right neighbor.

All of these performance issues were resolvable once the causes were iden-

---

[3]While this behavior is still a option, the typical implementation of the blocking send/receive protocol in current communication library implementations will try to allocate system memory to receive this unexpected message. This can also cause problems, as described in a later section.

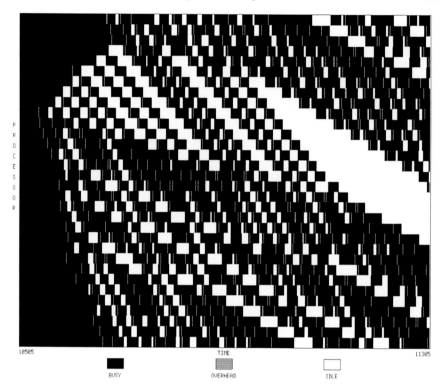

FIGURE 15.3: Gannt chart showing effect on ring algorithm of messages arriving at a processor from both left and right simultaneously. (From [380]. With permission.)

tified. See Figure 15.4 for the Gantt chart of the optimized version of the code.

In all cases performance was evaluated initially via strong scaling studies: per process wallclock times for performance sensitive phases of the code were measured for a number of fixed problem sizes and for a number of different processor counts. For issues (a) and (b), unexpectedly poor performance motivated the use of performance visualization to diagnose the performance problems. This expectation was based on an informal model of how performance should scale. The performance degradation resulting from issue (c) was not great enough to cause immediate suspicion that something was wrong. Subsequent fitting of the data to an algebraic performance model data was very accurate *if* communication costs were twice what was expected based on microbenchmarks results. Subsequent visualization of performance indicated that there was a performance problem. The source of the problem was identified via a discussion with the communication library developers. The workaround for (c) involved an explicit handshaking message at the begin-

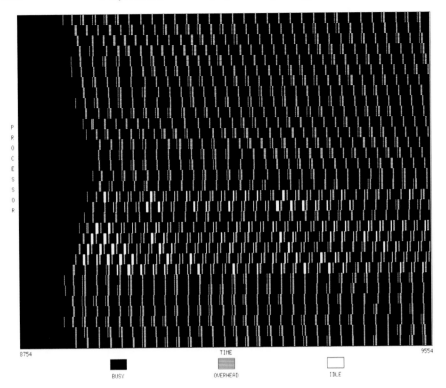

P
R
O
C
E
S
S
O
R

8754   TIME   9554

BUSY   OVERHEAD   IDLE

FIGURE 15.4: Gannt chart for optimized ring algorithm. (From [380]. With permission.)

ning of the ring algorithm and the use of a send command that did not utilize an internal handshaking message (i.e., a "ready send" protocol).

Performance modeling and performance visualization proved to be very important in this study. Such techniques are typically saved for the "hard" cases, and are not used in day-to-day CCSM performance engineering activities. In contrast, detailed performance scaling studies were and are used regularly, for performance debugging, tuning, and evaluation [378, 379].

15.5   Performance Portability: Supporting Options and Delaying Decisions

The study described in Section 15.4 was one of the first in a very active period of parallel algorithm design and experimentation within the CCSM development community. One recurring theme in subsequent CCSM work was the recognition that the communication protocol (transport layer, message passing library, one-sided versus two-sided communications, collective versus point-to-point commands, synchronous/blocking/nonblocking/... commands, pre-posting communication requests and/or delaying communication completion tests, aggregating or splitting communication requests, explicit handshaking messages and use of the "ready send" protocol, order of communication requests, etc.) can have a large impact on performance, and the best choice varies with parallel algorithm, problem size, number of processes, communication library, and target architecture [377]. Because of this, the communication protocol used varies throughout the CCSM component models, and some component models support compile- or runtime selection of the communication protocols to use for a given parallel algorithm within the component [185, 244, 381].

As the parallel algorithm research continued, a number of competing parallel algorithms were developed and the list of target platforms increased [382]. For fairness (and accuracy) in comparing both algorithms and platforms, each of the alternative algorithms needed to be optimized on the target platforms [136]. Algebraic performance models for some of the parallel algorithm alternatives [61, 137] were also developed that were sufficient to determine asymptotic behavior. However, predictions of performance for a particular problem size, target architecture, and implementation still required empirical data to determine constants used in the models. These studies also demonstrated clearly the utility of embedding a number of parallel implementations in the same model and the utility of an efficient experimental methodology for determining the optimal choices, as the best approach was a function of problem size, processor count, and target computer system [119, 136, 379, 383].

15.6 Case Study: Engineering Performance Portability into the Community Atmosphere Model

The design of CAM [99], the proposed successor to the CCM [188], began in the Spring of 2000. Atmospheric global circulation models, like CCM and CAM, are characterized typically by two computational phases: the dynamics, which advances the evolution equations for the atmospheric flow, and the physics, which approximates subgrid phenomena such as precipitation processes, clouds, long- and short-wave radiation, and turbulent mixing [99]. One of the decisions made early on in the design process was that CAM would support multiple dynamics solvers, henceforth referred to as *dynamical cores* or *dycores*, that share a common set of physics options. (The dycore is selected at compile-time.) As these dycores would not necessarily use the same computational grids nor support the same domain decompositions in their parallel implementations, it was decided to define an explicit interface between the dynamics and the physics. As a consequence, the physics data structures and parallelization strategies would no longer be required to be identical with those used in the dynamics. Instead a dynamics-physics coupler would be used to move data between data structures representing the dynamics state and the physics state.

In earlier work [120, 121, 136, 137], the goal had been to determine data structures and domain decompositions that work well with both the dynamics and the physics. By decoupling physics and dynamics data structures, a global design was no longer *necessarily* advantageous. Developers were thus free to investigate optimizing performance within the physics and dynamics independently, and to investigate different approaches to minimizing the overheads introduced by the decoupling of the physics and dynamics data structures. This activity also came at an interesting time architecturally. Vector systems such as the NEC SX-5 and SX-6 and the Cray X1 were possible production platforms, and there was strong motivation to support good performance on both vector and nonvector processor architectures. One effort at this time, described in [381], focused on optimizing performance in the physics.

The primary focus in [381] was the development and evaluation of a physics data structure that could be efficient for both vector and nonvector systems. All grid points in a three-dimensional grid with the same horizontal location are referred to as a *column*. The physics in CAM treats each column independently, and the processing of each column is very similar. Thus an index over the columns represents a potential direction for vectorization. In contrast, there is a tight coupling computationally between the vertical levels within a column, and the computational intensity is higher if one column is processed at a time. To support both scenarios, the index space representing the columns was "factored", with one factor of the index space assigned to the innermost index and the other assigned to the outermost index. This defines subsets of columns (referred to as *chunks*) that are processed together. Vectorization is applied within a chunk, while MPI and OpenMP parallelism are applied between chunks. Having larger numbers of columns in a chunk should improve vector performance, while having smaller numbers should improve nonvector processor performance.

Modern processor architectures are complex, however, and the above performance characterization of vector and nonvector processors is overly simplistic for use in algorithm optimization. Figure 15.5 describes an empirical evaluation of performance as a function of the number of columns assigned to a chunk for four computer systems. Runtimes are normalized with respect to the minimum observed runtime for each target system, and performance can not be compared between systems quantitatively in this figure. The Cray X1 vector architecture did in fact prefer large chunks (≥ 1026 in this study). Performance of the Power4 processor in the IBM p690 system was not very sensitive to chunk sizes within the range of 16 to 200, but the extremes (small and large) performed significantly worse. The Intel Itanium2 processor in

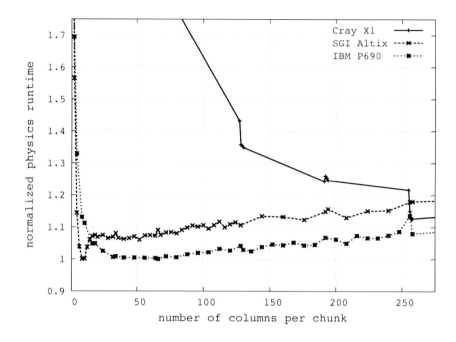

FIGURE 15.5: Sensitivity of physics performance to number of columns assigned to a chunk. (From [381].)

the SGI Altix showed a strong preference for chunks with 8 columns (for the given problem resolution). When CAM is ported now, similar studies are undertaken to determine the optimal chunk size on the new architecture.

Once the size of the chunks has been established, the assignment of columns to chunks and of chunks to processes needs to be determined. Each dycore has its own decomposition of columns among the processes. If the decomposition used by the dycore is retained by the physics, then the physics-dynamics coupling will not incur interprocess communication overhead. However, the cost of processing a column in the physics varies between columns and varies for a given column with simulation time. See Figure 15.6. The first graph describes how the cost for a column near the equator varies with simulation time for 24 hours for three different types of timesteps. (Not all physics calculations occur at the same time frequency.) Here runtime is normalized with respect to the average over all observations for a given type of timestep, and the relative cost of the different timesteps can not be determined from these data. The second graph describes how the cost of the most common type of timestep varies as a function of longitude for three different simulation times. Based on this analysis, a static load balancing scheme was developed.

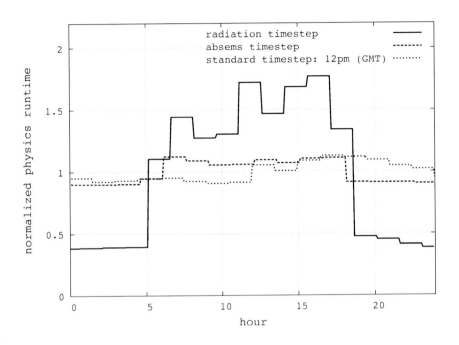

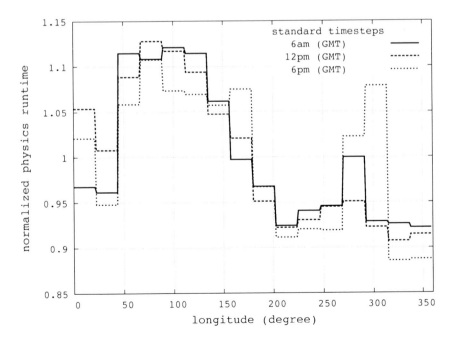

FIGURE 15.6: CAM physics load variability experiments. (From [381].)

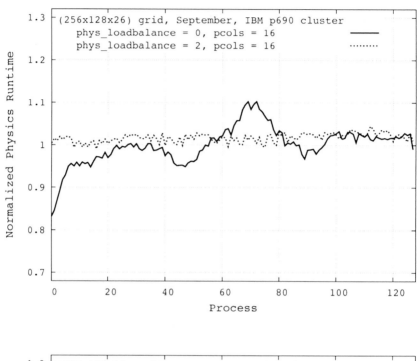

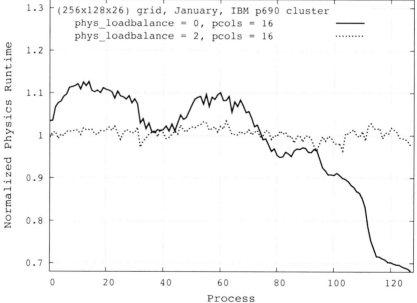

FIGURE 15.7: CAM physics load balancing experiments (128 processors on IBM p690 cluster, September and January). (From [381].)

Figure 15.7 demonstrates the efficacy of this load balancing scheme, graphing the cost of the physics over a simulation month as a function of process id both with (`phys_loadbalance=2`) and without (`phys_loadbalance=0`) load balancing. These results are from runs on an IBM p690 cluster using 128 processors, but similar results hold for the other architectures, including the Cray X1 vector system. The top graph is for the calendar month September and the bottom graph is for the calendar month January. Here the runtime is normalized against the average runtime when running without load balancing for a given simulation month, and performance with and without load balancing can be compared directly. This analysis did not take into account the cost of the interprocess communication required to remap the columns between the decompositions used in the dynamics and the physics. As this cost is sensitive to the architecture, problem size, and number of processes, the decision as to whether to use load balancing is determined on the basis of empirical experiments [379].

In summary, introducing flexibility into the design and using empirical studies for performance optimization was critical for supporting performance portability across these different platforms. This flexibility has continued to be very important for maintaining good performance when porting to new architectures and as the model evolves.

15.7   Case Study: Porting the Parallel Ocean Program to the Cray X1

In 2002–2003 the Parallel Ocean Program was ported to and optimized on the NEC SX-6, and then on the Earth Simulator [185]. A similar exercise was undertaken for the Cray X1 vector system in 2003–2004 [384]. This Cray X1 study was focused in particular on determining how the performance characteristics of the Earth Simulator and the X1 differed, and how to take this into account when optimizing POP performance. The first and third figures in Figure 15.8 are task graphs that document the impact of modifying approximately 700 lines of code (out of 45,000), decreasing the cost of *Task 1* by nearly a factor of 3. While there were some differences between the code modifications for the Earth Simulator and for the X1, they were on the order of tens of lines of code. Both the NEC and Cray compilers provided excellent feedback on where and why the code could not be vectorized. The visualization in Figure 15.8 was used simply to document the impact of vectorization. The second and fourth figures in Figure 15.8 are utilization count histograms that document the number of processes computing, communicating, or idle (waiting for communication) as a function of wallclock time, before and after vectorization respectively. This visualization indicates clearly the communication-bound nature of the run.

Another optimization, first identified and implemented in the port to the Earth Simulator, was to replace a series of halo updates, one per vertical level, with a single halo update for all vertical levels simultaneously.[4] The total volume of data moved was the same, but the number of requests, and message latency, was reduced to one per timestep. The top two histograms in Figure 15.9 show the impact of this optimization on the Cray X1. While still communication bound, execution is now 3 times faster. (Note that the top two histograms in Figure 15.9 document exactly half the amount of runtime as do the histograms in Figure 15.8.)

A communication benchmark that examines performance using different messaging layers and communication protocols [349] indicated that using Co-Array Fortran [262] one-sided messaging can be 5 times faster than using two-sided MPI messaging for latency-sensitive communication operations on the Cray X1. Based on this, Co-Array Fortran implementations of performance-sensitive halo updates and MPI collectives were introduced into POP. The performance impact of these modifications is displayed in the bottom two histograms in Figure 15.9. (The bottom two histograms document half the amount of runtime as the top two histograms in Figure 15.9.) No attempt was made to instrument the Co-Array communication, and the time spent in the Co-Array Fortran implementations of the halo update and MPI collectives is defined to be idle time. As can be seen, performance is doubled compared to the previous version, and the code is now no longer communication bound for this processor count.

This study occurred over a period of 12 months and included frequent communication with the Cray operating system and MPI library developers. As a consequence, the performance of the MPI collective operators were eventually improved to the point that the hand-tuned Co-Array Fortran implementations of the MPI collectives were no longer necessary. (The Co-Array implementations of the halo update were still important for performance.) Similarly, modifications to the operating system improved performance of both the MPI-only and MPI/Co-Array Fortran versions by over 80%. Figure 15.10 is a graph of performance (measured in simulated years per day, so larger is better) for the different versions of the code developed during the study, all collected on the same date. As can be seen, the optimizations were important even with the MPI library and operating system improvements. The four implementations with similar performance were all attempts to solve what were ultimately determined to be performance bottlenecks in the operating system. Moreover, performance of the best performer in Figure 15.10 was

---

[4]The halo for a subset of a computational grid is the region external to the subset from which data is required in the next step of the computation. For a parallel algorithm based on domain decomposition, a halo update is the interprocess communication required to acquire the halo data that resides in the memory space of other processes.

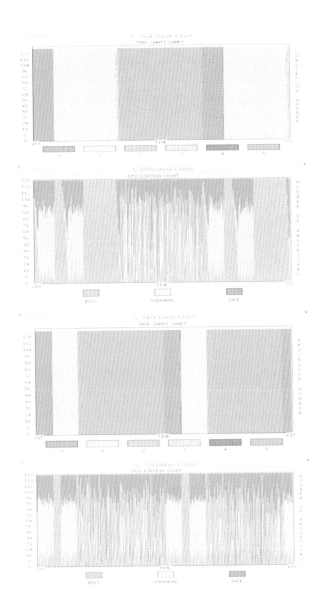

FIGURE 15.8: POP task graph and utilization count histogram on Cray X1: without and with vectorization. Horizontal axis is time. From [384]. (See color insert.)

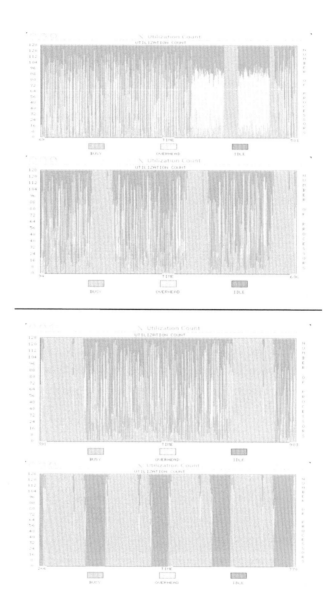

FIGURE 15.9: POP utilization count histograms on Cray X1: without and with MPI optimization (top) and Co-Array Fortran optimizations (bottom). Horizontal axis is time. From [384]. (See color insert.)

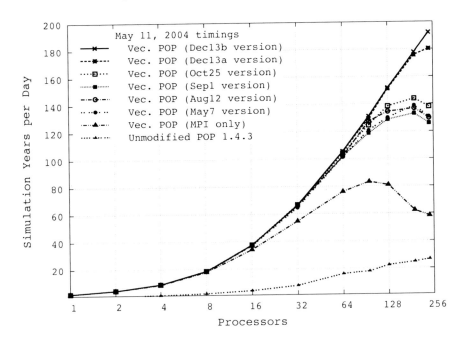

FIGURE 15.10: POP implementation performance comparison. (From [384].)

indistinguishable from the other 5 MPI/Co-Array Fortran implementations before the operating system optimizations were implemented.

In summary, this study demonstrated that kernel benchmarks can be critical in determining achievable performance and in evaluating alternative approaches. It also showed the advantage of using application-specific messaging layers for performance sensitive communication. With such a layer, communication can be implemented with whatever transport layer makes the most sense for a given architecture. Careful, detailed empirical studies also provide a mechanism for identifying system performance issues that can then be reported to computer center staff or to the system vendor. Finally, while not strictly required in this study, performance visualizations can be very useful in identifying and diagnosing performance problems.

15.8   Monitoring Performance Evolution

An important aspect of the performance engineering of the CCSM has been the monitoring of performance as the CCSM evolves, as the system software

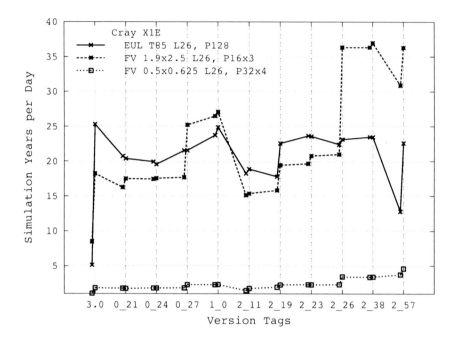

FIGURE 15.11: CAM Vector Performance History: March 2005 to March 2006. (From [376]. With permission.)

on the target platforms evolves, and as the production problem configurations change. The importance of this monitoring was made very clear during the period of time when vector architectures were being targeted. The underlying problem was that many of the contributors to the CCSM did not develop code on the target vector architectures, and, in consequence, their modifications often did not run efficiently on vector systems. Figure 15.11 is a graph of CAM performance on the Cray X1E for three different problem sizes and a fixed problem-size-dependent processor count over the period of March 2005 to March 2006. The X-axis is the CAM version number, where the labels are either major releases or versions checked-in by a subset of developers who were tasked to maintain performance on the X1E. Performance data (in simulated years per day) is plotted both for the labelled versions and for the immediately preceding versions. As can be seen, it was a constant battle to maintain good performance. See [374, 376] for additional details.

An analogous problem continues today with respect to scaling. Given the cost in turnaround time and computer allocation, most day-to-day development utilizes smaller problem sizes and runs with modest processor counts. In consequence, performance issues that arise at scale are sometimes not identified until attempting to use the code for large production runs. Examples

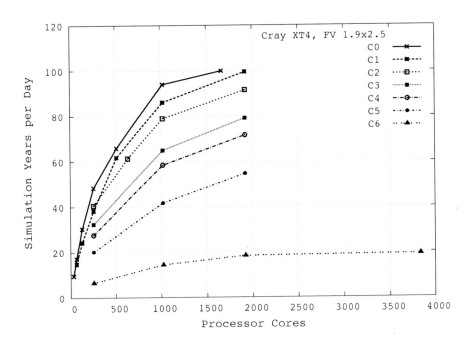

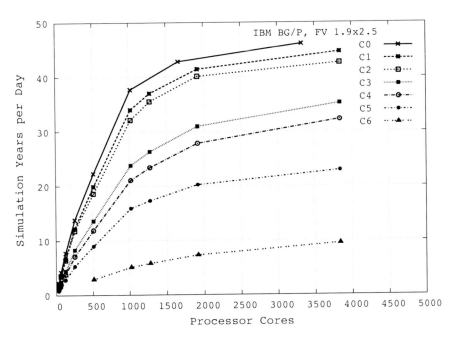

FIGURE 15.12: CAM performance for proposed physics modifications.

include modifications whose computational complexity increases faster than can be controlled by exploiting additional parallelism, sometimes because the complexity increases as a function of the number of processes itself. Other examples involve unexpected increases in communication overhead, sometimes due to changing performance characteristics of interprocess communication at scale.

The solution to the problems described above is in part education of the developers as to the performance issues. This requires an understanding by the performance engineers of the (ever changing) performance characteristics of the target platforms and the code, and the practical implications of these characteristics. However, careful monitoring of performance, both for current and future problem and computer configurations, continues to be critical. An example of both approaches was an evaluation in 2008 of the performance of a sequence of physics modifications under consideration for inclusion in the next generation version of CAM [375]. Figure 15.12 describes the impact on performance of these different options for a coarse grid resolution problem on a Cray XT4 and on an IBM BG/P. (Individual curves are truncated when performance no longer increases with processor core count.) Where before the dycore was the performance limiter, the new physics options increase significantly the importance of optimizing the performance of the physics.

15.9 Performance at Scale

Many of the performance issues of concern currently involve performance at scale, whether using ever increasing numbers of processor cores efficiently and/or maintaining good throughput rates as the problem size increases. For example, exploitable parallelism in CAM was extended recently by allowing different phases of the code to make use of different numbers of MPI processes [243], leaving some processes idle during some phases. This activity required a re-evaluation of MPI communication performance in the model, and the redesign of some communication logic used between these different phases [241].

One common theme at scale is the impact of MPI communication on performance and on memory requirements for large MPI process counts. Using the standard MPI communication protocols, including MPI collectives, often requires large system buffer space to hold necessary tables and to receive unexpected messages, that is messages that arrive before corresponding receive requests have been posted. On the Cray XT systems, these buffer sizes can be specified using environment variables, but this is not a robust solution as it is difficult to determine how much is sufficient *a priori*, and specifying too much system memory for MPI may prevent the user code from running, from lack of memory. System buffering of unexpected messages can be avoided by using communication protocols with additional flow control logic to eliminate these unexpected messages. For example, a process can delay sending a message until it has received a handshake message from the destination process indicating that the corresponding receive request has been posted. Introducing an explicit flow control option in each CCSM component and in the coupler has allowed us to address this issue without changing the default MPI buffer sizes (so far). However care needs to be taken in the design of these alternative algorithms so that performance is not degraded [241].

Another mitigation strategy is to use fewer MPI processes by exploiting shared-memory parallelism over the ever increasing number of processor cores available in an SMP node. Exploiting this approach within the CCSM was identified as a priority recently and, as mentioned earlier, all CCSM components can now use both MPI and OpenMP parallelism. Note that how OpenMP is introduced is also important. In CAM, the OpenMP parallelism was until recently applied to the same loops as the MPI parallelism. Thus, when MPI parallelism was fully exploited, there was no parallelism left to exploit with OpenMP. As part of the recent emphasis on exploiting OpenMP parallelism within the CCSM, OpenMP parallelism has now been applied to other loops as well, as an option. The best mix of MPI and OpenMP parallelism for a given processor count is still determined empirically.

A different approach to avoid unexpected messages and the consequent memory requirements is to use one-sided messaging, where messages are sent directly to memory controlled by a remote process. This can be implemented with MPI one-sided messaging or with messaging layers designed specifically for this operation, such as Co-Array Fortran and Unified Parallel C (UPC) [76]. The one-sided messaging approaches work best with hardware support in the interconnection network, which is missing in the current massively parallel systems (even though it was available in earlier systems such as the Cray X1 and the HP AlphaServer SC). Hardware support for one-sided messaging is expected to reappear in future systems, and it will be relatively simple to add support for one-sided messaging in the current CCSM messaging layer.

Another concern on certain massively parallel systems is (nonalgorithmic) performance variability occurring during or between runs. Performance variability can arise from performance inhomogeneity, either spatial from "slow" nodes or in the time-direction from operating system jitter. It can also arise from lack of sufficient isolation, so that your neighbor's job can consume a shared research such as network or I/O bandwidth that your job is attempting to use. The initial problem is identification that a performance problem is due to performance variability, and is not a performance problem intrinsic to the application code. The second step is the attribution, determining the source of the variability. What to do about this type of performance issue once it is identified is also problematic, but latency hiding techniques, in I/O for example, may also make code performance more robust in the presence of variability. So, performance portability techniques may need to include a degree of "performance variability tolerance" in the future.

With the increase in scale, the diagnosis of performance problems is becoming ever more difficult. In consequence, good performance tools and methodologies are increasingly important, but the functionality and robustness of the tools are also affected by scale. Building flexibility into the codes, in order to adapt easily to changing performance characteristics, continues to be an important technique, but progress in performance tools and, possibly, new programming paradigms will be needed as system, code, and problem size and complexity continue to grow.

15.10   Summary

To summarize, techniques that have been employed with good effect in the performance engineering of the CCSM include the following:

1. representative kernel codes for developing and examining new ideas and for narrowing the size of the space to be searched when determining optimal algorithm settings;

2. detailed performance scalability studies, tracking changes in performance of logical phases in application codes as process counts, problem sizes, and system architectures are varied;

3. performance modeling;

4. performance visualization;

5. support for a variety of messaging layers, communication algorithms, and communication algorithm implementations (protocols) for individual phases of a code, specified at compile-time or (preferably) runtime;

6. compile-time or runtime support for other algorithm options, such as data structure dimensions and loop bounds, load balancing, MPI vs. OpenMP parallelism, domain decomposition, in which the optimal choice is a function of the experiment particulars;

7. an efficient experimental methodology for determining the optimal settings for all of the compile-time and runtime options;

8. good communications with computer center and vendor staff.

For the most part these techniques have been employed without sophisticated performance engineering infrastructure. Recent improvements in performance modeling and autotuning should improve our ability to maintain and improve performance of the CCSM in the future, though the evolution of the model toward increasing complexity and the significantly different nature of proposed next generation high performance computing architectures (for example, many-core nodes with GPU accelerators) will place increasing demands on the CCSM performance engineering community.

15.11   Acknowledgments

Numerous individuals collaborated with the author in the work described herein, as documented in the bibliography, and thanks are due to all of them. Thanks also go to the greater CCSM development community, especially the staff of CCSM Software Engineering Group at NCAR [84] and the members of the CCSM Software Engineering Working Group [85].

This research used resources (Cray X1, X1E, XT3, XT4, and XT5) of the National Center for Computational Sciences at Oak Ridge National Laboratory, which is supported by the Office of Science of the U.S. Department of Energy under Contract No. DE-AC05-00OR22725. It used resources (IBM p690 cluster and BG/P and SGI Altix) of the Center for Computational Sciences, also at Oak Ridge National Laboratory. It used resources (IBM SP, IBM p575 cluster) of the National Energy Research Scientific Computing Center, which is supported by the Office of Science of the U.S. Department of Energy under Contract No. DE-AC03-76SF00098.

The work of Worley was supported by the Climate Change Research Division of the Office of Biological and Environmental Research and by the Office of Mathematical, Information, and Computational Sciences, both in the Office of Office of Science, U.S. Department of Energy, under Contract No. DE-AC05-00OR22725 with UT-Batelle, LLC. Accordingly, the U.S. Government retains a nonexclusive, royalty-free license to publish or reproduce the published form of this contribution, or allow others to do so, for U.S. Government purposes.

Chapter 16

Tuning an Electronic Structure Code

David H Bailey, Lin-Wang Wang, Hongzhang Shan

Lawrence Berkeley National Lab

Zhengji Zhao, Juan Meza, Erich Strohmaier

Lawrence Berkeley National Lab

Byounghak Lee

Texas State University, San Marcos

16.1   Introduction

Many material science and nanoscience computational problems require one to deal with large numbers of atoms – typically tens of thousands for best results. Such problems can be simulated accurately best by *ab initio* self-consistent methods. Some examples include quantum dots and wires, core/shell nanostructures, as well as more conventional systems such as semiconductor alloys.

For example, materials with separate electron states within the energy band gap ("mid-band-gap state" materials) have been proposed as solar cells [222]. Such devices possibly could achieve efficiencies from 40% to 63%, whereas solar cells in use today typically only achieve efficiencies in the 5% to 15% range [222]. One potential way to produce mid-band-gap state materials is to use certain semiconductor alloys, such as $ZnTe_{1-x}O_x$, where x denotes the fraction of oxygen atoms. For a small alloy percentage, ($\approx 3\%$), the oxygen states repulse the conduction band minimum (CBM) states of the ZnTe and form a separate band inside the energy band gap of ZnTe. Preliminary experimental results have shown that such mid-band-gap states do exist [389], but the characteristics of these mid-band-gap states are not known (i.e., whether they are spatially isolated or extended), nor is it known whether there is a clear energy band gap between the oxygen induced state and the CBM of ZnTe. If there is no gap, the electron from the ZnTe conduction band will be relaxed into the mid-band-gap states through phonon emission, which will render the material unusable for solar cell applications.

At the present time, accurate computational analysis of such materials is only possible by employing *ab initio* density functional theory (DFT) calculations [166, 196]. However, due to the small percentage of the oxygen atoms, large "supercells" containing thousands of atoms must be employed to properly describe the random distribution of these oxygen atoms. Due to this requirement for thousands of atoms, direct DFT methods are impractical. For example, using any of the classical $O(N^3)$ methods to simulate the 13,824-atom system that we will subsequently describe in this chapter would require four to six weeks of run time for a fully converged self-consistent result using a system with 20,000 processor cores. This reckoning generously presumes that such codes achieve a high fraction of peak and scale perfectly in performance to 20,000 cores.

In order to solve such problems on multi-peta- and exascale computer systems now on the drawing board, which will employ 1,000,000 or more processing cores, one of the present authors (Lin-Wang Wang) and some colleagues at Lawrence Berkeley National Laboratory have developed a new method whose computational cost increases nearly linearly with the number of atoms. For the 13,824-atom system mentioned above, for example, this method is roughly 400 times faster than conventional $O(N^3)$ methods (in terms of operation count), yet it yields essentially the same numerical results as the conventional methods. In addition, since it uses a divide-and-conquer scheme, this method is well-suited for very-large-scale parallel computer systems.

These techniques were implemented in a computer program that was named "LS3DF," an acronym for "linearly scaling, three-dimensional fragment" scheme. This program demonstrated well the potential for linear scaling (in terms of number of system size, i.e., number of atoms). However, the original LS3DF code ran at a performance rate that was a relatively low fraction of the peak performance on the numbers of processors being used. In addition, parallel scalability was limited to "only" about 1024 processing cores, which while very good is still far below the sizes of large state-of-the-art scientific computer systems available today.

This chapter gives some background on the LS3DF code, presents the results of some performance analysis that was done on the code, and then describes tuning that was done to improve this performance, which ultimately led to a code that not only runs extremely well but also won a coveted award granted by the Association of Computing Machinery (ACM), namely the ACM Gordon Bell Prize, in a special category for algorithm innovation.

Full details on the algorithm and methodology are presented in [350], from which this chapter is adapted. Among other things, this chapter includes new, significantly higher performance results than were available when [350] was published.

16.2 LS3DF Algorithm Description

A divide-and-conquer scheme is a natural approach for mapping the physical locality of a large problem to the architectural locality of a massively parallel computer. The method used in LS3DF is based on the observation that the total energy of a given system can be broken down into two parts: the electrostatic energy and the quantum mechanical energy (e.g, the kinetic energy and exchange correlation energy). While the electrostatic energy is long-range and must be solved via a global Poisson equation, the computationally expensive quantum mechanical energy is short-range [195] and can be solved locally. The idea is to divide the whole system into small fragments, calculate the quantum mechanical energies of these fragments, and then combine the separate fragment energies to obtain the energy of the whole system.

A critical issue in a divide-and-conquer scheme such as this is how to patch the fragments together. The core of the LS3DF algorithm is a novel patching scheme that cancels out the artificial boundary effects caused by the division of the system into smaller fragments. As a result of this cancellation, computed results are essentially the same as a direct calculation on the large system, which typically scales as $O(N^3)$, where N is the size of the system in atoms. In our method, once the fragment sizes are chosen to obtain a given numerical accuracy, the computational cost is proportional to the number of fragments. Hence, it is called the linearly scaling three-dimensional fragment (LS3DF) method. By using a small group of cores to solve the quantum mechanical part of each fragment independently, our method also scales well in performance with the number of processor cores. Only a small overhead is needed to patch the fragment charge densities into a global charge density, to solve the Poisson equation for the whole system, and to divide the global potential into fragment potentials. As a result, the scheme can be implemented on computer systems with hundreds of thousands of cores.

This divide-and-conquer scheme is illustrated in Figure 16.1, which depicts a two-dimensional system for clarity. In Figure 16.1, a periodic supercell is divided into $m_1 \times m_2$ small pieces. From each grid corner (i, j) one can define four fragments, with their sizes S equal to (in units of the smallest piece): $S = 1 \times 1$, 1×2, 2×1 and 2×2, respectively. Suppose one calculate the quantum energy $E_{i,j,S}$ and charge density $\rho_{i,j,S}$ of all of these fragments. Then the total quantum energy of the system can be calculated as $E = \sum_{i,j,S} \alpha_S E_{i,j,S}$, and the total charge density as $\rho(r) = \sum_{i,j,S} \alpha_S \rho_{i,j,S}(r)$. Here $\alpha_S = 1$ for the

$S = 1 \times 1$ and 2×2 fragments, and $\alpha_S = -1$ for the $S = 1 \times 2$ and 2×1 fragments. By allowing the usage of both positive and negative fragments in the above summation, the edge and corner effects between different fragments are canceled out, while one copy at the interior region of the fragment will be left to describe the original large system. This scheme can be extended to three dimensions in a straightforward way. The details of this method, as well as some of its novel features, are described in [352, 390].

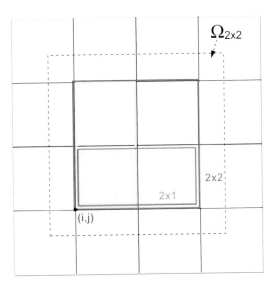

FIGURE 16.1: The division of space and fragment pieces from corner (i, j).

In the latest implementation of the LS3DF method, one starts with a 3D periodic supercell, and divides it into an $M = m_1 \times m_2 \times m_3$ grid. The atoms are assigned to fragments depending on their spatial locations. The artificially created surfaces of the fragments are passivated with hydrogen or partially charged pseudo-hydrogen atoms to fill the dangling bonds [351]. The wavefunctions of the fragments are described by planewaves within a periodic fragment box Ω_F (which is the square region plus a buffer region as shown by the dashed line in Figure 16.1 for a 2×2 fragment). Norm conserving pseudopotentials are used to describe the Hamiltonian. An all-band conjugate gradient method is used to solve the fragment wavefunctions [354].

The LS3DF method involves four important steps within each total potential self-consistent iteration, as illustrated in Figure 16.2. First, a total input potential $V_{in}^{tot}(r)$ (for the whole system) is provided. Secondly, the Gen_VF routine generates potentials for each fragment F, the potential $V_F(r) = V_{in}^{tot}(r) + \Delta V_F(r)$, $r \in \Omega_F$, where $\Delta V_F(r)$ is a fixed passivation potential for each fragment F which is only nonzero near its boundary [352]. Note that $V_F(r)$ is only defined in Ω_F. Third, PEtot_F solves Schrödinger's equation on each fragment for its wavefunctions $\psi_i^F(r)$. After the fragment wavefunctions $\psi_i^F(r)$ are solved for, the fragment charge density is computed, $\rho_F(r) = \sum_i |\psi_i^F(r)|^2$, and the charge density for the overall system is patched together $\rho_{tot}(r) = \sum_F \alpha_F \rho_F(r)$ by subroutine Gen_dens. In the final step carried out by subroutine GENPOT, a global Poisson equation is solved using fast Fourier transforms (FFTs) to obtain the global potential $V_{out}^{tot}(r)$. After potential mixing from previous iterations, the modified $V_{out}^{tot}(r)$ is used as the input for the next self-consistent iteration. Self-consistency is reached as $V_{out}^{tot}(r)$ approaches $V_{in}^{tot}(r)$ within a specified tolerance.

LS3DF

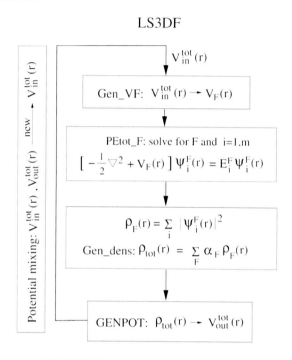

FIGURE 16.2: LS3DF flow chart.

16.3 LS3DF Code Optimizations

LS3DF was developed specifically for large-scale parallel computation, although it builds on the previous $O(N^3)$ DFT code PEtot [354] that has been in development for nearly ten years. The four major subroutines in LS3DF, namely Gen_VF, PEtot_F, Gen_dens, and GENPOT (Figure 16.2) were developed in steps. Initially, as a proof of concept, they were developed as separate executables using file I/O to pass the data between them. Later, these subroutines were integrated into a single executable. These codes are written in Fortran-90, using MPI for parallel computation.

In the most recent phase of the development, which yielded the code that we used in this study, we first focused on the PEtot_F subroutine, which dominates run time. Using the Integrated Performance Monitoring (IPM), developed by David Skinner of the Lawrence Berkeley National Laboratory [144], we found that the PEtot_F subroutine ran at about 15% of peak performance in the earlier phases of development. PEtot_F is derived from the PEtot code, which uses the same planewave q-space parallelization as other standard codes, such as PARATEC [86] and Qbox [156]. However, to save on memory requirements, it used a band-by-band algorithm that contributed to its relative low performance rate. Detailed profiling and analyses were carried out to increase the performance of PEtot_F. Closer analysis revealed that since PEtot_F solved for one electron wavefunction at a time ("band-by-band," i.e., solving for one eigenvalue/eigenvector at a time versus all at one time), the majority of operations (for nonlocal pseudopotential and wavefunction orthogonalization) were performed using BLAS-2 routines. The file I/O for communication also took a substantial amount of time.

Based on these analyses, we employed four major code optimizations to improve the performance and scalability of this code:

1. Within PEtot_F, solve for all the electron wavefunctions simultaneously (i.e., the all-band scheme), instead of one wavefunction at a time (i.e., the band-by-band scheme). With this approach, one can utilize BLAS-3 operations such as DGEMM, yielding higher performance than is possible with BLAS-2 routines. A typical matrix size for one of the fragments is 3000×200.

2. Implement a different algorithm for data communication between cores in Gen_VF and Gen_dens to allow better scaling for large numbers of cores.

3. Store the data in memory (through an LS3DF global module), and communicate with MPI calls, rather than store the data on disk and communicating via file I/O as in the earlier versions of the code. This change has resulted in a major improvement in scalability and overall performance.

4. Eliminate the setup overhead during each self-consistent call for PEtot_F and GENPOT through storage of the relevant variables in the LS3DF global module.

With regards to the first item, the original PEtot_F band-by-band algorithm has the advantage of requiring a rather modest amount of memory. But on the Cray XT4 system, with 2 GBbyte memory per core, we had enough memory to use the all-band method, which involves matrix-matrix multiplication that can be performed using BLAS-3 routines. This all-band optimization required changes in the data layout and rearrangements of the loop structures.

We have also implemented a new orthogonalization scheme for the all-band method. Instead of imposing the orthonormal condition for the wavefunction using a Gram-Schmidt scheme at each conjugate gradient step, we only impose the orthonormal condition after a few conjugate gradient steps by calculating an overlapping matrix. The use of the overlapping matrix instead of the direct band-by-band Gram-Schmidt algorithm also permits the use of BLAS-3 library routines such as DGEMM.

In the wake of this code optimization, the performance of the stand-alone PEtot code had increased from 15% of the theoretical peak to 56% for large system calculations, based on calculations performed on the "Franklin" system (a Cray XT4 system) at Lawrence Berkeley National Laboratory. This performance ratio is close to that of the best planewave codes, such as PARATEC [86] and Qbox [156]. The performance ratio of PEtot_F for the largest fragments was 45% on Franklin, which was slightly slower than the stand-alone code, probably due to the small size of the fragment. In a 2000-atom CdSe quantum rod sample problem, after the final code optimization, for 8000-core runs, the execution times for the four subroutines of the code were reduced to: Gen_VF 2.5 seconds (from the original 22 seconds), PEtot_F 60 seconds (from the original 170 seconds), Gen_dens 2.2 seconds (from the original 19 seconds), and GENPOT 0.4 seconds (from the original 22 seconds). These timings represented a factor of four overall improvement compared with the previous version. For Gen_VF and Gen_dens, the improvement was a factor of 10, while for GENPOT, the improvement was a factor of 50. The improvements for these three subroutines are critical for large-scale runs.

Shortly before the final completion of the original study, we were able to obtain access to the "Intrepid" system, which is a large Blue Gene/P system at Argonne National Laboratory. For one set of large runs on the Intrepid system (using 131,072 cores), we further improved Gen_VF and Gen_dens routines by employing point-to-point MPI isend and ireceive operations. As a result, on the Intrepid runs, these two routines together comprised less than 2% of the total run time. In particular, the breakdown for one self-consistent field (SCF) iteration is as follows: Gen_VF (0.37 sec.), PEtot_F (54.84 sec.), Gen_dens (0.56 sec.) and GENPOT (1.23 sec.). Since such a large percentage of the time now goes to PEtot_F (which has no inter-group communication), this bodes well for the future scalability of this code on petascale computer systems.

Although we were pleased with the performance and scalability of the resulting code, we made some additional improvements, including: (1) two-level parallelization in PEtot_F, to achieve greater parallelism; and (2) replacing DGEMM with a custom routine specialized for PEtot_F. The two-level parallelization code included parallelization over the plane wave basis set and on the wave function index i as indicated in Figure 16.2. This parallelization significantly increased the total concurrency in this application, thereby increasing the scalability of the code even further.

16.4   Test Systems

In order to test the scaling and floating-point operation performance of the LS3DF code, we set up a series of test problems involving $ZnTe_{1-x}O_x$ alloy systems. These alloy systems are in a distorted zinc blende crystal structure, with 3% of Te atoms being replaced by oxygen atoms. Although the LS3DF method can be used to calculate the force and relax the atomic position, for these particular systems, we found that the atomic relaxation can be described accurately by the classical valence force field (VFF) method [292]. These alloy systems are characterized by the sizes of their periodic supercells. The size of a supercell can be described as $m_1 \times m_2 \times m_3$ in the unit of the cubic eight-atom zinc blend unit cell. Thus, the total number of atoms is $8m_1m_2m_3$. We used a nonlocal norm-conserving pseudopotential to describe the Hamiltonian, and a planewave basis function with a 50 Ryd energy cutoff to describe the wavefunction. The real space grid for each eight-atom unit cell is $40 \times 40 \times 40$. The d-state electrons in Zn atom are not included in the valence electron calculation. Thus in average, there are four valence electrons per atom.

We found that for the fragment calculations, a reciprocal q-space implementation of the nonlocal potential is faster than a real-space implementation. Thus, we used a q-space nonlocal Kleinman-Bylander projector for the nonlocal potential calculation [269]. The accuracy of LS3DF, as compared with the equivalent DFT computation, increases sharply with the fragment size. For the LS3DF calculation, they used the eight-atom cubic cell as the smallest fragment size, as shown in Figure 16.1. Using this fragment size, the LS3DF results are very close to direct DFT calculated results. For example, the total energy differed by only a few meV per atom, and the atomic forces differed by only 10^{-5} atomic units [352]. In short, the LS3DF and the direct DFT results are essentially the same.

To test the weak scaling of the LS3DF code, we chose alloy supercells of dimensions $m_1 \times m_2 \times m_3$, namely $3 \times 3 \times 3$, $4 \times 4 \times 4$, $5 \times 5 \times 5$, $6 \times 6 \times 6$, $8 \times 6 \times 9$, $8 \times 8 \times 8$, $10 \times 10 \times 8$ and $12 \times 12 \times 12$. These problems correspond to 216, 512, 1000, 1728, 3456, 4096, 6400, and 13824 atoms, respectively. To study the physics of the oxygen induced states, large supercells are needed to properly describe the atomic configuration due to the small oxygen percentages (e.g., 3%) used in laboratory experiments.

For the $3 \times 3 \times 3$ system, we also calculated the full system with a direct local density approximation (LDA) method (using PEtot). Since the band gap and eigenenergy differences between the direct LDA method and the LS3DF method are about 2 meV (for the LS3DF method, researchers took its converged potential, then calculated the eigenenergy of the full system). Since the energy gaps being studied investigate are few tenths of an eV to a few eV, the LS3DF method is numerically accurate enough. For example, in a related study, we used LS3DF to calculate thousand-atom quantum rods and their dipole moments [352]. These calculated dipole moments differed from the direct LDA results by less than 1%.

For the runs done on the Intrepid system, we modified the above parameters as follows. First of all, we employed a cutoff value of 40 Ryd instead of 50, and employed a real-space grid of size $32 \times 32 \times 32$. These changes were made to adjust for the smaller amount of main memory per core on the Blue Gene/P system. Secondly, we generated larger problem sizes, including alloy supercells of dimensions $4 \times 4 \times 4$, $8 \times 4 \times 4$, $8 \times 8 \times 4$, $8 \times 8 \times 8$, $16 \times 8 \times 8$ and $16 \times 16 \times 8$.

16.5   Performance Results and Analysis

To assess the optimizations implemented in LS3DF and to demonstrate the benefits of this new code, we conducted several computational experiments. First of all, we executed LS3DF with a constant problem size across the currently available range of concurrency (i.e., "strong scaling"), and for a variety of problem sizes (i.e., "weak scaling"). We then compared code performance to other important DFT codes such as PARATEC and VASP.

Part of these benchmark runs were performed on the Franklin system at the National Energy Research Scientific Computing (NERSC) Center. At the time the initial study done, as reported in [350], the Franklin system was a Cray XT4 system with 9660 compute nodes, each of which had two 2.6 GHz AMD Opteron cores and 4 GByte main memory. The entire system thus had a theoretical peak performance rate of 101Tflop/s. Subsequently the Franklin system was upgraded to 9572 nodes, each of which had four 2.3 GHz quad-core Opterons and 8 Gbyte memory. Thus the upgraded system had a theoretical peak performance rate of 352 Tflop/s.

The second set of runs were performed on the Jaguar system at ORNL. At the time the initial study was done, Jaguar had 7832 XT4 compute nodes, each with a quad-core 2.1 GHz AMD Opteron processor and 8 GByte of memory. The theoretical peak performance rate of this system was approximately 263 Tflop/s. Subsequently this system was upgraded to 37,538 nodes (150,152 cores) and a theoretical peak performance of 1.381 Pflop/s.

The third set of runs were performed on the Intrepid system at ANL, a Blue Gene/P system. At the time the initial study was done, Intrepid featured 40,960 nodes and 163,840 cores, with a theoretical peak performance rate of 556 Tflop/s (and this system was used for all runs mentioned here).

Performance results are given in Table 16.1. Figures are listed in separate sections of the table for runs on Franklin, Jaguar, and Intrepid. Tflop/s figure in the table for various problem sizes were calculated based on operation counts measured on the Franklin system using the CrayPat tool [14]. For the very largest problems, which could not be run on Franklin, we estimated the operation count based on the number of fragments and the small-problem operation counts. This estimation scheme was found to be consistently within 1% of the actual operation count for those problems that could be run on Franklin. Note that the Tflop/s figures reported in Table 16.1 were calculated based on wall-clock time. On Franklin, these figures are roughly 10% lower than the Tflop/s figures reported by the CrayPat tool, which uses user time rather than wall-clock time, but for consistency across all systems, the wall-clock reckoning was used.

To evaluate the strong scaling behavior of LS3DF, we chose a medium-sized problem with 3456 atoms and a fragment grid size of $m_1 \times m_2 \times m_3 = 8 \times 6 \times 9$. In this test, we used $N_p = 40$, where N_p is the number of cores within each group used in PEtot_F. The value of 40 was determined by the parallel efficiency for each group. Experience has shown (for example, Table 16.1, third test case for the Jaguar system) that when the value of N_p is increased beyond 40, the scaling within each group drops off, which drives the overall efficiency down. We increased the number of groups N_g from 27 to 432. This change represents a 16-fold range of concurrency levels from 1080 cores to 17,280 cores on the Franklin system. To estimate the run time, two SCF iterations of LS3DF were executed and the times for the second iteration were recorded, since this is the iteration which are iterated several dozen times for a converged calculation (the first iteration has some small additional overhead due to array and index setups).

Figure 16.3 shows the speedup of LS3DF and the PEtot_F component for the range of cores evaluated on Franklin. Speedup and parallel efficiency figure for the 17,280-core runs (using the 1080-core run as baseline), were 15.3 and 95.8% for the PEtot_F portion, and 13.8 and 86.3% for LS3DF, both of which are excellent. Overall, LS3DF achieved a performance rate of 31.35 Tflop/s on 17,280 cores. All computations are performed on 64-bit floating-point data.

We analyzed the results of our strong scaling experiment with Amdahl's Law:

$$P_p = P_s \left(\frac{n}{1 + (n-1)\alpha} \right). \tag{16.1}$$

Here P_p is parallel performance, P_s is the serial performance, n is the number of cores, and α is the fraction of serial work in the code. In particular, we employed least-squares fitting to determine the parameters P_s and α. The resulting formula fits our performance data extremely well, with an average absolute relative deviation of the fitting, namely $\sum_n |(P_{fitted}/P_{measured} - 1)|/n$, of only 0.26% and a single maximal deviation of 0.48%. Fitted values for the single core performance are 2.39 Gflop/s for the effective single core performance and hypothetical fractions of the remaining serial work components of 1/362,000 for PEtot_F and 1/101,000 overall for LS3DF.

Figure 16.4 shows computational efficiencies for a variety of problems and different code execution parameters for runs on Franklin. Overall, the excellent scalability demonstrated in the strong scaling experiment is confirmed. The small variations in code performance for a given concurrency level appear to

TABLE 16.1: Summary of test results. "% peak" is the fraction of the peak performance for the number of cores used.

sys. size	atoms	cores	N_p	Tflop/s	% peak
Franklin					
$3 \times 3 \times 3$	216	270	10	0.57	40.4%
$3 \times 3 \times 3$	216	540	20	1.14	40.8%
$3 \times 3 \times 3$	216	1080	40	2.27	40.5%
$4 \times 4 \times 4$	512	1280	20	2.64	39.6%
$5 \times 5 \times 5$	1000	2500	20	5.15	39.6%
$6 \times 6 \times 6$	1728	4320	20	8.72	38.8%
$8 \times 6 \times 9$	3456	1080	40	2.28	40.5%
$8 \times 6 \times 9$	3456	2160	40	4.51	40.2%
$8 \times 6 \times 9$	3456	4320	40	8.88	39.5%
$8 \times 6 \times 9$	3456	8640	40	17.04	37.9%
$8 \times 6 \times 9$	3456	17280	40	31.35	34.9%
$8 \times 8 \times 8$	4096	2560	20	5.46	41.0%
$8 \times 8 \times 8$	4096	10240	20	19.72	37.0%
$10 \times 10 \times 8$	6400	2000	20	4.18	40.2%
$10 \times 10 \times 8$	6400	16000	20	29.52	35.5%
$12 \times 12 \times 12$	13824	17280	10	32.17	35.8%
$8 \times 8 \times 8$	4096	8192	20	30.5	40.4%
$16 \times 8 \times 8$	8192	16384	20	60.6	40.2%
$16 \times 16 \times 8$	16384	36864	20	135.2	40.0%
Jaguar					
$8 \times 8 \times 6$	3072	7680	20	17.3	26.8%
$8 \times 8 \times 6$	3072	15360	40	33.0	25.6%
$8 \times 8 \times 6$	3072	30720	80	53.8	20.9%
$8 \times 6 \times 9$	3456	17280	40	36.5	25.2%
$16 \times 8 \times 6$	6144	15360	20	33.6	26.0%
$16 \times 12 \times 8$	12288	30720	20	60.3	23.4%
$4 \times 4 \times 4$	512	2048	20	7.79	41.3%
$8 \times 8 \times 8$	4096	16384	20	58.66	38.9%
$16 \times 16 \times 8$	16384	65536	20	197.86	32.8%
$16 \times 16 \times 18$	36864	147456	20	442.32	32.6%
Intrepid					
$4 \times 4 \times 4$	512	4096	64	4.4	31.6%
$8 \times 4 \times 4$	1024	8192	64	8.8	31.5%
$8 \times 8 \times 4$	2048	16384	64	17.5	31.4%
$8 \times 8 \times 8$	4096	32768	64	34.5	31.1%
$16 \times 8 \times 8$	8192	65536	64	60.2	27.1%
$16 \times 16 \times 8$	16384	131072	64	107.5	24.2%
$16 \times 16 \times 10$	20480	163840	64	224.0	40.0%

depend, to first order, on code execution parameters such as the size N_p of processor groups working on individual fragments. Notice that for a given concurrency, the computational efficiency is almost independent of the size

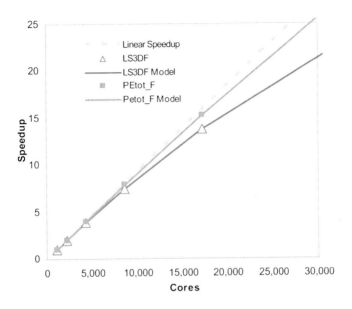

FIGURE 16.3: Strong scaling speedups for LS3DF and PEtot_F. The curves are models based on Amdahl's Law.

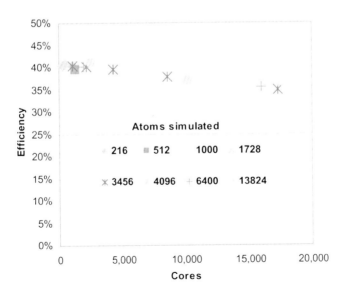

FIGURE 16.4: Computational efficiency for different runs on the NERSC Franklin system.

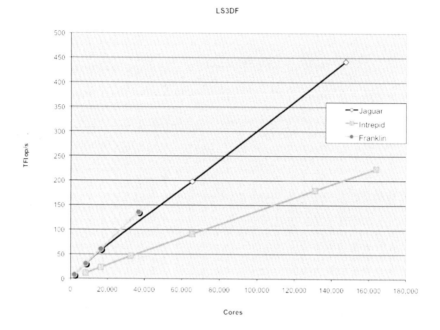

FIGURE 16.5: Weak scaling floating point operation rates on different machines.

of the physical system studied. The slight drop of efficiency for very high concurrency is again mostly due to Gen_VF and Gen_dens.

Figure 16.5 shows the total flop/s rate for weak scaling runs (with constant number of atoms to number of cores ratio) on different machines. The fairly straight lines suggest that the code is well poised for future petascale computer systems.

On the sustained test run on Franklin, convergence on the $8 \times 6 \times 9$ system was achieved with 60 iterations, using 17,280 cores of Franklin. This run required one hour of run time, or in other words one minute per iteration. The total sustained performance rate for that run was 31.4 Tflop/s, which is 35% of the theoretical peak rate (89.9 Tflop/s) on 17,280 cores.

16.6 Science Results

As mentioned above, we used LS3DF to achieve fully converged results for the $m_1 \times m_2 \times m_3 = 8 \times 6 \times 9$ system. This physical system has 3456 atoms,

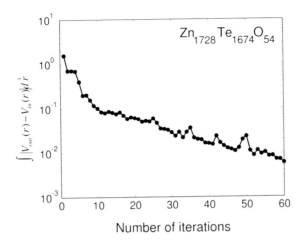

FIGURE 16.6: LS3DF convergence: Input and output potential difference as a function of self-consistent iteration steps.

and requires 60 SCF iterations (the outer loop in Figure 16.2) to achieve full convergence. Using 17,280 processors, the run requires one minute for each iteration, and thus one hour for the entire calculation. The SCF convergence of the system can be measured by the difference between the input $V_{in}(r)$ and output $V_{out}(r)$ potentials, as shown in Figure 16.2. We have chosen to use the absolute difference rather than the relative difference because $V_{in}(r)$ can have an arbitrary shift, and because the absolute difference can be directly related to the energy difference. The difference $\int |V_{in}(r) - V_{out}(r)| d^3 r$ as a function of self-consistent iteration is shown in Figure 16.6 (in atomic units). As one can see, overall this difference decays steadily. However, there are a few cases where this difference jumps. This is typical in the potential mixing method, since there is no guarantee that this difference will decrease at every step. Overall, the convergence rate is satisfactory, and the final 10^{-2} potential difference is comparable to the criterion used in our nanosystem dipole moment calculations. In addition, because the charge density response to a potential change in LS3DF is similar to a direct LDA method, and we are using the same charge mixing scheme, we would expect that the LS3DF method will have similar convergence properties as the direct LDA method, and should therefore converge for all systems with a band gap.

The converged potential $V(r)$ was then used to solve the Schrödinger equation for the whole system for only the band edge states. This was done using our folded spectrum method (FSM) [353]. Since not all the occupied eigenstates are calculated, the FSM method scales linearly with the size of the system. Overall this step does not take much time and it can be considered as a fast post-process of the LS3DF calculations. There is a well-known LDA band

(a) Bottom of conduction band state   (b) Top of O band state

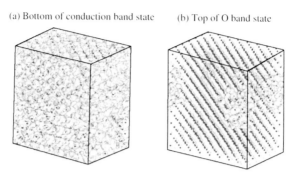

FIGURE 16.7: Isosurface plots of the electron wavefunction squares for the bottom of conduction band (a), and top of oxygen-induced band (b). The small dots are Zn, Te or oxygen atoms.

gap error that can be corrected using a modified nonlocal pseudopotential for the s, p, d states [292]. The calculated CBM state is shown in Figure 16.7(a), while the highest oxygen induced states is shown in Figure 16.7(b). Between the CBM and the highest oxygen induced state, there is a 0.2 eV band gap. This should prevent the electron in CBM from falling down to the oxygen induced states. Thus, our simulations predict that the $ZnTe_{0.97}O_{0.03}$ alloy could be used for solar cell applications.

One interesting point is that the oxygen induced states form a very broad band (0.7 eV) inside the band gap of ZnTe. As a result, its theoretical maximum efficiency of a solar cell made from this alloy will be smaller than the 63% estimated based on a narrow mid-band-gap [222]. Also, as shown in Figure 16.7(b), the oxygen induced states can cluster among a few oxygen atoms. Such a clustering is more localized in the high energy states than in the lower energy states within the oxygen induced band, which will significantly reduce the electron mobility (i.e., conductivity) in those states.

16.7   Summary

In summary, we have developed and deployed a new approach to the problem of large-scale *ab initio* electronic structure calculations. Our approach targets systems with a band gap and that require highly accurate planewave based calculations. It can be applied to nanostructures, defects, dislocations, grain boundaries, alloys, and large organic molecules. It simultaneously addresses the two critical issues for any large-scale computer simulation: the scaling of the total computational cost relative to the size of the physical problem, and the parallel scaling of the computation to very large numbers of processor cores.

On one of the test computations, for the $m_1 \times m_2 \times m_3 = 16 \times 16 \times 8$ ZnTeO system (16,384 atoms), we achieved 60.6 Tflop/s on 36,864 cores of the Franklin system. On a $16 \times 16 \times 10$ problem (20,480 atoms), we achieved 224 Tflop/s on 163,840 cores of the Intrepid system. On a $16 \times \times 16 \times 18$ system (36,864 atoms), we achieved 442 Tflop/s. Performance figures (and all others reported in this chapter) are for 64-bit floating-point operations. Virtually perfect linear weak scaling was achieved on all three systems.

More importantly, our simulations have yielded substantive scientific results. First, we found that there is a 0.2 eV band gap between the ZnTe conduction band and the oxygen induced band, implying that this alloy can be used for solar cell applications. Secondly, we found that the oxygen induced states form a very broad band (0.7 eV) inside the band gap of ZnTe. As a result, simulations predict that the theoretical maximum efficiency of solar cells made from this alloy will most likely be lower than the 63% figures estimated based on a narrow mid-band-gap. Also, as shown in Fig. 16.7(b), the oxygen induced states can cluster among a few oxygen atoms. Such a clustering will have an impact on the mobility of the electrons in those states.

Lastly, based on performance analysis, we have found that there is no intrinsic obstacle to scaling the LS3DF code to run on over 1,000,000 processing cores, with over 1 Pflop/s sustained performance.

16.8 Acknowledgments

This work was sponsored by the Performance Engineering Research Institute, which is supported by the SciDAC Program of the U.S. Department of Energy, and by the Director, Office of Science, Basic Energy Sciences, Division of Material Science and Engineering, and the Advanced Scientific Computing Research Office, of the U.S. Dept. of Energy under Contract No. DE-AC02-05CH11231. It used resources at the National Energy Research Scientific Computing Center (NERSC), the National Center for Computational Sciences (NCCS) and the Advanced Leadership Computing Facility (ALCF).

Bibliography

[1] ATLAS Frequently Asked Questions.
http://math-atlas.sourceforge.net/faq.html

[2] BLAS: Basic linear algebra subprograms.
http://www.netlib.org/blas

[3] CactusEinstein toolkit home page.
http://www.cactuscode.org/Community/NumericalRelativity

[4] GEO 600.

[5] Gnu standard: Formatting error messages.
http://www.gnu.org/prep/standards/html_node/Errors.html

[6] Kranc: Automated code generation.

[7] LIGO: Laser Interferometer Gravitational wave Observatory.

[8] LISA: Laser Interferometer Space Antenna.

[9] Mesh refinement with Carpet.

[10] Netlib repository.
http://www.netlib.org

[11] Queen Bee, the core supercomputer of LONI.

[12] Sun Constellation Linux Cluster: Ranger.

[13] Top500 Supercomputer Sites.
http://www.top500.org

[14] Optimizing applications on the Cray X1TM system, 2009. http://docs.cray.com/books/S-2315-50/html-S-2315-50/z1055157958smg.html

[15] *ROSE Web Reference*, 2010.
http://www.rosecompiler.org

[16] *SciDAC Performance Engineering Research Institute*, 2010.
http://www.peri-scidac.org/perci

[17] D. Abramson, A. Lewis, T. Peachey, and C. Fletcher. An automatic design optimization tool and its application to computational fluid dynamics. In *Proceedings of the ACM/IEEE Conference on Supercomputing (SC01)*, pages 25–25, New York, NY, 2001. ACM.

[18] E.M. Abreu. Gordon Moore sees another decade for Moore's Law. February 11 2003.
http://www.cnn.com/2003/TECH/biztechj/02/11/moore.law.reut

[19] M.F. Adams. A distributed memory unstructured Gauss–Seidel algorithm for multi-grid smoothers. In *ACM/IEEE Proceedings of SC2001: High Performance Networking and Computing*, Denver, CO, November 2001.

[20] M.F. Adams, H.H. Bayraktar, T.M. Keaveny, and P. Papadopoulos. Ultrascalable implicit finite element analyses in solid mechanics with over a half a billion degrees of freedom. In *ACM/IEEE Proceedings of SC2004: High Performance Networking and Computing*, 2004.

[21] M.F. Adams, M. Brezina, J. J. Hu, and R.S. Tuminaro. Parallel multigrid smoothing: Polynomial versus Gauss–Seidel. *Journal of Computational Physics*, 188(2):593–610, 2003.

[22] L. Adhianto, S. Banerjee, M. Fagan, M. Krentel, G. Marin, J. Mellor-Crummey, and N.R. Tallent. HPCToolkit: Tools for performance analysis of optimized parallel programs. *Concurrency and Computation: Practice and Experience*, 2010. http://dx.doi.org/10.1002/cpe.1553

[23] L. Adhianto, J. Mellor-Crummey, and N.R. Tallent. Effectively presenting call path profiles of application performance. In *Proceedings of the 2010 Workshop on Parallel Software Tools and Tool Infrastructures, held in conjunction with the 2010 International Conference on Parallel Processing*, 2010.

[24] N.R. Adiga, et al. An overview of the BlueGene/L supercomputer. November 2002.

[25] Advanced Micro Devices. AMD CodeAnalyst performance analyzer. http://developer.amd.com/cpu/codeanalyst

[26] M. Alcubierre, B. Brügmann, P. Diener, M. Koppitz, D. Pollney, E. Seidel, and R. Takahashi. Gauge conditions for long-term numerical black hole evolutions without excision. *Physical Review D*, 67:084023, 2003.

[27] M. Alcubierre, B. Brügmann, T. Dramlitsch, J.A. Font, P. Papadopoulos, E. Seidel, N. Stergioulas, and R. Takahashi. Towards a stable numerical evolution of strongly gravitating systems in general relativity: The conformal treatments. *Physical Review D*, 62:044034, 2000.

[28] A.S. Almgren, J.B. Bell, P. Colella, L.H. Howell, and M. Welcome. A conservative adaptive projection method for the variable density incompressible Navier-Stokes equations. 142:1–46, 1998.

[29] Alpaca: Cactus tools for application-level profiling and correctness analysis. http://www.cct.lsu.edu/~eschnett/Alpaca

[30] AMD. *Instruction-Based Sampling: A New Performance Analysis Technique for AMD Family 10h Processors*, 2007.

[31] G. Ammons, T. Ball, and J.R. Larus. Exploiting hardware performance counters with flow and context sensitive profiling. In *SIGPLAN Conference on Programming Language Design and Implementation*, pages 85–96, New York, NY, USA, 1997. ACM.

[32] J.M. Anderson, L.M. Berc, J. Dean, S. Ghemawat, M.R. Henzinger, S-T A. Leung, R.L. Sites, M.T. Vandevoorde, C.A. Waldspurger, and W.E. Weihl. Continuous profiling: Where have all the cycles gone? *ACM Transactions on Computer Systems*, 15(4):357–390, 1997.

[33] Astrophysics Simulation Collaboratory (ASC) home page.

[34] E. Ayguade, R.M. Badia, F.D. Igual, J. Labarta, R. Mayo, and E.S. Quintana-Orti. An extension of the StarSs programming model for platforms with multiple GPUs. In *Procs. of the 15th international Euro-Par Conference (Euro-Par 2009)*, pages 851–862. Spinger, 2009.

[35] R. Azimi, M. Stumm, and R. Wisniewski. Online performance analysis by statistical sampling of microprocessor performance counters. June 2005.

[36] L. Bachega, S. Chatterjee, K. Dockser, J. Gunnels, M. Gupta, F. Gustavson, C. Lapkowski, G. Liu, M. Mendell, C. Wait, and T.J.C. Ward. A high-performance SIMD floating point unit design for BlueGene/L: Architecture, compilation, and algorithm design. September 2004.

[37] R. Badia, J. Labarta, R. Sirvent, J.M. Perez, J.M. .Cela, and R. Grima. Programming grid applications with grid superscalar. *Journal of Grid Computing*, 1(2):151–170, 2003.

[38] D. Bailey, E. Barszcz, J. Barton, D. Browning, R. Carter, L. Dagum, R. Fatoohi, S. Fineberg, P. Frederickson, T. Lasinski, R. Schreiber, H. Simon, V. Venkatakrishman, and S. Weeratunga. The NAS parallel benchmarks. *International Journal of Supercomputer Applications*, 5:66–73, 1991.

[39] D.H. Bailey. Twelve ways to fool the masses when giving performance results on parallel computers. *Supercomputing Review*, pages 54–55, August 1991.

[40] D.H. Bailey. Misleading performance reporting in the supercomputing field. *Scientific Programming*, 1:141–151, 1992.

[41] D.H. Bailey. Little's law and high performance computing, 1997. http://crd.lbl.gov/~dhbailey/dhbpapers/little.pdf

[42] D.H. Bailey and A.S. Snavely. Performance modeling: Understanding the present and predicting the future, 2005.

[43] S. Balay, K. Buschelman, V. Eijkhout, W.D. Gropp, D. Kaushik, M.G. Knepley, L.C. McInnes, B.F. Smith, and H. Zhang. PETSc users manual. Technical Report ANL-95/11 - Revision 3.0.0, Argonne National Laboratory, 2008.

[44] M.M. Baskaran, N. Vydyanathan, U. Bonkhugula, J. Ramanujam, A. Rountev, and P. Sadayappan. Compiler-assisted dynamic scheduling for effective parallelization of loop nests on multicore processors. In *14th ACM SIGPLAN Symposium on Principles and Practice of Parallel Programming*, Raleigh, North Carolina, February 2009.

[45] H.H. Bayraktar, M.F. Adams, P.F. Hoffmann, D.C. Lee, A. Gupta., P. Papadopoulos, and T.M. Keaveny. Micromechanics of the human vertebral body. In *Transactions of the Orthopaedic Research Society*, volume 29, page 1129, San Francisco, 2004.

[46] D. Becker, R. Rabenseifner, and F. Wolf. Timestamp synchronization for event traces of large-scale message-passing applications. In *Proceedings of 14th European PVM and MPI Conference (EuroPVM/MPI)*, pages 315–325, Paris, 2007.

[47] G. Bell. Bell's law for the birth and death of computer classes. *Communications of the ACM*, 5(1):86–94, January 2008.

[48] J. Bell, M. Berger, J. Saltzman, and M. Welcome. A three-dimensional adaptive mesh refinement for hyperbolic conservation laws. 15(1):127–138, 1994.

[49] R. Bell, A. Malony, and S. Shende. A portable, extensible, and scalable tool for parallel performance profile analysis. In *Proceedings of European Conference on Parallel Computing*, 2003.

[50] P. Bellens, J.M. Perez, R.M. Badia, and J. Labarta. CellSs: A programming model for the Cell BE architecture. In *Proceedings of the 2006 ACM/IEEE Conference on Supercomputing (SC06)*, 2006.

[51] M.J. Berger and P. Colella. Local adaptive mesh refinement for shock hydrodynamics. *Journal of Computational Physics*, 82(1):64–84, May 1989.

[52] M.J. Berger and J. Oliger. Adaptive mesh refinement for hyperbolic partial differential equations. *Journal of Computational Physics*, 53:484–512, 1984.

[53] C. Bernard, C. DeTar, S. Gottlieb, U.M. Heller, J. Hetrick, N. Ishizuka, L. Kärkkäinen, S.R. Lantz, K. Rummukainen, R. Sugar, D. Toussaint, and M. Wingate. Lattice QCD on the IBM scalable POWERParallel systems SP2. In *ACM/IEEE Proceedings of SC 1995: High Performance Networking and Computing*, San Diego, California, November 1995.

[54] D.E. Bernholdt, B.A. Allan, R. Armstrong, F. Bertrand, K. Chiu, T.L. Dahlgren, K. Damevski, W.R. Elwasif, T.G.W. Epperly, M. Govindaraju, D.S. Katz, J.A. Kohl, M. Krishnan, G. Kumfert, J.W. Larson, S. Lefantzi, M.J. Lewis, A.D. Malony, L.C. McInnes, J. Nieplocha, B. Norris, S.G. Parker, J. Ray, S. Shende, T.L. Windus, and S. Zhou. A component architecture for high-performance scientific computing. *Intl. Journal of High-Performance Computing Applications*, ACTS Collection Special Issue, 2005.

[55] J. Bilmes, K. Asanovic, C-W Chin, and J. Demmel. Optimizing matrix multiply using PHiPAC: a portable, high-performance, ANSI C coding methodology. In *International Conference on Supercomputing*, pages 340–347, Vienna, Austria, 1997.

[56] D. Biskamp. *Magnetohydrodynamic Turbulence*. Cambridge University Press, U.K., 2003.

[57] C.M. Bitz and W.H. Lipscomb. An energy-conserving thermodynamic model of sea ice. *Journal of Geophysical Research*, 104:15669–15677, 1999.

[58] L.S. Blackford, J. Choi, A. Cleary, E. DAzevedo, J. Demmel, I. Dhillon, J. Dongarra, S. Hammarling, G. Henry, A. Petitet, K. Stanley, D. Walke, and R.C. Whaley. *ScaLAPACK Users Guide*. SIAM, Philadelphia, 1997.

[59] W. Blume, R. Doallo, R. Eigenmann, J. Grout, J. Hoeflinger, T. Lawrence, J. Lee, D. Padua, Y. Paek, B. Pottenger, L. Rauchwerger, and P. Tu. Parallel programming with polaris. *Computer*, 29(12), December 1996.

[60] U. Bondhugula, A. Hartono, J. Ramanujam, and P. Sadayappan. A practical automatic polyhedral parallelizer and locality optimizer. In *Proceedings of ACM SIGPLAN Conference on Programming Language Design and Implementation*, June 2008.

[61] J. Brehm, P.H. Worley, and M. Madhukar. Performance modeling for SPMD message-passing programs. *Concurrency: Practice and Experience*, 10(5):333–357, 1998.

[62] D. Brown, P. Diener, O. Sarbach, E. Schnetter, and M. Tiglio. Turduckening black holes: An analytical and computational study. *Physical Review D (submitted)*, 2008.

[63] P.N. Brown, R.D. Falgout, and J.E. Jones. Semicoarsening multigrid on distributed memory machines. *SIAM Journal on Scientific Computing*, 21(5):1823–1834, 2000.

[64] S. Browne, J. Dongarra, N. Garner, G. Ho, and P. Mucci. A portable programming interface for performance evaluation on modern processors. *The International Journal of High Performance Computing Applications*, 14(4):189–204, 2000.

[65] S. Browne, J. Dongarra, N. Garner, G. Ho, and P. Mucci. A portable programming interface for performance evaluation on modern processors. *International Journal of High Performance Computing Applications*, 14(3):189–204, 2000.

[66] H. Brunst, A.D. Malony, S. Shende, and R. Bell. Online remote trace analysis of parallel applications on high-performance clusters. In *Proceedings of the ISHPC Conference (LNCS 2858)*, pages 440–449. Springer, 2003.

[67] H. Brunst and W.E. Nagel. Scalable performance analysis of parallel systems: Concepts and experiences. In *Parallel Computing: Software, Alghorithms, Architectures Applications*, pages 737–744, 2003.

[68] H. Brunst, W.E. Nagel, and A.D. Malony. A distributed performance analysis architecture for clusters. In *Proceedings of the IEEE International Conference on Cluster Computing (Cluster 2003)*, pages 73–83. IEEE Computer Society, 2003.

[69] B. Buck and J.K. Hollingsworth. An API for runtime code patching. *The International Journal of High Performance Computing Applications*, 14(4):317–329, Winter 2000.

[70] M. Burtscher, B.D. Kim, J. Diamond, J. McCalpin, L. Koesterke, and J. Browne. Perfexpert: An automated HPC performance measurement and analysis tool with optimization recommendations. In *Proceedings of ACM/IEEE Conference on Supercomputing (SC10)*, New York, NY, November 2010. ACM.

[71] A. Buttari, J. Langou, J. Kurzak, and J. Dongarra. A class of parallel tiled linear algebra algorithms for multicore architectures. *Parallel Computing*, 35(1):38–53, 2009.

[72] Cactus computational toolkit home page. http://www.cactuscode.org

[73] D. Callahan, J. Cocke, and K. Kennedy. Estimating interlock and improving balance for pipelined architectures. *Journal of Parallel and Distributed Computing*, 5(4):334–358, 1988.

[74] K. Camarda, Y. He, and K.A. Bishop. A parallel chemical reactor simulation using Cactus. In *Proceedings of Linux Clusters: The HPC Revolution, NCSA*, 2001.

[75] R. Car and M. Parrinello. *Physics Review Letters*, 55, 2471, 1985.

[76] W.W. Carlson, J.M. Draper, D.E. Culler, K. Yelick, E. Brooks, and K. Warren. Introduction to upc and language specification. Technical Report CCS-TR-99-157, Center for Computing Sciences, 17100 Science Dr., Bowie, MD 20715, May 1999.

[77] S. Carr and K. Kennedy. Improving the ratio of memory operations to floating-point operations in loops. *ACM Transactions on Programming Languages and Systems*, 16(6):1768–1810, 1994.

[78] L. Carrington, A. Snavely, X. Gao, and N. Wolter. A performance prediction framework for scientific applications. *ICCS Workshop on Performance Modeling and Analysis (PMA03)*, June 2003.

[79] L. Carrington, N. Wolter, A. Snavely, and C.B. Lee. Applying an automated framework to produce accurate blind performance predictions of full-scale HPC applications. *DoD Users Group Conference (UGC2004)*, June 2004.

[80] M. Casas, R. Badia, and J. Labarta. Automatic analysis of speedup of MPI applications. In *Proceedings of the 22nd ACM International Conference on Supercomputing (ICS)*, pages 349–358, 2008.

[81] M. Casas, R.M. Badia, and J. Labarta. Automatic structure extraction from MPI applications tracefiles. In *European Conference on Parallel Computing*, pages 3–12, 2007.

[82] M. Casas, H. Servat, R.M. Badia, and J. Labarta. Analyzing the temporal behavior of application using spectral analysis. In *Research Report UPC-RR-CAP-2009-14*, 2009.

[83] C. Cascaval, E. Duesterwald, P.F. Sweeney, and R.W. Wisniewski. Multiple page size modeling and optimization. *Parallel Architectures and Compilation Techniques, 2005. PACT 2005. 14th International Conference on*, pages 339–349, 17-21 September 2005.

[84] CCSM Software Engineering Group.
http://www.ccsm.ucar.edu/cseg

[85] CCSM Software Engineering Working Group.
http://www.ccsm.ucar.edu/csm/working_groups/Software

[86] National Energy Research Scientific Computing Center. Parallel total energy code, 2009.

[87] C. Chen. *Model-Guided Empirical Optimization for Memory Hierarchy*. PhD thesis, University of Southern California, 2007.

[88] C. Chen, J. Chame, and M. Hall. Combining models and guided empirical search to optimize for multiple levels of the memory hierarchy. March 2005.

[89] C. Chen, J. Chame, and M. Hall. CHiLL: A framework for composing high-level loop transformations. Technical Report 08-897, University of Southern California, June 2008.

[90] C. Chen, J. Chame, and M.W. Hall. Combining models and guided empirical search to optimize for multiple levels of the memory hierarchy. In *Proceedings of the International Symposium on Code Generation and Optimization*, March 2005.

[91] D. Chen, N. Vachharajani, R. Hundt, S.W. Liao, V. Ramasamy, P. Yuan, W. Chen, and W. Zheng. Taming hardware event samples for FDO compilation. pages 42–53, April 2010.

[92] J. Choi and J.J. Dongarra. Scalable linear algebra software libraries for distributed memory concurrent computers. In *FTDCS '95: Proceedings of the 5th IEEE Workshop on Future Trends of Distributed Computing Systems*, page 170, Washington, DC, USA, 1995. IEEE Computer Society.

[93] I-H Chung and J.K. Hollingsworth. Using Information from Prior Runs to Improve Automated Tuning Systems. In *Proceedings of the 2004 ACM/IEEE conference on Supercomputing (SC04)*, page 30, Washington, DC, USA, 2004. IEEE Computer Society.

[94] I.-H. Chung and J.K. Hollingsworth. A case study using automatic performance tuning for large-scale scientific programs. In *High Performance Distributed Computing, 2006 15th IEEE International Symposium on High Performance Distributed Computing*, pages 45–56, 2006.

[95] C. Coarfa, J. Mellor-Crummey, N. Froyd, and Y. Dotsenko. Scalability analysis of SPMD codes using expectations. In *ICS '07: Proceedings of the 21st annual International Conference on Supercomputing*, pages 13–22, New York, NY, 2007. ACM.

[96] P. Colella. Multidimensional Upwind Methods for Hyperbolic Conservation Laws. *Journal of Computational Physics*, 87:171–200, 1990.

[97] P. Colella and M.D. Sekora. A limiter for PPM that preserves accuracy at smooth extrema. *Journal of Computational Physics*, submitted.

[98] W.D. Collins, C.M. Bitz, M.L. Blackmon, G.B. Bonan, C.S. Bretherton, J.A. Carton, P. Chang, S.C. Doney, J.H. Hack, T.B. Henderson, J.T. Kiehl, W.G. Large, D.S. McKenna, B.D. Santer, and R.D. Smith. The community climate system model version 3 (CCSM3). *Journal of Climate*, 19(11):2122–2143, 2006.

[99] W.D. Collins and P.J. Rasch, et al. Description of the NCAR community atmosphere model (CAM 3.0). NCAR Tech Note NCAR/TN-464+STR, National Center for Atmospheric Research, Boulder, CO 80307, 2004.

[100] W.D. Collins, et al. The formulation and atmospheric simulation of the community atmosphere model: CAM3. *Journal of Climate*, 2005.

[101] Community Climate System Model.
http://www.ccsm.ucar.edu

[102] K.D. Cooper, D. Subramanian, and L. Torczon. Adaptive optimizing compilers for the 21st century. *The Journal of Supercomputing*, 23(1):7–22, August 2002.

[103] C. Ţăpuş, I-H Chung, and J.K. Hollingsworth. Active harmony: Towards automated performance tuning. In *Proceedings of the ACM/IEEE Conference on Supercomputing (SC02)*, pages 1–11, Los Alamitos, CA, USA, 2002. IEEE Computer Society Press.

[104] D. Culler, J.P. Singh, and A. Gupta. *Parallel Computer Architecture: A Hardware/Software Approach*. Morgan Kaufmann, San Francisco, 1999.

[105] A.N. Cutler. A history of the speed of light. 2001.
http://www.sigma-engineering.co.uk/light/lightindex.shtml

[106] L. Dagum and R. Menon. OpenMP: an industry-standard API for shared-memory programming. *IEEE Computational Science and Engineering*, 5(1):46–55, January/March 1998.

[107] A. Danalis, K. Kim, L. Pollock, and M. Swany. Transformations to parallel codes for communication-computation overlap. In *Proceedings of IEEE/ACM Conference on Supercomputing (SC05)*, November 2005.

[108] K. Datta, M. Murphy, V. Volkov, S. Williams, J. Carter, L. Oliker, D. Patterson, J. Shalf, and K. Yelick. Stencil computation optimization and autotuning on state-of-the-art multicore architectures. In *Proceedings of ACM/IEEE Conference on Supercomputing (SC08)*, 2008.

[109] K. Davis, A. Hoisie, G. Johnson, D. Kerbyson, M. Lang, S. Pakin, and F. Petrini. A performance and scalability analysis of the bluegene/l architecture.

[110] J. Dean, J.E. Hicks, C.A. Waldspurger, W.E. Weihl, and G. Chrysos. ProfileMe: Hardware support for instruction-level profiling on out-of-order processors. In *MICRO 30: Proceedings of the 30th annual ACM/IEEE International Symposium on Microarchitecture*, pages 292–302, Washington, DC, 1997. IEEE Computer Society.

[111] J. Demmel, J. Dongarra, V. Eijkhout, E. Fuentes, A. Petitet, R. Vuduc, C. Whaley, and K. Yelick. Self adapting linear algebra algorithms and software. *Proceedings of the IEEE*, 93(2), 2005. Special issue on Program Generation, Optimization, and Adaptation.

[112] J.W. Demmel. *Applied Numerical Linear Algebra*. Society for Industrial and Applied Mathematics, Philadephia, PA, 1997.

[113] R. Desikan, D. Burger, S. Keckler, and T. Austin. Sim-alpha: a validated, execution-driven Alpha 21264 simulator. Technical Report TR-01-23, Department of Computer Sciences, The University of Texas at Austin, 2001.

[114] R.E. Dickinson, K.W. Oleson, G. Bonan, F. Hoffman, P. Thornton, M. Vertenstein, Z-L Yang, and X. Zeng. The Community Land Model and its climate statistics as a component of the Climate System Model. *Journal of Climate*, 19(11):2032–2324, 2006.

[115] P. Diener, E.N. Dorband, E. Schnetter, and M. Tiglio. Optimized high-order derivative and dissipation operators satisfying summation by parts, and applications in three-dimensional multi-block evolutions. *Journal of Scientific Computing*, 32:109–145, 2007.

[116] F. Dijkstra and A. van der Steen. Integration of two ocean models.

[117] S. Donadio, J. Brodman, T. Roeder, K. Yotov, D. Barthou, A. Cohen, M.J. Garzarán, D. Padua, and K. Pingali. A language for the compact representation of multiple program versions. In *Proceedings of the 18th International Workshop on Languages and Compilers for Parallel Computing*, October 2005.

[118] J. Dongarra, A.D. Malony, S. Moore, P. Mucci, and S. Shende. Performance instrumentation and measurement for terascale systems. In *Proceedings of the ICCS 2003 Conference (LNCS 2660)*, pages 53–62, 2003.

[119] J.B. Drake, I.T. Foster, J.J. Hack, J.G. Michalakes, B.D. Semeraro, B. Tonen, D.L. Williamson, and P.H. Worley. PCCM2: A GCM adapted for scalable parallel computer. In *Fifth Symposium on Global Change Studies*, pages 91–98. American Meteorological Society, Boston, 1994.

[120] J.B. Drake, I.T. Foster, J.G. Michalakes, B. Toonen, and P.H. Worley. Design and performance of a scalable parallel community climate model. *Parallel Computing*, 21(10):1571–1591, 1995.

[121] J.B. Drake, S. Hammond, R. James, and P.H. Worley. Performance tuning and evaluation of a parallel community climate model. In *Proceedings of 1999 ACM/IEEE Conference on Supercomputing (SC99)*, page 34, New York, NY, USA, 1999. ACM.

[122] J.B. Drake, P.W. Jones, and G. Carr. Overview of the software design of the Community Climate System Model. *International Journal of High Performance Computing Applications*, 19(3):177–186, Fall 2005.

[123] J.B. Drake, P.W. Jones, and G.R. Carr, Jr. Special issue on climate modeling. *International Journal of High Performance Computing Applications*, 19(3), August 2005.

[124] J.B. Drake, P.W. Jones, M. Vertenstein, J.B. White III, and P.H. Worley. Software design for petascale climate science. In D.A. Bader, editor, *Petascale Computing: Algorithms and Applications*, chapter 7, pages 125–146. Chapman & Hall/CRC, New York, NY, 2008.

[125] P.J. Drongowski. Instruction-based sampling: A new performance analysis technique for AMD family 10h processors, November 2007.
http://developer.amd.com/Assets/AMD_IBS_paper_EN.pdf

[126] A. Dubey, L.B. Reid, and R. Fisher. Introduction to FLASH 3.0, with application to supersonic turbulence. *Physica Scripta*, 132:014046, 2008.

[127] J.K. Dukowicz, R.D. Smith, and R.C. Malone. A reformulation and implementation of the Bryan-Cox-Semtner ocean model. *Journal of Atmospheric and Oceanic Technology*, 10:195–208, 1993.

[128] J.W. Eaton. Octave home page.
http://www.octave.org

[129] S. Eranian. Perfmon2: A flexible performance monitoring interface for Linux. pages 269–288, July 2006.

[130] M. Ester, H.P. Kriegel, J. Sander, and X. Xu. A density-based algorithm for discovering clusters in large spatial databases with noise. In *Proceedings of the Second International Conference on Knowledge Discovery and Data Mining*, pages 226–231, 1996.

[131] FEAP.

[132] A. Federova. *Operating System Scheduling for Chip Multithreaded Processors*. PhD thesis, Harvard University, September 2006.

[133] W. Feng. The importance of being low power in high-performance computing. *CTWatch Quarterly*, 1(3):12–20, 2005.

[134] Solaris memory placement optimization and sun fireservers.
http://www.sun.com/software/solaris/performance.jsp, March 2003.

[135] P.F. Fischer, G.W. Kruse, and F. Loth. Spectral element methods for transitional flows in complex geometries. *Journal of Scientific Computing*, 17, 2002.

[136] I.T. Foster, B. Toonen, and P.H. Worley. Performance of parallel computers for spectral atmospheric models. *Journal of Atmospheric and Oceanic Technology*, 13(5):1031–1045, 1996.

[137] I.T. Foster and P.H. Worley. Parallel algorithms for the spectral transform method. 18(3):806–837, May 1997.

[138] F. Freitag, J. Caubet, M. Farreras, T. Cortes, and J. Labarta. Exploring the predictability of MPI messages. In *Proceedings of the 17th IEEE International Parallel and Distributed Processing Symposium (IPDPS03)*, pages 46–55, 2003.

[139] M. Frigo. A fast Fourier transform compiler. In *Proceedings of ACM SIGPLAN Conference on Programming Language Design and Implementation*, May 1999.

[140] M. Frigo and S.G. Johnson. FFTW: An adaptive software architecture for the FFT. In *Proceedings of 1998 IEEE Intl. Conf. Acoustics Speech and Signal Processing*, volume 3, pages 1381–1384. IEEE, 1998.

[141] M. Frigo and S.G. Johnson. FFTW for version 3.0, 2003.
http://www.fftw.org/fftw3.pdf

[142] M. Frigo, C.E. Leiserson, and K.H. Randall. The implementation of the Cilk-5 multithreaded language. In *Proceedings of the 1998 ACM SIGPLAN Conference on Programming Language Design and Implementation*, pages 212–223, Montreal, Quebec, Canada, June 1998.

[143] N. Froyd, J. Mellor-Crummey, and R. Fowler. Low-overhead call path profiling of unmodified, optimized code. In *Proceedings of 19th International Conference on Supercomputing*, pages 81–90, New York, NY, 2005. ACM Press.

[144] K. Frlinger and D. Skinner. Capturing and visualizing event flow graphs of mpi applications. *Proceedings of the Workshop on Productivity and Performance (PROPER 2009)*, August 2009.

[145] T. Gamblin, B.R. de Supinski, M. Schulz, R. Fowler, and D.A. Reed. Scalable load-balance measurement for SPMD codes. In *Proceedings of ACM/IEEE Conference on Supercomputing (SC08)*, pages 1–12, Piscataway, NJ, 2008. IEEE Press.

[146] M. Garcia, J. Corbalan, and J. Labarta. LeWI: A runtime balancing algorithm for nested parallelism. In *Proceedings of the International Conference on Parallel Processing (ICPP'09)*, 2009.

[147] M. Geimer, B. Kuhlmann, F. Pulatova, F. Wolf, and B.J.N. Wylie. Scalable collation and presentation of call-path profile data with cube. In *Parallel Computing: Architectures, Algorithms and Applications: Proceedings of Parallel Computing (ParCo07)*, volume 15, pages 645–652, Julich (Germany), 2007.

[148] M. Geimer, S. Shende, A. Malony, and F. Wolf. A generic and configurable source-code instrumentation component. In G. Allen, J. Nabrzyski, E. Seidel, G. van Albada, J. Dongarra, and P. Sloot, editors, *International Conference on Computational Science (ICCS)*, volume 5545 of *Lecture Notes in Computer Science*, pages 696–705, Baton Rouge, LA, May 2009. Springer.

[149] S. Girbal, N. Vasilache, C. Bastoul, A. Cohen, D. Parello, M. Sigler, and O. Temam. Semi-automatic composition of loop transformations for deep parallelism and memory hierarchies. *International Journal of Parallel Programming*, 34(3):261–317, June 2006.

[150] J. Gonzalez, J. Gimenez, and J. Labarta. Automatic detection of parallel applications computation phases. In *Proceedings of the 23rd IEEE International Parallel and Distributed Processing Symposium (IPDPS09)*, 2009.

[151] T. Goodale, G. Allen, G. Lanfermann, J. Massó, T. Radke, E. Seidel, and J. Shalf. The Cactus framework and toolkit: Design and applications. In *Vector and Parallel Processing – VECPAR'2002, 5th International Conference, Lecture Notes in Computer Science*, Berlin, 2003. Springer.

[152] S. Graham, P. Kessler, and M. McKusick. gprof: A call graph execution profiler. *SIGPLAN '82 Symposium on Compiler Construction*, pages 120–126, June 1982.

[153] J.A. Gunnels, R.A. Van De Geijn, and G.M. Henry. FLAME: Formal linear algebra methods environment. *ACM Transactions on Mathematical Software*, 27, 2001.

[154] D. Gunter, K. Huck, K. Karavanic, J. May, A. Malony, K. Mohror, S. Moore, A. Morris, S. Shende, V. Taylor, X. Wu, and Y. Zhang. Performance database technology for SciDAC applications. 2007.

[155] F. Gygi. Architecture of Qbox: A scalable first-principles molecular dynamics code. *IBM Journal of Research and Development*, 52, January/March 2008.

[156] F. Gygi, E. Draeger, B.R. de Supinski, R.K. Yates, F. Franchetti, S. Kral, J. Lorenz, C.W. Überhuber, J.A. Gunnels, and J.C. Sexton. Large-scale first-principles molecular dynamics simulations on the BlueGene/L platform using the Qbox code. In *Proceedings of ACM/IEEE Conference on Supercomputing (SC05)*, 2005.

[157] F. Gygi, E.W. Draeger, M. Schulz, B.R. de Supinski, J.A. Gunnels, V. Austel, J.C. Sexton, F. Franchetti, S. Kral, and C.W. Überhuber. Large-scale electronic structure calculations of high-z metals on the BlueGene/L Platform. In *Proceedings of ACM/IEEE Conference on Supercomputing (SC06)*, November 2006.

[158] J.J. Hack, B.A. Boville, B.P. Briegleb, J.T. Kiehland, P.J. Rasch, and D.L. Williamson. Description of the NCAR community climate model (CCM2). NCAR Tech. Note NCAR/TN–382+STR, National Center for Atmospheric Research, Boulder, CO, 1992.

[159] M. Hall, J. Chame, J. Shin, C. Chen, G. Rudy, and M.M. Khan. Loop transformation recipes for code generation and auto-tuning. In *LCPC*, October, 2009.

[160] M. Hall, D. Padua, and K. Pingali. Compiler research: The next fifty years. *Communications of the ACM*, February 2009.

[161] M.W. Hall, J.M. Anderson, S.P. Amarasinghe, B.R. Murphy, S. Liao, E. Bugnion, and M.S. Lam. Maximizing multiprocessor performance with the SUIF compiler. *IEEE Computer*, 29(12):84–89, December 1996.

[162] A. Hartono, B. Norris, and P. Sadayappan. Annotation-based empirical performance tuning using Orio. In *Proceedings of the 23rd International Parallel and Distributed Processing Symposium*, May 2009.

[163] A. Hartono and S. Ponnuswamy. Annotation-based empirical performance tuning using Orio. In *23rd IEEE International Parallel and Distributed Processing Symposium (IPDPS) Rome, Italy*, May 2009.

[164] J.L. Hennessy and D.A. Patterson. *Computer Architecture: A Quantitative Approach*. Morgan Kaufmann, San Francisco, 2006.

[165] M.D. Hill and A.J. Smith. Evaluating Associativity in CPU Caches. *IEEE Transactions on Computers*, 38(12):1612–1630, 1989.

[166] P. Hohenberg and W. Kohn. Inhomogeneous electron gas. *Physical Review*, 136:B864, 1964.

[167] A. Hoisie, O. Lubeck, and H. Wasserman. Performance and scalability analysis of teraflop-scale parallel architectures using multidimensional wavefront applications. *International Journal of High Performance Computing Applications*, 14:330–346, 2000.

[168] J.K. Hollingsworth and P.J. Keleher. Prediction and adaptation in Active Harmony. *Cluster Computing*, 2(3):195–205, 1999.

[169] J.K. Hollingsworth, B.P. Miller, and J. Cargille. Dynamic program instrumentation for scalable performance tools. In *1994 Scalable High Performance Computing Conference*, pages 841–850, Knoxville, TN, May 1994.

[170] R. Hooke and T.A. Jeeves. Direct search solution of numerical and statistical problems. *Journal of the ACM*, 8(2):212–229, 1961.

[171] HPC challenge benchmark.
http://icl.cs.utk.edu/hpcc/index.html

[172] K. Huck and A. Malony. PerfExplorer: A performance data mining framework for large-scale parallel computing. In *Proceedings of ACM/IEEE Conference on Supercomputing (SC05)*, 2005.

[173] K. Huck, A. Malony, S. Shende, and A. Morris. Knowledge Support and Automation for Performance Analysis with PerfExplorer 2.0. *The Journal of Scientific Programming*, 16(2-3):123–134, 2008. (special issue on Large-Scale Programming Tools and Environments).

[174] K.A. Huck., A.D. Malony, and A. Morris. Design and implementation of a parallel performance data management framework. In *Proceedings of the 2005 International Conference on Parallel Processing (ICPP05)*, pages 473–482, Washington, DC, USA, 2005. IEEE Computer Society.

[175] E.C. Hunke and J.K. Dukowicz. An elastic-viscous-plastic model for sea ice dynamics. *Journal of Physical Oceanography*, 27:1849–1867, 1997.

[176] S. Hunold and T. Rauber. Automatic tuning of PDGEMM towards optimal performance. In *Proceedings European Conference on Parallel Computing*, August 2005.

[177] S. Husa, I. Hinder, and C. Lechner. Kranc: A Mathematica application to generate numerical codes for tensorial evolution equations. *Computer Physics Communications*, 174:983–1004, 2006.

[178] R. Ierusalimschy, L.H. de Figueiredo, and W.C. Filho. Lua an extensible extension language. *Software: Practice and Experience*, 26:635–652, June 1996.

[179] Intel Corporation. Intel 64 and IA-32 architectures software developers manualvolume 3b: System programming guide, part 2, number 253669-032us, September 2009.
http://www.intel.com/Assets/PDF/manual/253669.pdf

[180] E. Ipek, B.R. de Supinski, M. Schulz, and S.A. McKee. An approach to performance prediction for parallel applications. In *Euro-Par 2005 Parallel Processing*, pages 196–205, 2005.

[181] ITER: International thermonuclear experimental reactor.

[182] M. Itzkowitz and Y. Maruyama. HPC profiling with the Sun Studio(TM) performance tools. In *Third Parallel Tools Workshop*, Dresden, Germany, September 2009.

[183] E. Jaeger, L. Berry, and J. Myra, et al. Sheared poloidal flow driven by mode conversion in tokamak plasmas. *Physical Review Letters*, 90, 2003.

[184] J.A. Joines and C.R. Houck. On the use of non-stationary penalty functions to solve nonlinear constrained optimization problems with GA's. pages 579–584 vol.2, June 1994.

[185] P.W. Jones, P.H. Worley, Y. Yoshida, J.B. White III, and J. Levesque. Practical performance portability in the Parallel Ocean Program (POP). *Concurrency and Computation: Practice and Experience*, 17(10):1317–1327, 2005.

[186] G. Karypis and V. Kumar. Parallel multilevel k-way partitioning scheme for irregular graphs. *ACM/IEEE Proceedings of SC1996: High Performance Networking and Computing*, 1996.

[187] D.J. Kerbyson, H.J. Alme, A. Hoisie, F. Petrini, H.J. Wasserman, and M. Gittings. Predictive performance and scalability modeling of a large-scale application. In *Proceedings of ACM/IEEE Conference on Supercomputing (SC01)*, pages 37–37, New York, NY, USA, 2001. ACM.

[188] J.T. Kiehl, J.J. Hack, G. Bonan, B.A. Boville, D.L. Williamson, and P.J. Rasch. The National Center for Atmospheric Research Community Climate Model: CCM3. *Journal of Climate*, 11:1131–1149, 1998.

[189] J.G. Kim and H.W. Park. Advanced simulation technique for modeling multiphase fluid flow in porous media. In *Computational Science and Its Applications - Iccsa 2004, LNCS 2004, by A. Lagana et. al.*, pages 1–9, 2004.

[190] T. Kisuki, P.M.W. Knijnenburg, and M.F.P. O'Boyle. Combined selection of tile sizes and unroll factors using iterative compilation. In *PACT '00: Proceedings of the 2000 International Conference on Parallel Architectures and Compilation Techniques*, Washington, DC, USA, 2000. IEEE Computer Society.

[191] A. Knüpfer, R. Brendel, H. Brunst, H. Mix, and W.E. Nagel. Introducing the Open Trace Format (OTF). In *Proceedings of the 6th International Conference on Computational Science*, volume 3992 of *Springer Lecture Notes in Computer Science*, pages 526–533, Reading, UK, May 2006.

[192] A. Knupfer and W.E. Nagel. Construction and compression of complete call graphs for post-mortem program trace analysis. In *Proceedings of the International Conference on Parallel Processing (ICPP)*, pages 165–172, 2005.

[193] S-H Ko, K.W. Cho, Y.D. Song, Y.G. Kim, J-S Na, and C. Kim. *Development of Cactus driver for CFD analyses in the grid computing environment*, pages 771–777. Springer, 2005.

[194] S. Kohn, G. Kumfert, J. Painter, and C. Ribbens. Divorcing language dependencies from a scientific software library. In *Proceedings of the 10th SIAM Conference on Parallel Processing*, 2001.

[195] W. Kohn. Density functional and density matrix method scaling linearly with the number of atoms. *Physical Review Letters*, 76(17):3168–3171, 1996.

[196] W. Kohn and L.J. Sham. Self-consistent equations including exchange and correlation effects. *Physical Review*, 140:A1133, 1965.

[197] T.G. Kolda, R.M. Lewis, and V. Torczon. Optimization by direct search: New perspectives on some classical and modern methods. *SIAM Review*, 45(3):385–482, 2004.

[198] M. Kotschenreuther, G. Rewoldt, and W.M. Tang. Comparison of initial value and eigenvalue codes for kinetic toroidal plasma instabilities. *Computer Physics Communications*, 88:128–140, August 1995.

[199] Kranc: Automated code generation.
http://www.cct.lsu.edu/~eschnett/Kranc

[200] A.S. Kronfeld. Quantum chromodynamics with advanced computing. *Journal of Physics: Conference Series*, 125:012067, 2008.

[201] R. Kufrin. PerfSuite: An accessible, open source performance analysis environment for Linux. In *Sixth International Conference on Linux Clusters (LCI)*, 2005.

[202] P. Kulkarni, W. Zhao, H. Moon, K. Cho, D. Whalley, J. Davidson, M. Bailey, Y. Paek, and K. Gallivan. Finding effective optimization phase sequences. *SIGPLAN Not.*, 38(7):12–23, 2003.

[203] J. Labarta, J. Gimenez, E. Martinez, P. Gonzalez, H. Servat, G. Llort, and X. Aguilar. Scalability of tracing and visualization tools. In *Parallel Computing 2005*, Malaga, 2005.

[204] J. Labarta, S. Girona, V. Pillet, T. Cortes, and L. Gregoris. Dip: A parallel program development environment. In *Proceedings of 2nd International EuroPar Conference (EuroPar 96)*, Lyon (France), August 1996.

[205] J.C. Lagarias, J.A. Reeds, M.H. Wright, and P.E. Wright. Convergence properties of the Nelder-Mead simplex algorithm in low dimensions. *SIAM Journal on Optimization*, 9:112–147, 1998.

[206] J.R. Larus and T. Ball. Rewriting executable files to measure program behavior. *Software Practice and Experience*, 24(2):197–218, 1994.

[207] E.D. Lazowska, J. Zahorjan, G.S. Graham, and K.C. Sevcik. *Quantitative System Performance: Computer System Analysis Using Queueing Network Models*. Prentice-Hall, Inc., Upper Saddle River, NJ, USA, 1984.

[208] C. Lechner, D. Alic, and S. Husa. From tensor equations to numerical code — computer algebra tools for numerical relativity. In *SYNASC 2004 — 6th International Symposium on Symbolic and Numeric Algorithms for Scientific Computing*, Timisoara, Romania, 2004.

[209] B.C. Lee, D.M. Brooks, B.R. de Supinski, M. Schulz, K. Singh, and S.A. McKee. Methods of inference and learning for performance modeling of parallel applications. In *PPoPP '07: Proceedings of the 12th ACM SIGPLAN Symposium on Principles and Practice of Parallel Programming*, pages 249–258, New York, NY, 2007. ACM.

[210] W.W. Lee. Gyrokinetic particle simulation model. *Journal of Computational Physics*, 72:243–269, 1987.

[211] Y. Lee and M. Hall. A code isolator: Isolating code fragments from large programs. In *Proceedings of the Seventeenth Workshop on Languages, Compilers for Parallel Computing (LCPC'04)*, September 2004.

[212] M. Legendre. Dyninst as a binary rewriter. In *Paradyn/Dyninst week*, 2009. http://www.dyninst.org/pdWeek09/slides/legendre-binrewriter.pdf

[213] J. Levon and P. Elie. Oprofile: A system profiler for Linux.
http://oprofile.sourceforge.net

[214] Y. Li, J. Dongarra, and S. Tomov. A note on auto-tuning GEMM for GPUs. In *9th International Conference on Computation Science (ICCS'09)*, Baton Rouge, LA, May 2009.

[215] C. Liao, D.J. Quinlan, R. Vuduc, and T. Panas. Effective source-to-source outlining to support whole program empirical optimization. In *Proceedings of the 22nd International Workshop on Languages and Compilers for Parallel Computing (LCPC09)*, October 2009.

[216] Z. Lin, S. Ethier, T.S. Hahm, and W.M. Tang. Size scaling of turbulent transport in magnetically confined plasmas. *Physical Review Letters*, 88, 2002.

[217] Z. Lin, T.S. Hahm, W.W. Lee, W.M. Tang, and R.B. White. Turbulent transport reduction by zonal flows: Massively parallel simulations. *Science*, 281(5384):1835–1837, September 1998.

[218] K.A. Lindlan, J. Cuny, A.D. Malony, S. Shende, B. Mohr, R. Rivenburgh, and C. Rasmussen. A tool framework for static and dynamic analysis of object-oriented software with templates. In *Proceedings of ACM/IEEE Conference on Supercomputing (SC2000)*, 2000.

[219] X.S. Liu, X.H. Zhang, K.K. Sekhon, M.F. Adam, D.J. McMahon, E. Shane, J.P. Bilezikian, and X.E. Guo. High-resolution peripheral quantitative computed tomography can assess microstructural and mechanical properties of human distal tibial bone. *Journal of Bone and Mineral Research*, in press.

[220] G. Llort, J. Gonzalez, H. Servat, J. Gimenez, and J. Labarta. On-line detection of large-scale parallel application's structure. In *IPDPS 2010*, April 2010.

[221] C.K. Luk, R. Cohn, R. Muth, H. Patil, A. Klauser, G. Lowney, S. Wallace, V.J. Reddi, and K. Hazelwood. Pin: Building customized program analysis tools with dynamic instrumentation. In *Proceedings of Programming Language Design and Implementation (PLDI)*, pages 191–200, 2005.

[222] A. Luque and A. Marti. Increasing the efficiency of ideal solar cells by photon induced tansitions at intermediate lavels. *Physical Review Letters*, 78:5014, 1997.

[223] A. Macnab, G. Vahala, L. Vahala, and P. Pavlo. Lattice boltzmann model for dissipative MHD. In *29th EPS Conference on Controlled Fusion and Plasma Physics*, volume 26B, Montreux, Switzerland, June 17-21, 2002.

[224] S. Major, D. Rideout, and S. Surya. Spatial hypersurfaces in causal set cosmology. *Classical Quantum Gravity*, 23:4743–4752, Jun 2006.

[225] A. Malony and S. Shende. *Performance technology for complex parallel and distributed systems*, pages 37–46. Kluwer, Norwell, MA, 2000.

[226] A. Malony, S. Shende, and A. Morris. Phase-based parallel performance profiling. In *ParCo 2005: Parallel Computing 2005*, Malaga, Spain, September 2005.

[227] G. Marin and J. Mellor-Crummey. Crossarchitecture performance predictions for scientific applications using parameterized models. In *Proceedings of the Joint International Conference on Measurement and Modeling of Computer Systems (SIGMETRIC 2004)*, pages 2–13, New York, NY, 2004.

[228] V. Marjanovic, J. Labarta, E. Ayguad, and M. Valero. Effective communication and computation overlap with hybrid MPI/SMPSs. In *Poster at PPoPP 2010*, 2010.

[229] J. Markoff. Measuring how fast computers really are. *New York Times*, page 14F, September 1991.

[230] S. Mayanglambam, A. Malony, and M. Sottile. Performance Measurement of Applications with GPU Acceleration using CUDA. In *Parallel Computing (ParCo)*, 2009. to appear.

[231] P.E. McKenney. Differential profiling. *Software: Practice and Experience*, 29(3):219–234, 1998.

[232] K.I.M. McKinnon. Convergence of the Nelder–Mead simplex method to a nonstationary point. *SIAM Journal on Optimization*, 9(1):148–158, 1998.

[233] McLachlan, a public BSSN code.

[234] J. Mellor-Crummey. Harnessing the power of emerging petascale platforms. *Journal of Physics: Conference Series*, 78, June 2007.

[235] J. Mellor-Crummey, R.J. Fowler, G. Marin, and N. Tallent. HPCView: A tool for top-down analysis of node performance. *Journal of Supercomputing*, 23(1):81–104, 2002.

[236] A. Mericas. Performance monitoring on the POWER5 microprocessor. In L.K. John and L. Eeckhout, editors, *Performance Evaluation and Benchmarking*, pages 247–266. CRC PRESS, 2006.

[237] A. Mericas, et al. CPI analysis on POWER5, Part 2: Introducing the CPI breakdown model.
https://www.ibm.com/developerworks/library/pa-cpipower2

[238] Message Passsing Interface Forum. MPI: A Message Passing Interface Standard. *International Journal of Supercomputer Applications (Special Issue on MPI)*, 8(3/4), 1994.

[239] C. Mikenberg and G. Rodriguez. Tracedriven cosimulation of highperformance computing systems using omnet++. In *2nd International Workshop on OMNeT++, in conjunction with the 2nd International Conference on Simulation Tools and Techniques (SIMUTools'09)*, 2009.

[240] G.H. Miller and P. Colella. A Conservative Three-Dimensional Eulerian Method for Coupled Solid-Fluid Shock Capturing. *Journal of Computational Physics*, 183:26–82, 2002.

[241] R. Mills, F. Hoffman, P. Worley, K. Perumalla, A. Mirin, G. Hammond, and B. Smith. Coping at the user-level with resource limitations in the Cray message passing poolkit MPI at scale: How not to spend your summer vacation. In R. Winget and K. Winget, editor, *Proceedings of the 51st Cray User Group Conference, May 4-7, 2009*, Eagan, MN, 2009. Cray User Group, Inc.

[242] F. Miniati and P. Colella. Block structured adaptive mesh and time refinement for hybrid, hyperbolic + n-body systems. *Journal of Computational Physics*, 227:400–430, 2007.

[243] A. Mirin and P. Worley. Extending scalability of the Community Atmosphere Model. *Journal of Physics: Conference Series*, 78, 2007. doi: 10.1088/1742-6596/78/1/012082

[244] A.A. Mirin and W.B. Sawyer. A scalable implemenation of a finite-volume dynamical core in the Community Atmosphere Model. *International Journal of High Performance Computing Applications*, 19(3), August 2005.

[245] B. Mohr, A.D. Malony, S. Shende, and F. Wolf. Towards a performance tool interface for OpenMP: An approach based on directive rewriting. In *Proceedings of Third European Workshop on OpenMP*.

[246] B. Mohr and F. Wolf. KOJAK – A tool set for automatic performance analysis of parallel programs. In *Procs. of the International Conference on Parallel and Distributed Computing (Euro-Par 2003). (Lecture notes in computer science; 2790)*, pages 1301–1304, August 2003.

[247] G.E. Moore. Cramming more components onto integrated circuits. *Electronics*, 38(8), April 1965.

[248] A. Morris, W. Spear, A. Malony, and S. Shende. Observing performance dynamics using parallel profile snapshots. In *EuroPar 2008*, volume LNCS 5168, pages 162–171, Canary Island, Spain, August 2008. Springer.

[249] D. Mosberger-Tang. libunwind.
http://www.nongnu.org/libunwind

[250] T. Mytkowicz, A. Diwan, M. Hauswirth, and P.F. Sweeney. Producing wrong data without doing anything obviously wrong! In *Proceedings of the 14th International Conference on Architectural Support for Programming Languages and Operating Systems*, pages 265–276, New York, NY, USA, 2009. ACM.

[251] W. E. Nagel, A. Arnold, M. Weber, H-C. Hoppe, and K. Solchenbach. VAMPIR: Visualization and Analysis of MPI Resources. *Supercomputer*, 12(1):69–80, 1996.

[252] W.E. Nagel, A. Arnold, M. Weber, H.C. Hoppe, and K. Solchenbach. VAMPIR: Visualization and analysis of MPI resources. *The International Journal of Supercomputer Applications and High Performance Computing*, 11(2):144–159, 1997.

[253] A. Nataraj, A. Malony, A. Morris, D. Arnold, and B. Miller. TAUoverMRNet (ToM): A framework for scalable parallel performance monitoring. In *International Workshop on Scalable Tools for High-End Computing (STHEC '08)*, 2008.

[254] A. Nataraj, A.D. Malony, S. Shende, and A. Morris. Integrated parallel performance views. *Cluster Computing*, 11(1):57–73, 2008.

[255] A. Nataraj, A. Morris, A.D. Malony, M. Sottile, and P. Beckman. The ghost in the machine: Observing the effects of kernel operation on parallel application performance. In *Proceedings of 2007 ACM/IEEE Conference on Supercomputing (SC2007)*, Reno, Nevada, November 10–16 2007.

[256] A. Nataraj, M. Sottile, A. Morris, A.D. Malony, and S. Shende. TAUoverSupermon: Low-overhead online parallel performance monitoring. In *Europar'07: European Conference on Parallel Processing*, 2007.

[257] National Center for Supercomputing Applications. Blue Waters hardware. http://www.ncsa.illinois.edu/BlueWaters/hardware.html

[258] J.A. Nelder and R. Mead. A simplex method for function minimization. *Computer Journal*, 7:308–313, 1965.

[259] Y.L. Nelson, B. Bansal, M. Hall, A. Nakano, and K. Lerman. Model-guided performance tuning of parameter values: A case study with molecular dynamics visualization. *IEEE International Symposium on Parallel and Distributed Processing (IPDPS 2008)*, April 2008.

[260] O.Y. Nickolayev, P.C. Roth, and D.A. Reed. Real-time statistical clustering for event trace reduction. In *Proceedings of the 2008 ACM/IEEE conference on Supercomputing (SC08)*, pages 1–12, 2008.

[261] M. Noeth, P. Ratn, F. Mueller, M. Schulz, and B. de Supinski. Scalatrace: Scalable compression and replay of communication traces in high performance computing. *Journal of Parallel and Distributed Computing*, 69(8):969–710, Aug 2009.

[262] R.W. Numrich and J.K. Reid. Co-Array Fortran for parallel programming. *ACM Fortran Forum*, 17(2):1–31, 1998.

[263] Office of Science, U.S. Department of Energy. A science-based case for large-scale simulation. http://www.pnl.gov/scales, July 30 2003.

[264] L. Oliker, A. Canning, J. Carter, J. Shalf, and S. Ethier. Scientific computations on modern parallel vector systems. In *Proceedings of ACM/IEEE Conference on Supercomputing (SC04)*, page 10, Washington, DC, USA, 2004. IEEE Computer Society.

[265] L. Oliker, A. Canning, and J. Carter, et al. Scientific application performance on candidate petascale platforms.

[266] M. Olszewski, J. Ansel, and S. Amarasinghe. Kendo: Efficient deterministic multi-threading in software. March 2009.

[267] M. Parrinello. From silicon to RNA: The coming of age of first-principle molecular dynamics. *Solid State Communications*, 103, 107, 1997.

[268] D.A. Patterson and J.L. Hennessy. *Computer Organization and Design: The Hardware/Software Interface*. Morgan Kaufmann, San Francisco, 2008.

[269] M.C. Payne, M.P. Teter, D.C. Allan, T.A. Arias, and J.D. Joannopoulos. Iterative minimization techniques for ab initio total-energy calculations: Molecular dynamics and conjugate gradients. *Reviews of Modern Physics*, 64:1045, 1992.

[270] E. Perelman, G. Hamerly, M.V. Biesbrouck, T. Sherwood, and B. Calder. Using simpoint for accurate and efficient simulation. *ACM SIGMETRICS Performance Evaluation Review*, 31:318–319, 2003.

[271] SciDAC Performance Engineering Research Institute (PERI).

[272] PETSc: Portable, extensible toolkit for scientific computation.

[273] S. Phillips. Victoria Falls: Scaling highly-threaded processor cores. In *HotChips 19*, 2007.

[274] P. Pierce. The NX message passing interface. *Parallel Computing*, 20(4):463–480, April 1994.

[275] V. Pillet, J. Labarta, T. Cortes, and S. Girona. PARAVER: A tool to visualise and analyze parallel code. In *Proceedings of WoTUG-18: Transputer and occam Developments*, volume 44, pages 17–31, Amsterdam, 1995. IOS Press.

[276] V. Pillet, J. Labarta, T. Cortes, and S. Girona. PARAVER: A tool to visualize and analyze parallel code. Technical Report UPC-CEPBA 95-3, European Center for Parallelism of Barcelona (CEPBA), Universitat Politècnica de Catalunya (UPC), 1995.
http://tinyurl.com/paraver95

[277] S. Pinker. *The Blank Slate: The Modern Denial of Human Nature*. Viking, New York, 2002.

[278] PLASMA project.
http://icl.cs.utk.edu/plasma

[279] S. Pop, A. Cohen, C. Bastoul, S. Girbal, G. Silber, and N. Vasilache. Graphite: Polyhedral analyses and optimizations for gcc. In *Proceedings of the 2006 GCC Developers Summit*, page 2006, 2006.

[280] A. Qasem, G. Jin, and J. Mellor-Crummey. Improving performance with integrated program transformations. Technical Report TR03-419, Rice University, October 2003.

[281] A. Qasem and K. Kennedy. Profitable loop fusion and tiling using model-driven empirical search. In *Proceedings of the 2006 ACM International Conference on Supercomputing*, June 2006.

[282] Coefficient of determination.
mathbits.com/mathbits/tisection/statistics2/correlation.htm

[283] P. Ratn, F. Mueller, M. Schulz, and B. de Supinski. Preserving time in large-scale communication traces. In *International Conference on Supercomputing*, pages 46–55, June 2008.

[284] Rice University. HPCToolkit performance tools.
http://hpctoolkit.org

[285] D. Rideout and S. Zohren. Evidence for an entropy bound from fundamentally discrete gravity. *Classical Quantum Gravity*, 2006.

[286] P.C. Roth, D.C. Arnold, and B.P. Miller. Mrnet: A software-based multicast/reduction network for scalable tools. In *International Conference on Supercomputing*, pages 21–36. IEEE Computer Society, 2003.

[287] G. Rudy. CUDA-CHiLL: A programming language interface for GPGPU optimizations and code generation. Master's thesis, May 2010.

[288] V. Salapura, K. Ganesan, A. Gara, M. Gschwind, J. Sexton, and R. Walkup. Next-generation performance counters: Towards monitoring over a thousand concurrent events. Technical Report RC24351 W0709-061, IBM Research Division, 2007.

[289] E. Schnetter. Multi-physics coupling of Einstein and hydrodynamics evolution: A case study of the Einstein Toolkit. CBHPC 2008 (Component-Based High Performance Computing) (accepted), 2008.

[290] E. Schnetter, P. Diener, E.N. Dorband, and M. Tiglio. A multi-block infrastructure for three-dimensional time-dependent numerical relativity. *Classical Quantum Gravity*, 23:S553–S578, 2006.

[291] E. Schnetter, S.H. Hawley, and I. Hawke. Evolutions in 3D numerical relativity using fixed mesh refinement. *Classical and Quantum Gravity*, 21:1465–1488, 2004.

[292] J. Schrier, D.O. Demchenko, L.-W. Wang, and A.P. Alivisatos. Optical properties of zno/zns and zno/znte heterostructures for photovoltaic applications. *NanoLett.*, 7:2377, 2007.

[293] M. Schulz and B.R. de Supinski. A flexible and dynamic infrastructure for MPI tool interoperability. In *Proceedings of ICPP 2006*, pages 193–202, 2006.

[294] M. Schulz and B.R. de Supinski. $p^n MPI$ tools: A whole lot greater than the sum of their parts. In *Proceedings of SC07*, 2007.

[295] National Science and Technology Council Committee on Technology High-End Computing Revitalization Task Force. Report of the High-End Computing Revitalization Task Force (HECRTF). 2004.

[296] S. Seidl. VTF3 – A fast Vampir trace file low-level management library. Technical Report ZHR-R-0304, Dresden University of Technology, Center for High-Performance Computing, Nov 2003.

[297] H. Servat, G. Llort, J. Gimenez, and J. Labarta. Detailed performance analysis using coarse grain sampling. In *2nd Workshop on Productivity and Performance (PROPER 2009)*, 2009.

[298] S. Shende. *The Role of Instrumentation and Mapping in Performance Measurement*. PhD thesis, University of Oregon, August 2001.

[299] S. Shende, A. Malony, and A. Morris. *Optimization of Instrumentation in Parallel Performance Evaluation Tools*, volume 4699 of *LNCS*, pages 440–449. Springer, 2008.

[300] S. Shende and A.D. Malony. The TAU parallel performance system. *The International Journal of High Performance Computing Applications*, 20(2):287–331, Summer 2006.

[301] S. Shende, A.D. Malony, J. Cuny, K. Lindlan, P. Beckman, and S. Karmesin. Portable Profiling and Tracing for Parallel Scientific Applications using C++. In *Proceedings of the SIGMETRICS Symposium onParallel and Distributed Tools, SPDT'98*, pages 134–145, 1998.

[302] S. Shende, A.D. Malony, C. Rasmussen, and M. Sottile. A Performance Interface for Component-Based Applications. In *Proceedings of International Workshop on Performance Modeling, Evaluation and Optimization, International Parallel and Distributed Processing Symposium*, 2003.

[303] J. Shin, M.W. Hall, J. Chame, C. Chen, P. Fischer, and P.D. Hovland. Autotuning and specialization: Speeding up Nek5000 with compiler technology. In *Proceedings of the International Conference on Supercomputing*, June 2010.

[304] J. Shin, M.W. Hall, J. Chame, C. Chen, and P.D. Hovland. Autotuning and specialization: Speeding up matrix multiply for small matrices with compiler technology. In *The Fourth International Workshop on Automatic Performance Tuning*, October 2009.

[305] K. Singh, M. Bhadauria, and S.A. McKee. Real time power estimation of multi-cores via performance counters. *Proceedings of Workshop on Design, Architecture and Simulation of Chip Multi-Processors*, November 2008.

[306] K. Singh, E. Ipek, S.A. McKee, B.R. de Supinski, M. Schulz, and R. Caruana. Predicting parallel application performance via machine learning approaches. *Concurrency And Computation: Practice and Experience*, 19(17):2219–2235, 2007.

[307] D. Skinner. Performance monitoring of parallel scientific applications. Technical Report LBNL-5503, Lawrence Berkeley National Laboratory, 2005.

[308] A. Sloss, D. Symes, and C. Wright. *ARM System Developer's Guide: Designing and Optimizing System Software.* Morgan Kaufmann Publishers Inc., San Francisco, CA, 2004.

[309] A.J. Smith. A comparative study of set associative memory mapping algorithms and their use for cache and main memory. *IEEE Transactions on Software Engineering*, (2):121–130.

[310] R.D. Smith, J.K. Dukowicz, and R.C. Malone. Parallel ocean general circulation modeling. *Phys. D*, 60(1-4):38–61, 1992.

[311] A. Snavely, L. Carrington, N. Wolter, J. Labarta, R. Badia, and A. Purkayastha. A framework for application performance modeling and prediction. In *Proceedings of ACM/IEEE Conference on Supercomputing (SC02)*, 2002.

[312] A. Snavely, X. Gao, C. Lee, N. Wolter, J. Labarta, J. Gimenez, and P. Jones. Performance modeling of HPC applications. *Proceedings of the Parallel Computing Conference 2003*, October 2003.

[313] F. Song, F. Wolf, N. Bhatia, J. Dongarra, and S. Moore. An Algebra for Cross-Experiment Performance Analysis. In *Proceedings of International Conference on Parallel Processing (ICPP-04)*, August 2004.

[314] SPIRAL project.
       http://www.spiral.net

[315] B. Sprunt. Pentium 4 performance-monitoring features. *IEEE Micro*, 22(4):72–82, 2002.

[316] A. Srivastava and A. Eustace. Atom: A system for buiding customized porgram analysis tools. In *Proceedings of of the SIGPLAN 94 Conf. on Porgramming Language Design and Implementation*, pages 196–205, Orlando, FL, June 1994.

[317] STREAM: Sustainable memory bandwidth in high performance computers.
       http://www.cs.virginia.edu/stream

[318] E. Strohmaier and H. Shan. Architecture independent performance characterization and benchmarking for scientific applications. In *International Symposium on Modeling, Analysis and Simulation of Computer and telecommunication Systems*, October 2004.

[319] E. Strohmaier and H. Shan. Apex-MAP: A global data access benchmark to analyze HPC systems and parallel programming paradigms. In *Proceedings of 2005 ACM/IEEE Conference on Supercomputing (SC05)*, 2005.

[320] R. Subramanya and R. Reddy. Sandia DNS code for 3D compressible flows - Final Report. Technical Report PSC-Sandia-FR-3.0, Pittsburgh Supercomputing Center, PA, 2000.

[321] Sun Microsystems. Sun Studio Performance Analyzer. http://developers.sun.com/sunstudio/overview/topics/analyzing.jsp 2009.

[322] V. Tabatabaee, A. Tiwari, and J.K. Hollingsworth. Parallel Parameter Tuning for Applications with Performance Variability. In *SC '05: Proceedings of the 2005 ACM/IEEE conference on Supercomputing*, page 57, Washington, DC, 2005. IEEE Computer Society.

[323] B. Talbot, S. Zhou, and G. Higgins. Review of the Cactus framework: Software engineering support of the third round of scientific grand challenge investigations, task 4 report - earth system modeling framework survey.

[324] N. Tallent, J. Mellor-Crummey, L. Adhianto, M. Fagan, and M. Krentel. Diagnosing performance bottlenecks in emerging petascale applications. In *Proceedings of ACM/IEEE Conference on Supercomputing (SC09)*, pages 1–11, New York, NY, USA, 2009. ACM.

[325] N.R. Tallent, L. Adhianto, and J. Mellor-Crummey. Scalable identification of load imbalance in parallel executions using call-path profiles. In *Proceedings of ACM/IEEE Conference on Supercomputing (SC10)*, New York, NY, November 2010. ACM.

[326] N.R. Tallent and J. Mellor-Crummey. Effective performance measurement and analysis of multithreaded applications. In *Proceedings of the 14th ACM SIGPLAN Symposium on Principles and Practice of Parallel Programming*, pages 229–240, New York, NY, USA, 2009. ACM.

[327] N.R. Tallent, J. Mellor-Crummey, and M.W. Fagan. Binary analysis for measurement and attribution of program performance. In *Proceedings of the 2009 ACM SIGPLAN Conference on Programming Language Design and Implementation*, pages 441–452, New York, NY, USA, 2009. ACM.

[328] N.R. Tallent, J. Mellor-Crummey, and A. Porterfield. Analyzing lock contention in multithreaded applications. In *Proceedings of the 15th ACM SIGPLAN Symposium on Principles and Practice of Parallel Programming*, 2010.

[329] J. Tao, G. Allen, I. Hinder, E. Schnetter, and Y. Zlochower. XiRel: Standard benchmarks for numerical relativity codes using Cactus and Carpet. Technical Report CCT-TR-2008-5, Louisiana State University, 2008.

[330] V. Taylor, X. Wu, and R. Stevens. Prophesy: An infrastructure for performance analysis and modeling of parallel and grid applications. *SIGMETRICS Perform. Eval. Rev.*, 30(4):13–18, 2003.

[331] The Parallel Ocean Program.
http://climate.lanl.gov/Models/POP

[332] The R Foundation for Statistical Computing. R project for statistical computing.
http://www.r-project.org, 2007.

[333] K. Thompson and D.M. Ritchie. Unix programmers manual, sixth edition, May 1975.

[334] K.S. Thorne. Gravitational Radiation – a New Window Onto the Universe. (Karl Schwarzschild Lecture 1996). *Reviews of Modern Astronomy*, 10:1–28, 1997.

[335] M.M. Tikir, L. Carrington, E. Strohmaier, and A. Snavely. A genetic algorithm approach to modeling the performance of memory-bound computations. In *Proceedings of ACM/IEEE Conference on Supercomputing (SC07)*, 2007.

[336] F.X. Timmes and F.D. Swesty. The accuracy, consistency, and speed of an electron-positron equation of state based on table interpolation of the helmholtz free energy. *Astrophysical Journal, Supplement*, 126:501–516, 2000.

[337] A. Tiwari, C. Chen, J. Chame, M. Hall, and J.K. Hollingsworth. A scalable autotuning framework for compiler optimization. In *Proceedings of the 24th International Parallel and Distributed Processing Symposium*, April 2009.

[338] S. Tomov, J. Dongarra, and M. Baboulin. Towards dense linear algebra for hybrid GPU accelerated manycore systems. Technical Report UT-CS-08-632, University of Tennessee, 2008. LAPACK Working Note 210.

[339] University of Oregon. TAU Portable Profiling.
http://tau.uoregon.edu

[340] University of Oregon. TAU Portal.
http://tau.nic.uoregon.edu

[341] J. Vetter. Dynamic statistical profiling of communication activity in distributed applications. In *Proceedings of the 2002 ACM SIGMETRICS International Conference on Measurement and Modeling of Computer Systems*, pages 240–250, New York, NY, USA, 2002. ACM.

[342] J. Vetter and C. Chambreau. mpiP: Lightweight, scalable MPI profiling.
http://www.llnl.gov/CASC/mpip

[343] J.S. Vetter and C. Chambreau. mpiP: Lightweight, scalable MPI profiling, April 2005. http://www.llnl.gov/CASC/mpip

[344] V. Volkov and J. Demmel. Benchmarking GPUs to tune dense linear algebra. In *Supercomputing 08*. IEEE, 2008. to appear.

[345] M.J. Voss and R. Eigenmann. ADAPT: Automated de-coupled adaptive program transformation. *Parallel Processing, 2000. Proceedings. 2000 International Conference on*, 2000.

[346] R. Vuduc, J. Demmel, and K. Yelick. OSKI: A library of automatically tuned sparse matrix kernels. In *Proceedings of SciDAC 2005, Journal of Physics: Conference Series*. Institute of Physics Publishing, June 2005.

[347] R. Vuduc, J. Demmel, and K. Yelick. OSKI: A library of automatically tuned sparse matrix kernels. *Journal of Physics: Conference Series*, 16:521–530, June 2005.

[348] D.W. Walker, P.H. Worley, and J.B. Drake. Parallelizing the spectral transform method. Part II. *Concurrency: Practice and Experience*, 4(7):509–531, October 1992.

[349] A.J. Wallcraft. SPMD OpenMP vs MPI for ocean models. In *Proceedings of the First European Workshop on OpenMP*, Lund, Sweden, 1999. Lund University. http://www.it.lth.se/ewomp99

[350] L.-W. Wang, B. Lee, H. Shan, Z. Zhao, J. Meza, E. Strohmaier, and D. Bailey. Linearly scaling 3D fragment method for large-scale electronic structure calculations. *Proceedings of ACM/IEEE Conference on Supercomputing (SC08)*, 2008.

[351] L.-W. Wang and J. Li. First-principles thousand-atoms quantum dot calculations. *Physical Review B*, 69:153302, 2004.

[352] L.-W. Wang, Z. Zhao, and J. Meza. Linear scaling three-dimensional fragment method for large-scale electronic structure calculations. *Physical Review B*, 77:165113, 2008.

[353] L.-W. Wang and A. Zunger. Solving Schrodinger's equation around a desired energy: Application to silicon quantum dots. *Journal of Chemical Physics*, 100:2394, 1994.

[354] L.W. Wang. Parallel planewave pseudopotential ab initio package, 2004. http://hpcrd.lbl.gov/~linwang/PEtot/PEtot.html

[355] T.A. Weaver, G.B. Zimmerman, and S.E. Woosley. Presupernova evolution of massive stars. 225:1021–1029, 1978.

[356] V.M. Weaver and S.A. McKee. Can hardware performance counters be trusted? pages 141–150, September 2008.

[357] J. Weinberg, M.O. McCracken, E. Strohmaier, and A. Snavely. Quantifying locality in the memory access patterns of HPC applications. *Proceedings of ACM/IEEE Conference on Supercomputing (SC05)*, pages 50–61, Nov. 2005.

[358] R.C. Whaley. Atlas version 3.8: Status and overview. In *International Workshop on Automatic Performance Tuning (iWAPT07)*, Tokyo, Japan, September 2007.

[359] R.C. Whaley and J. Dongarra. Automatically tuned linear algebra software. In *Proceedings of Supercomputing '98*, November 1998.

[360] R.C. Whaley and J.J. Dongarra. Automatically tuned linear algebra software. In *SuperComputing*, 1998.

[361] R.C. Whaley, A. Petitet, and J. Dongarra. Automated empirical optimization of software and the ATLAS project. *Parallel Computing*, 27(1-2):3–35, 2001.

[362] S. Williams. *Auto-tuning Performance on Multicore Computers*. PhD thesis, EECS Department, University of California, Berkeley, Dec 2008.

[363] S. Williams, J. Carter, L. Oliker, J. Shalf, and K. Yelick. Lattice Boltzmann simulation optimization on leading multicore platforms. In *Interational Conference on Parallel and Distributed Computing Systems (IPDPS)*, Miami, FL, 2008.

[364] S. Williams, J. Carter, L. Oliker, J. Shalf, and K. Yelick. Lattice Boltzmann simulation optimization on leading multicore platforms. *Journal of Parallel and Distributed Computing*, 69(9):762–777, 2009.

[365] S. Williams, L. Oliker, R. Vuduc, J. Shalf, K. Yelick, and J. Demmel. Optimization of sparse matrix-vector multiplication on emerging multicore platforms. In *Proceedings of ACM/IEEE Conference on Supercomputing (SC07)*, 2007.

[366] S. Williams, L. Oliker, R. Vuduc, J. Shalf, K. Yelick, and J. Demmel. Optimization of sparse matrix-vector multiplication on emerging multicore platforms. *Parallel Computing - Special Issue on Revolutionary Technologies for Acceleration of Emerging Petascale Applications*, 35(3):178–194, 2008.

[367] S. Williams, D. Patterson, L. Oliker, J. Shalf, and K. Yelick. The roofline model: A pedagogical tool for auto-tuning kernels on multicore architectures. In *IEEE HotChips Symposium on High-Performance Chips (HotChips 2008)*, August 2008.

[368] S. Williams, A. Watterman, and D. Patterson. Roofline: An insightful visual performance model for floating-point programs and multicore architectures. *Communications of the ACM*, April 2009.

[369] D. L. Williamson. Description of NCAR Community Climate Model (CCM0B). NCAR Tech. Note NCAR/TN-210+STR, NTIS PB83 231068, National Center for Atmospheric Research, Boulder, Colo., 1983.

[370] D.L. Williamson, J.T. Kiehl, V. Ramanathan, R.E. Dickinson, and J.J. Hack. Description of NCAR community climate model (CCM1). NCAR Tech. Note NCAR/TN-285+STR, NTIS PB87–203782/AS, June 1987.

[371] I. Witten and E. Frank. *Data Mining: Practical Machine Learning Tools and Techniques*. Morgan Kaufmann, 2005.

[372] F. Wolf, B. Mohr, J. Dongarra, and S. Moore. Efficient pattern search in large traces through successive refinement. In *Proceedings of the European Conference on Parallel Computing (EuroPar 2004, LNCS 3149)*, pages 47–54. Springer, 2004.

[373] F. Wolf, B. Wylie, E. Ábrahám, D. Becker, W. Frings, K. Fürlinger, M. Geimer, M. Hermanns, B. Mohr, S. Moore, M. Pfeifer, and Z. Szebenyi. Usage of the SCALASCA toolset for scalable performance analysis of large-scale parallel applications. In *Proceedings of the 2nd HLRS Parallel Tools Workshop*, pages 157–167, Stuttgart, Germany, July 2008. Springer. ISBN 978-3-540-68561-6.

[374] P. Worley. Performance of the Community Atmosphere Model on the Cray X1E and XT3. In R. Winget and K. Winget, editor, *Proceedings of the 48th Cray User Group Conference, May 8-11, 2006*, Eagan, MN, 2006. Cray User Group, Inc.

[375] P. Worley and A. Mirin. Performance Results for the new CAM Benchmark Suite, June 2008. Poster Presentation at the 13th Annual CCSM Workshop, June 17-19, 2008, Breckenridge, CO.

[376] P. Worley, A. Mirin, J. Drake, and W. Sawyer. Performance engineering in the community atmosphere model. *Journal of Physics: Conference Series*, 46:356–362, 2006. doi: 10.1088/1742-6596/46/1/050

[377] P.H. Worley. MPI performance evaluation and characterization using a compact application benchmark code. In *Proceedings of the Second MPI Developers Conference and Users' Meeting*, pages 170–177. IEEE Computer Society Press, Los Alamitos, CA, 1996.

[378] P.H. Worley. Scaling the unscalable: A case study on the AlphaServer SC. In *Proceedings of ACM/IEEE Conference on Supercomputing (SC02)*. 2002.

[379] P.H. Worley. Benchmarking using the Community Atmosphere Model. In *Proceedings of the 2006 SPEC Benchmark Workshop, January 23, 2006*, Warrenton, VA, 2006. The Standard Performance Evaluation Corp.

[380] P.H. Worley and J.B. Drake. Parallelizing the spectral transform method. *Concurrency: Practice and Experience*, 4(4):269–291, June 1992.

[381] P.H. Worley and J.B. Drake. Performance portability in the physical parameterizations of the Community Atmosphere Model. *International Journal of High Performance Computing Applications*, 19(3):1–15, August 2005.

[382] P.H. Worley and I.T. Foster. Parallel spectral transform shallow water model: a runtime–tunable parallel benchmark code. In J. J. Dongarra and D. W. Walker, editors, *Proceedings of the Scalable High Performance Computing Conference*, pages 207–214. IEEE Computer Society Press, Los Alamitos, CA, 1994.

[383] P.H. Worley, I.T. Foster, and B. Toonen. Algorithm comparison and benchmarking using a parallel spectral transform shallow water model. In G.-R. Hoffman and N. Kreitz, editors, *Coming of Age: Proceedings of the Sixth ECMWF Workshop on Use of Parallel Processors in Meteorology*, pages 277–289. World Scientific Publishing Co. Pte. Ltd., Singapore, 1995.

[384] P.H. Worley and J. Levesque. The performance evolution of the Parallel Ocean Program on the Cray X1. In R. Winget and K. Winget, editor, *Proceedings of the 46th Cray User Group Conference, May 17-21, 2004*, Eagan, MN, 2004. Cray User Group, Inc.

[385] C.E. Wu, A. Bolmarcich, M. Snir, D. Wootton, F. Parpia, A. Chan, E. Lusk, and W. Gropp. From trace generation to visualization: A performance framework for distributed parallel systems. In *Proceedings of ACM/IEEE Conference on Supercomputing (SC00)*, November 2000.

[386] J. Xiong, J. Johnson, R. Johnson, and D. Padua. SPL: A language and compiler for DSP algorithms. In *Proceedings of ACM SIGPLAN Conference on Programming Language Design and Implementation*, June 2001.

[387] Q. Yi, K. Seymour, H. You, R. Vuduc, and D. Quinlan. POET: parameterized optimizations for empirical tuning. In *Proceedings of the 21st International Parallel and Distributed Processing Symposium*, March 2007.

[388] K. Yotov, X. Li, G. Ren, M.J. Garzarán, D. Padua, K. Pingali, and P. Stodghill. Is search really necessary to generate high-performance BLAS? *Proceedings of the IEEE*, 93(2):358–386, 2005.

[389] K.M. Yu, W. Walukiewicz, J. Wu, W. Shan, J.W. Beeman, M.A. Scarpulla, O.D. Dubon, and P. Becta. Diluted ii-vi oxide semiconductors with multiple band gaps. *Physical Review Letters*, 91:246403, 2003.

[390] Z. Zhao, J. Meza, and L.-W. Wang. A divide and conquer linear scaling three dimensional fragment method for large scale electronic structure calculations. *Journal of Physics: Condensed Matter*, 20(294203), 2008.

[391] H. Zima, M. Hall, C. Chen, and J. Chame. Model-guided autotuning of high-productivity languages for petascale computing. In *Proceedings of the Symposium on High Performance Distributed Computing*, May 2009.

[392] B. Zink, E. Schnetter, and M. Tiglio. Multipatch methods in general relativistic astrophysics – hydrodynamical flows on fixed backgrounds. *Physical Review D*, 77:103015, 2008.

Index